Food for the Future

Food for the Future
Conditions and Contradictions
of Sustainability

Edited by

Patricia Allen
University of California, Santa Cruz

JOHN WILEY & SONS, INC.
New York / Chichester / Brisbane / Toronto / Singapore

Copyright © 1993 by John Wiley & Sons, Inc.

Library of Congress Cataloging in Publication Data:
Food for the future : conditions and contradictions of sustainability
  / [edited by] Patricia Allen.
      p.   cm.
      "A Wiley Interscience publication."
      ISBN 0-471-58082-1 (pbk.)
      1. Agriculture—Economic aspects.   2. Food supply.   3. Sustainable
  agriculture—Economic aspects.   I. Allen, Patricia, 1954–   .
  HD1415.F633   1993
  338.1'9—dc20   92-41780

Printed in the United States of America

10 9 8 7 6 5 4 3 2 1

Toward that time when children are no longer hungry, and parents listen to contented laughter instead of anguished cries.

Contributors

Patricia Allen
Agroecology Program
University of California
Santa Cruz, California

Miguel A. Altieri
Division of Biological Control
University of California
Berkeley, California
            and
Latin American Consortium on
    Agroecology and Development
Santiago, Chile

Frederick H. Buttel
Department of Rural Sociology
University of Wisconsin
Madison, Wisconsin

Katherine L. Clancy
Department of Nutrition and Food
    Management
Syracuse University
Syracuse, New York

Kenneth A. Dahlberg
Department of Political Science
Western Michigan University
Kalamazoo, Michigan

Harriet Friedmann
Department of Sociology
University of Toronto
Toronto, Ontario Canada

David Goodman
Board of Environmental Studies
University of California
Santa Cruz, California

Kathleen Merrigan
U.S. Senate Committee on
    Agriculture, Nutrition, and
    Forestry
Washington, D.C.

James O'Connor
Boards of Sociology and Economics
University of California
Santa Cruz, California

Michael Redclift
Dept. of Agriculture, Horticulture,
    and the Environment
Wye College
University of London
London UK

Tom Regan
*Department of Philosophy and
   Religion*
North Carolina State University
Raleigh, North Carolina

Carolyn Sachs
*Department of Agricultural
   Economics and Rural Sociology*
Pennsylvania State University
University Park, Pennsylvania

Neill Schaller
Institute for Alternative Agriculture
Greenbelt, Maryland

Lori Ann Thrupp
*Center for International
   Development and Environment*
World Resources Institute
Washington, D.C.

Garth Youngberg
Institute for Alternative Agriculture
Greenbelt, Maryland

Preface

Today environmental issues and social and economic justice are at the top of the global agenda for survival. The 1992 "Earth Summit" and Global Forum in Brazil brought together more heads of state and nongovernmental organizations from the North and the South than any other meeting in history. They raised global environmental and economic issues that have never before been deemed so urgent. The sustainability of food and agriculture was regarded as an issue of unparalleled importance.

This is because agriculture is central to human well being, because it is fundamental to environmental issues, and because it is in crisis. Food production and distribution are the cornerstones of human subsistence. Unless we develop a sustainable food and agriculture system, life itself is at risk. Agriculture also has far-reaching environmental effects. Food production, distribution, and consumption form the most basic intersection between society and nature, a fact that has generated intensive discussions and dynamic developments relevant to all "green" issues. Yet despite global awareness of the issues, profound systemic and social crises continue to confront the world in food and agriculture. Resources are depleted, toxins enter the food chain, and people go hungry and starve. Such enormous ecological and social problems require immediate, concerted efforts toward transformation to a new form of agriculture, one that ensures our ability to provide ourselves with food both now and into the future.

This transformation is the objective of those working toward agricultural sustainability. In recent times interest in sustainable agriculture has exploded into people's awareness, into theoretical dialogue, and into the efforts of government, universities, and nonprofit organizations. Ten years ago, discussion of sustainable agriculture was considered anathema in food and agriculture circles; today it has become imperative in those same circles. Traditional agricultural universities are developing sustainable agriculture research and education programs. The U.S. Department of Agriculture now operates a sustainable agriculture program and international agencies such as U.S. AID and the U.N. FAO are developing their own. The range of efforts to transform agriculture under the banner of sustainability is vast and includes people representing many different perspectives and interests: small and large farmers, the affluent and the impoverished, development planners and community activists, Western scientists and teachers of traditional knowledge, environmentalists and agribusiness people, national policy makers, and bioregionalists.

This increase in interest and the diversity of sustainable agriculture proponents are both major achievements of, and challenges for, sustainable agriculture. The widespread adoption of the term "sustainable agriculture" has produced a new imperative—one that sustainable agriculture not be subverted and subordinated to the priorities of conventional agriculture. Avoiding this requires analyzing root causes of sustainability problems, asking questions and designing solutions at the multiple levels of the food and agriculture system, and implementing plans for effective transformations.

This book was written for those concerned about creating a sustainable food and agriculture system. I first decided that a book such as this was needed while organizing symposia and seminars on agricultural sustainability over the past several years at the University of California Santa Cruz Agroecology Program. In developing these projects and talking with so many of the people involved in sustainable agriculture, again and again it became apparent that sustainability was conceptualized largely in terms of farm-level agricultural practices and the natural sciences. Working at these levels is crucial, but it is not enough, since it leaves many fundamental questions unasked and therefore unanswered. For example, what do we want to sustain—food production, groundwater levels, profits, existing gender relations? Who should benefit from sustainable agriculture—family farmers, transnational food industries, the hungry? What types of economic structures will facilitate the development of a sustainable agriculture—free market, planning, a combination of both? Despite the present urgency of agricultural sustainability problems, there are surprisingly few studies that address such questions concerning what sustainability means and how we can achieve it.

This book takes up these questions and examines the critical social, political, economic, and ethical aspects of sustainability in food and agriculture. The authors address theoretical and practical aspects of a transformation to sustainability—aspects that have been obscured by the current emphasis on production per se. They discuss new approaches to understanding and developing sustainability, the limitations and future potential of sustainable food and agriculture systems, and ways in which we can work together from different positions to achieve sustainability. The contributors to this book share a wealth of experience and scholarship. My only regret is that the voices of those most at risk from an unsustainable agriculture—the hungry, the impoverished, and the oppressed—are not represented in the first person.

My hope is that this book will help expand the parameters of theory and practice in sustainable agriculture in ways that will make it unthinkable to exclude social issues from its agenda in the future. I envision a sustainable agriculture in which we assign commensurate importance to production and consumption; our research equally emphasizes social and natural science; we consider the needs of both farm owners and farmworkers; we are compelled as much by social problems such as hunger, land tenure, poverty, gender, and culture as we are by environmental problems; and we are as concerned with global sustainability as we are with that in our own communities.

A book such as this is the result of many people's efforts. I appreciate the work and cooperation of all of the authors and thank them for their intelligence, diligence, and patience with my editorial requests. The book would not have been possible without the inspired and strategic contributions of Valerie Kuletz, who also managed the project while I cared for my newborn daughter. Debra Van Dusen was invaluable in helping to conceptualize and organize programs and projects that contributed to the very early ideas for the book; in the final stages she provided essential editorial suggestions for my chapters. The perspective of the Agroecology Program at the University of California, Santa Cruz is that sustainability can never be achieved without solving both ecological and social problems in food and agriculture. I value the support of the Agroecology Program, for my work on the social aspects of sustainability. The confidence and insights of James O'Connor have been integral to this project and have meant a great deal to me. I also appreciate generous financial assistance from the Educational Foundations of America. No project such as this is brought to completion without exacting a price from those to whom one is closest. As ever, I cherish the love and forbearance of family and friends, especially David, Kitt, Paul, and my parents, who provided limitless support and essential wisdom.

PATRICIA ALLEN

Santa Cruz, California
January 1993

Contents

Connecting the Social and the Ecological in Sustainable Agriculture

Patricia Allen

Although the goal of agriculture is first and ultimately sustaining human life, agricultural sustainability has been constructed almost exclusively in the discourse and domain of nature and the natural sciences. The effort toward a sustainable reconstruction of agriculture has privileged environmental priorities and natural science approaches while ignoring social priorities and approaches, despite the fact that social and ecological problems are inseparably connected in food and agriculture systems. Unless we closely examine people's relations with each other, in addition to those between people and nature, we foreclose our ability to bring about the deep structural changes on which sustainable agriculture ultimately depends. This book concentrates on the need to integrate the "social" with the "natural" in sustainable agriculture.

Critics of conventional agriculture in the United States have developed alternative ideas and practices known collectively as "sustainable agriculture." Sustainability proponents have called attention to agricultural resource issues, placed agricultural sustainability on public research and policy agendas, increased demand for pesticide-free food, and developed conservation-ori-

Food for the Future: Conditions and Contradictions of Sustainability, edited by Patricia Allen.
ISBN: 0-471-58082-1 © 1993 John Wiley & Sons, Inc.

ented agricultural techniques. Yet, while the sustainable agriculture movement has effectively demonstrated conventional agriculture's problematic
treatment of the environment, too often this has been at the expense of
attending to equally pressing social problems. As Carolyn Sachs and I discuss
in Part II of this volume, these approaches do not question inequities such
as hunger, poverty, racial oppression, or gender subordination that many
experience in current agrarian structures (e.g., family farms, rural communities, wage labor).

In the past decade "sustainability" has become a central agricultural symbol, moving from a fringe concern to one that is becoming institutionalized
(Buttel and Gillespie, 1988). New organizations have emerged to advocate
sustainability platforms and established institutions have adopted the mantle
of sustainable agriculture. The concept has attracted farmers, consumers,
environmentalists, and agricultural experts alike.[1] Since agricultural sustainability is increasingly embraced as a goal in agriculture (see Youngberg
et al., this volume), yet has accomplished relatively little of major significance
(Buttel, this volume), it is critical that we widen our definition and practice
of this concept, for it has great potential as a transformational tool. A reformulation of its theory and practice is essential to prevent sustainable agriculture
from reproducing the ecological and social problems of current food and
agriculture systems, since agricultural sustainability is a socially constructed,
ideologically based discourse that has as its root a social concept and problem. The purpose of this chapter is to demonstrate the social basis of sustainable agriculture and why social approaches are required for achieving sustainability.

AGRICULTURE: A COMPOSITION OF THE NATURAL AND THE SOCIAL

Agriculture does not exist and cannot function except at the intersection of
society and nature. The editor of the *Journal of Sustainable Agriculture* stresses
the connection between agriculture and society when he writes that in sustainable agriculture not only agriculture but also society needs to be sustained
(Poincelot, 1990). Agriculture is a self-conscious, human productive activity
that has always been socially organized and becomes more so as it develops.
The understanding that human relations with nature are always mediated
through social institutions and systems is central to developing a sustainable
food and agriculture system. Sustainability problems do not arise from only
the interface between society and external nature but also from contradictions
within society itself. In addition, what are often construed as "natural" or
environmental problems (e.g., floods obliterating housing) actually result

from social factors (e.g., state policies encouraging flood plain development). In sustainable agriculture it is important to understand that we are working in a situation in which both nature and society have been developed, produced, and reproduced by the ideas and activities of human beings.

The Production of Nature

In our Western conception of nature, the environment is usually considered to be a physical space and set of laws that exist and operate external to and independent of us. While there is a "nature" that preexisted human beings and we are subject to natural laws such as gravity, nature in sustainable agriculture is a humanly reconstructed nature. In his historical treatment of agriculture and the environment Redclift (1987) regards the environment not as an external form, but as a process that is the result of relationships between physical space, natural resources, and economic forces. Smith (1984) shows that humans produce their means of existence from nature and at the same time generate additional needs that require further production, leading to further divisions of social labor. While hunters and gatherers appropriated from nature long before they began to farm, with agriculture humans began to produce their means of subsistence through their interaction with nature. People's relations with nature then become mediated through the social institutions designed to regulate production and distribution of surplus. Nature is thus a dialectical process of transformation between humans and nature—nature is thus "produced" by humans as well as producing itself.

We apply human labor and appropriate from nature in order to transform natural materials into forms that are useful for us as humans. The reproduction of nature through agriculture is the basis for human subsistence and the development of society. As Harriet Friedmann (this volume) points out, "From the first domestication of plants and animals, humans irreversibly posed for themselves the problem of creating social relations through which to act in concert on nature." Similarly, both sustainability and nonsustainability are material conditions constructed by humans in terms of both their causes and consequences, and therefore their solutions. Elements of the biosphere (soil, water, energy) become "resources" only when people define, use, and exchange them as such. As Michael Redclift (this volume) reminds us, the sustainability of the resource base makes little sense if it is separated from the human agents who manage the environment. Agricultural sustainability problems are defined and have meaning for us only insofar as they affect the very social activities of production and survival. Sustainability problems therefore cannot be correctly understood as purely or even mainly "natural."

The Production of Society

Society is also produced. Like nature, it is a product of human relations, definitions, priorities, and uses. The individual *produces* society through relationships with other individuals (Sayer, 1987) and through structural activity. An example is the economy. As Friedmann notes, the economy is a distinctly social creation, governed by rules of contract, property, and entitlements. Yet it has become so opaque—so much of an objectively perceived entity—that it is difficult to see that malleable human ideas and institutions shape it, support it, and can change it. We "naturalize" such social relationships—they appear natural, external, and eternal to us. When social relations are naturalized, they appear immutable, a false perception that limits our ability to effect change.

In sustainable agriculture there is a similar tendency to naturalize the social, even though current social relations in agriculture have no exclusively natural basis. Rather, they have developed and changed historically as the products of class and power laden human choices and social institutions. An example is the way in which the present forms of property ownership and agricultural markets are taken as givens instead of social constructions. As Friedmann points out, current relationships reflect the more or less conscious activities of people responding to their experiences. When people and their activities become more conscious, they can weigh alternative futures differently. In today's context of Eastern European disintegration, the Third World debt crisis, and the U.S. recession, analysts and politicians speak of how society must adjust to supposed economic "laws" such as competition and productivity. But such fatalistic approaches cannot resolve the enormous socioeconomic pressures facing us. Instead, we must seek to shape the economy to fit social needs, which will involve substantial rethinking. Only with such a fundamental reframing is it possible to ask, as James O'Connor (this volume) does, whether current political economic structures are compatible with sustainability.

Developing agricultural sustainability requires that we think more creatively about the prerequisites for sustainability and barriers necessary to overcome. As Redclift discusses, the conflicts over environment are not only about the way people seek to dominate nature, but also (perhaps more importantly) about the way people dominate each other. Thus, the agenda for sustainable agriculture cannot be simply about external nature or the environment; it must also address human needs and social relations. If we avoid naturalizing what is primarily social and instead recognize the role of human intervention in nature and the historical generation of social structures, it becomes axiomatic that sustainability problems can be transcended rather than accepted as inevitable. This requires a commitment to transform our relations not only with external nature, but with other human beings.

THE NATURE-BASED APPROACH OF SUSTAINABLE AGRICULTURE

The genesis and early development of the discourse on sustainability show the extent to which it has been constructed more in the realm of "nature"—resources and environment than in that of social relations. People have been concerned with resource problems in agriculture from the time of the enclosures of the 1700s through the New Deal farm policies of the 1930s and up until today. According to Ruttan (1992) contemporary concerns with natural resources and environmental change represent the third wave of social interest in U.S. sustainability issues since World War II. The first of these, in the late 1940s and early 1950s, questioned the adequacy of natural resources to sustain growth. The second, in the late 1960s and early 1970s, added concerns about the externalities of commodity production (e.g., industrial pollution and pesticides in food). Agricultural sustainability first emerged in its present form during the energy crisis of the 1970s, causing people to question the energy intensifies of industrialized agriculture and reconsider the pesticide concerns raised by Rachel Carson in *Silent Spring* (1962). Interest in and activities around sustainable agriculture grew in the early 1980s, primarily in response to concerns about resource depletion (such as groundwater overdrafts and soil erosion and salinization), environmental contamination (such as nitrates in groundwater), direct and indirect pesticide poisoning, and diminishing marginal yield increases in response to additional inputs. In Ruttan's view, the third wave of sustainability concern began in the mid-1980s and focuses on issues such as global warming, ozone depletion, and acid rain that are unarguably transnational in scope.

Anyone reviewing the literature on sustainable agriculture will immediately realize the emphasis on "natural" constraints to sustainability. For example, Crews et al. (1991:146) state that although the profitability of sustainable agricultural systems is constrained by the social structure of agriculture, "sustainability itself is constrained solely by the ecological conditions of agriculture." Sustainable agriculture is seen primarily as a natural/technical process, rather than a social one; a relation between people and nature versus a relation between people *and people* and nature. While current efforts in sustainable agriculture address environmental problems caused by conventional approaches to agriculture, they have generally ignored social relations problems in agriculture. Hamlin (1991:508–509) writes, "A peculiar feature of the consideration of alternative agricultures (and of Green politics) is the appeal to nature." This is reflected in the popularity of Wes Jackson's ideas. One of the initial and most influential sustainable agriculture proponents and researchers, Jackson bases his approach to sustainable agriculture on "nature as analogy" (Jackson, 1990).

The focus on nature in sustainable agriculture, though incomplete, certainly has an important and highly pragmatic basis. As Goodman and Redclift (1990: 202) state, "No area of concern demonstrates the difficulty of managing the contradictions of the food system as clearly as the environment." Even in its industrialized form, agriculture remains dependent on natural resources and processes such as soil, water, and weather. For some, agriculture is a special case of the intersection of production and nature (see, for example, Mann, 1990 and Goodman et al., 1987), since it is an "eco-regulatory" activity[2] which depends upon the primary appropriation of nature (Benton, 1989). Thus ecological sustainability problems come into particularly sharp focus in agriculture, even though all material production, including industry, is nature dependent.

The environmental problems to which agricultural sustainability proponents have called our attention are indeed serious. The destruction of ecological conditions of production in agriculture, a deterioration that has accelerated in past decades, threatens our future ability to produce food. Resource degradation has already severely reduced or destroyed agricultural productivity in a number of areas, due to groundwater depletion, soil salinization, and unmanageable pest problems caused by pesticide use (Lockeretz, 1989). Nonetheless, strategies to resolve what appear to be purely environmental problems require the inclusion of social factors if they are to succeed. As Smith (1984: 18) points out, "The relation with nature is an historical product, and even to posit nature as external to society (a primary methodological axiom of positivist 'science,' for example) is literally absurd since the very act of positing nature requires entering a certain relation *with* nature."

NEED FOR INCLUDING THE SOCIAL IN SUSTAINABLE AGRICULTURE

Since we cannot neglect that nature is socially constructed or overlook that the goals of sustainable agriculture are ultimately for human benefit, sustainable agriculture needs to include a social dimension. Dominant approaches to sustainable agriculture are partial and contain contradictions that limit their effectiveness (as discussed by myself and Carolyn Sachs). We can expand these horizons by framing sustainability issues in two broad dimensions: *degradation* (ecological and social) and *temporality* (present and future). As Lori Ann Thrupp (this volume) points out, we face both the erosion of natural resource conditions *and* human dignity and livelihoods. Thrupp sees the causes of environmental and human degradation as not only ecological and technical, but, more importantly, arising from socioeconomic and political factors.

While insufficient attention to environmental conditions in agriculture has led to ecological degradation, our failure to solve social problems has led to both human and ecological degradation. In terms of temporality, ecological destruction is a problem primarily (though not exclusively) for the long term and for future generations, while social problems constitute immediate crises for the present generation as well as prospective crises for future generations. Most U.S. sustainable agriculture proponents address only a portion of these concerns because they work primarily to conserve the ecological conditions needed to provide for the long term, but not also the social conditions needed to provide for both the present and the long term. The need to expand our focus is embedded in O'Connor's discussion of conditions of production. Of the three conditions of production, external nature or environment is only one; the other two critical conditions are human labor power and infrastructure/space. Sustaining production and human life depends on all three.

How can we give greater priority to the social conditions that affect not only the lives of present and future people but the ecological conditions on which they depend? A beginning is to define sustainability in social as well as ecological terms. Such a definition is not without precedent. The Food and Agriculture Organization of the United Nations (1989) recognizes the centrality of human needs in sustainable agriculture when it states the sustainable agriculture should "satisfy changing human needs while maintaining or enhancing the quality of the environment and conserving natural resources." And for Harwood (1990:4), a sustainable agriculture is "an agriculture that can evolve indefinitely toward greater human utility, greater efficiency of resource use, and a balance with the environment that is favorable both to humans and to most other species."

These conceptualizations carry important political implications, particularly in terms of how they can affect the life possibilities for those traditionally underprivileged in the global food and agriculture system. This is addressed in the Bruntland Report on sustainable development. In this influential document, as David Goodman (this volume) notes, technical criteria are not allowed to "silence a preeminently political question: sustainable development for whom?" Here the emphasis is on meeting basic needs such as access to resources required for day-to-day survival, needs that are not met for one-fifth of the world's people. Altieri, Redclift, and Thrupp also demonstrate that the principal social problem of sustainability is poverty and argue that greater equity or reduction of poverty must be achieved before the question of environmental quality can be fully addressed. As Goodman explains, for example, while rural poverty is the proximate cause of environmental problems such as desertification and deforestation, this poverty is caused by political economic structures that encourage land concentration, undermine traditional resource management systems, privatize common property resources, and subsidize unsustainable technologies. He cites the example of Northeast Brazil, where "the high unequal distribution of property rights and land-use changes associated with heavily subsidized agricultural modernization programs are the principal causes of rural poverty." In addition, as Thrupp discusses, the effects of resource degradation accrue disproportionately to the poor, to women, and to racial minorities.

Achieving the goal of environmental preservation, even in the United States, is not possible without transforming social institutions and policies. Soil erosion, for example, is a "natural" process but is greatly accelerated by continuous, intensive cultivation practices encouraged by agricultural policies. Similarly, declining water tables, common in many agricultural regions, are caused by extensive irrigation, also encouraged by agricultural investment and tax policies. And increased application of pesticides carries the seed of intensifying future needs for more chemical toxins as pests develop resistance to standard preparations.

While meeting human needs requires the preservation of the environment, the inverse is also absolutely necessary. Ecological sustainability cannot be achieved in the absence of equitable control and distribution of resources. According to Richard Jolly of the United Nations Children's Fund, "Unless we focus on the basic human needs of those in absolute poverty, sustainable development isn't going to work" (Myers, 1989). Sustainable agriculture, therefore, must be based on fulfillment first and foremost of basic human needs, both for generations to come and for those generations living now. These needs include consumption (food, water, fuel), protection (clothing, shelter), and regeneration (dignity, self-determination, freedom from exploitation) (Allen and Sachs, 1992). Working to meet these needs requires reframing our concept of sustainability to include a social dimension and a concomitant expansion of our approach to sustainable agriculture research.

THE SCIENCE OF SUSTAINABILITY AND THE ROLE OF EPISTEMOLOGY

Located as it has been in the natural sciences, the study of sustainable agriculture has overlooked important social factors. As Frederick H. Buttel (this volume) discusses, sustainable agriculture has premised its appeals and program on natural science: the "impetus for and legitimacy of sustainability come from natural science, natural scientists, and natural science data." The overwhelming majority of research dollars for sustainable agriculture is spent on natural sciences and focused on farm-level projects and production innovations, while a minuscule amount is devoted to the social constraints and possibilities. This natural science approach is applied even when social goals are sought. As noted by Hamlin (1991:508–509), "One cannot help but be struck by the degree to which proposed social changes are sanctioned in appeals to biology, toxicology and the earth sciences, rather than notions of justice, effective government or progress."

Epistemology, the process by which people come to have knowledge of their world, is central to solving agricultural problems because it defines problems and the kinds of questions asked, determines which are important and how they are approached, and forms the foundation out of which solutions and strategies are devised. As O'Neill (1986:91) observes, "In adopting certain categories for social inquiry we also adopt a certain view of the social world, of its problem areas and of its fixed points, of the actions it makes available and ways in which their results are constrained." Redclift also emphasizes the intimate connection between knowledge and power.

The knowledge system and epistemology of sustainability determines what actions will be undertaken on its behalf. In traditional scientific approaches such actions will reflect the separation of nature and society: "It is commonplace that science treats nature as external in the sense that scientific method and procedure dictates an absolute abstraction both from the social context of the events and objects under scrutiny and from the social context of the scientific activity itself" (Smith, 1984:4). This has important consequences since it tends to naturalize what is social and sets tragically inaccurate limits on what is possible. While agriculture's predominating natural science epistemology is essential to preserve or reconstruct ecological conditions of production in agriculture, by itself it is insufficient. In sustainability work we must be aware of the ideological constructions and choices made within science—the "objective reality" we create subjectively.

Still, sustainable agriculture researchers have broadened the agricultural science agenda considerably. They reject the view of most of the agricultural sciences (inherited from Francis Bacon) that nature is an object solely to be controlled and manipulated to serve human needs, and they focus more on integrated farm systems than do their conventional agriculture counterparts. Core sustainable agriculture premises differ from those of conventional agriculture in their recognition of nature-mediated unintended consequences such as soil erosion and pest resistance to pesticides, which call for a different configuration of agricultural strategies. Agroecological approaches—ecological cornerstones of sustainability—include principles of diversity, adaptability, durability, and symbiosis and offer a substantial improvement on reductionist approaches to dominating nature.

Sustainable agriculture researchers tend to share much with their conventional agriculture counterparts, however, including faith in the truth and objectivity of Western science, the use of standard agronomic and economic categories and measures of success, and a tendency to go directly from problem description to prescription without the intermediate step of comprehensive explanation or analysis. Sustainable agriculture premises mirror those of conventional agriculture when they exclude humans and social institutions from analysis. In their models for developing a sustainable agriculture, many programs and activities treat social relations as a constant and interaction with nature as the only manipulable variable (Allen, 1991). While some agroecologists study agricultural systems from an ecological *and* socioeconomic perspective (as discussed by Miguel Altieri, this volume), they often do this by attempting to generalize ecological principles to human systems. Many ecologists situate the human species as only one among many in the totality of nature rather recognizing humans' pivotal role in agriculture and the powerful impact of human intentionality and social institutions. Through their labor, which produces both nature and society, humans are absolutely at the center of nature (as they can know it).

Overlooking the centrality of human action has led sustainability advocates and researchers to define problems mainly or solely in terms of nature and environment, and has resulted in a sense that sustainability problems can best be addressed through the traditional Western epistemological tools of logical positivism, reductionism, and neoclassical economics. These focus on the descriptive (rather than the prescriptive) and on observable "things," leading to a tendency to reify social relations in food and agriculture, that is, to treat them as if they had acquired a fixed quality and were features of an external, natural world rather than a social world. Framing sustainable agriculture in a natural science discourse that excludes social relations not only ignores social problems, but leaves unexamined the degree to which environmental problems have social causes. By its emphasis on the natural sciences and farm-level production, the sustainable agriculture movement has embraced an epistemology that is at once too abstract and too positivistic. It is overly abstract in that it does not give sufficient emphasis to the actions of real people producing their world. It is too positivistic in that it focuses on what *is* without considering what *should* be (much less understanding the production of what is).

What is critically needed is to explore options for the natural and the social together—options currently invisible in the dominant discourse on sustainability. We need new epistemological and research approaches that investigate the natural processes of ecology *combined with* the social relations compatible with sustainability. This will involve, for instance, examining not only techniques for reducing soil erosion or water depletion, but also new forms of social organization that alternative agricultural practices presuppose. External nature must figure actively in the social sciences and humans must figure actively in the natural sciences. In most discussions of sustainable agriculture, as in those of conventional agriculture, the primacy of the social over the natural has been misleadingly inverted. The central questions of sustainable agriculture must be not only how we can produce adequate food and fiber in environmentally sound ways, but also as Smith (1984:63) notes, "*how* we produce nature and *who* controls this production of nature" as well as who benefits in terms of wealth, income, working conditions, and consumption. We must ask if sustainability is even possible, much less desirable, without the elimination of patriarchy, racism, and class exploitation—all of which maintain systems of power that reinforce the contradictory social relations on which nonsustainable food and agriculture systems are based.

NEW CONCEPTUAL APPROACHES

The concept of sustainable agriculture is meaningful to the extent that it provides a potential for guiding action. A more detailed and practical theory of sustainability must include methods for determining which problems are most immediate and which actions can be taken to solve them. A new vision for a future food and agriculture system is needed, one that includes new approaches. As Redclift points out, Western science is not a universal epistemology and achieving sustainability requires using many types of cognitive maps, including not only "scientific reasoning" but also "indigenous knowledge."

We must enlarge the scope and importance of values in sustainability. In discussions of sustainability, this is usually limited to "rural" or "traditional" values felt to be embodied in family farming. Tom Regan (this volume) broadens our horizons on this issue by exploring what the sustainability movement can gain from moral philosophy, in particular from feminist ethics and animal rights philosophy, and shows that this expanded moral understanding sets the stage for a broad-scale movement toward sustainability. In examining intensive animal agriculture, for example, he argues against anthropocentrism, and states that we have responsibilities not only to disenfranchised humans but also to other species.

New forms and units of analysis are also required. Sustainable agriculture has tended to focus on only one aspect of the food chain—farm-level production, defined in technical terms—and generally overlooks the equally important components of distribution, exchange, and consumption, all of which are primarily socially determined aspects. Political ecology—a promising and powerful new conceptual framework for studying and solving agricultural sustainability problems—is introduced by Lori Ann Thrupp with a compelling example of land degradation and agriculture in Central America. Kenneth A. Dahlberg (this volume) illustrates why we must go beyond the narrow focus on production agriculture to a broad analysis of complete food systems: production processes and inputs; food processing, preparation, and preservation; distribution, use, and consumption; recycling and disposal of food wastes; and infrastructure, marketing transportation, storage, government services, and policies. He points out that these operate at levels ranging from the household to the international and therefore need to be addressed and resolved at these various levels.

Our approach to sustainable agriculture must be both historical and global since, as Friedmann discusses, the appropriation of nature and its transformation have for several centuries been conducted on a world scale. Today the food and agriculture system is truly global, both in environmental and political economic terms.[3] Acid rain produced in one region affects crop growth in another; ozone depletion creates global, not only regional, climate changes, and agriculture in one region is dependent on genetic resources and market pressures from others. This internationalization of agriculture, which has developed around the global movement of capital, labor, and biological resources (Redclift, 1987), has been the basis of privilege for some and impoverishment for others and it has led to uneven sustainability problems. The North-led Green Revolution, for example, has produced enormous ecological problems in the South including micronutrient deficiencies, nitrate pollution, water logging, desertification, groundwater depletion, and pesticide pollution (Shiva, 1988). Redclift (1987:32) states that "The illusory pursuit of 'food security' in North America and Western Europe has helped to produce regional structures which are a major impediment to greater self-sufficiency in food production in the South. At the same time the absence of 'environmental security' in the South represents an enormous threat to the achievement of real food security in developing countries." Concomitantly, the erosion of food self-sufficiency in the South has led to increased grain production in the North (de Janvry and Le Veen, 1986), often accomplished via environmentally destructive farming practices.

Despite the tendency to think about agricultural sustainability in sectoral, national, or regional terms, it is both limiting and unrealistic to choose nation, state, or "community" boundaries as the essential units of analysis because the logic of sustainability cannot be understood apart from world forces. Friedmann illustrates the conditions under the postwar food regime in which national agricultures have been destroyed, and describes how transnationals and agrifood capital now control agriculture as they disconnect production from consumption at a local level and relink them globally through international buying and selling. National or regional separatism ignores the historical, transnational nature of sustainability issues and problems. Not only are Northern priorities significant in determining the forms of resource exploitation in the South, but as Redclift (1987:82) notes, "In an important sense the life-styles of the rich, industrialized nations were forged from the convergence of three factors in the global development of agriculture: the movement of crops, the control over trade and the (usually) forcible relocation of labour." Thus we cannot consider sustainability to be the exclusive purview of individual nations. Further, greater inequalities exist *within* nations than between nations. According to the World Bank (1986), incomes and food consumption vary more within households and regions than among countries. Wealth and privilege, poverty and oppression, and the suffering caused by nonsustainability accrue less along national lines than along those of class, race, and gender.

THE AIM OF THIS BOOK

Conventional agriculture has produced not only nutrition, but malnutrition. Where it has produced abundance, it has often done so at the cost of environmental destruction. Pesticides produce pests, irrigation produces groundwater depletion, cultivation produces soil erosion, increased yields produce hunger. The food and agriculture crisis marches on as soils degrade and people starve day after day. While the enormity of the suffering this crisis engenders is staggering, it also provides us with the opportunity to develop concepts and strategies to transcend this situation, both for the present and for the future. The still-emerging discourse on sustainability presents an opening for a fundamental "paradigm shift" in the way we think about and practice agriculture. The purpose of this book is to help overcome the duality of the natural and social in sustainable agriculture and to demonstrate the essential contributions of new social approaches to this critical field.

To a large degree, the natural science focus and the deemphasizing of social issues in sustainable agriculture have contributed to its current popular acceptance. Sustainable agriculture has been tolerated, and even embraced, to the degree that it provides limited opposition to the institutional and ideological formulations and foundations of the dominant agricultural culture. But while a nonoppositional approach may be responsible for the acceptance of sustainable agriculture efforts, it also limits our ability to produce truly efficacious or far-reaching solutions. Under the pressure of the current sustainability movement, certain features of the food and agriculture system may change, but its central formation, laws of motion, and winners and losers will remain unaltered. That is, the conditions of agricultural production will be maintained for those who currently possess them rather than extending new possibilities to those desperately at risk—the landless, the hungry, the poor—mostly women and children.

Agriculture is an historical, dialectical process of natural and social transformation. In sustainable agriculture we must struggle against the temptation to project "societal laws" as "natural laws." We must not let present epistemologies falsely set limits on human possibilities or deify the "natural" as a model for social processes. At the same time we must not overlook the natural—all forms of life are dependent on naturally given conditions and we must be willing to recognize nature-imposed limits. The crucial distinction that we must make, however, is not so much between what is natural and what is social, but rather between what can and what cannot be reconfigured and improved.

It is imperative that we make the most pressing social problems a visible part of the debate on agricultural sustainability. Sustainability discourse must not be silent about human miseries, however distant they may be. The boundaries of farms, communities, or nations should not be the boundaries of ethical concern, and therefore should not set the parameters for problem analysis and solution. There are no obvious solutions to the urgent and complex problems faced in agriculture. There are no models, no social and economic systems yet developed that provide all of the bases for an agriculture that nourishes everyone and at the same time preserves resources for future generations.

Thus, today is a critical time to reconsider our sense of agricultural sustainability and to transform the premises and ideology of sustainability discourse, since this creates the framework for actions that can make sustainability a reality. The basis for this change should be a concept of sustainability that proscribes the exploitation of people as well as that of nature, one that combines the approaches of the social and natural sciences as well as alternative epistemologies. This concept would create a framework for working toward emancipatory social strategies while at the same time building on the work already done to learn about nature-constrained boundaries on social possibilities. Such changes are necessary to redirect sustainable agriculture from a set of narrowly defined practices that will benefit some to a transformation that can improve the life possibilities of all of us, especially those traditionally condemned to suffer the most.

NOTES

1. Established environmental groups such as the Sierra Club, the Natural Resources Defense Council, and the Environmental Defense Fund have been active in the analysis and development of sustainable agriculture. It was the National Resources Defense Council who gained national attention for pesticides in food with the publication of its report on alar in apples. Grower groups such as the California Certified Organic Farmers, the New England Organic Farmers Association, and the Biodynamic Agriculture Association have

long been active in sustainable agriculture; today, traditional farmer organizations such as the National Farmers Union increasingly show interest. Major agricultural universities such as the University of California and Ohio State University, along with the U.S. Department of Agriculture, have established sustainable agriculture research and education programs. U.S.-based organizations that focus on sustainability at an international level include the Committee for Agricultural Sustainability in Developing Countries, the International Alliance for Sustainable Agriculture, the World Resources Institute, and the Pesticide Action Network. Numerous nongovernmental organizations work on sustainability at national or local levels: the Committee for Sustainable Agriculture, the California Action Network, the Land Institute, the Center for Rural Affairs, the Institute for Alternative Agriculture, the Rodale Research Institute, and the Center for Science in the Public Interest, to name a few.

2. Benton characterizes "eco-regulatory" practices as those for which (1) the subject of labor is the conditions for the growth and development of the product, not the transformation of a raw material into a product; (2) labor is applied to regulating and reproducing, rather than transforming; (3) the time and space characteristics of the labor activity are shaped by organic developmental processes; and (4) nature-given conditions are both conditions of the labor process and subjects of labor.

3. To some degree, agriculture has always been international, a condition greatly accelerated under colonialism. This has involved First World investments in plantations or agribusiness contracts with Third World producers, investments by First World food processors and distributors in retail marketing in order to expand consumption of processed foods, and the opening of new markets for agricultural exports, primarily for U.S. cereals (de Janvry and Le Veen, 1986).

REFERENCES

Allen, P. 1991. Sustainable Agriculture at the Crossroads. *Capitalism, Nature, Socialism* 2(3):20–28.

Allen, P., and C. Sachs. 1991. The Social Side of Sustainability: Class, Gender, and Ethnicity. *Science as Culture* 2(13):569–590.

Allen, P., and C. Sachs. 1992. The Poverty of Sustainability: An Analysis of Current Discourse. *Agriculture and Human Values* 9(4):30–37.

Altieri, M. 1988. Beyond Agroecology: Making Sustainable Agriculture Part of a Political Agenda. *American Journal of Alternative Agriculture* 3:1142–1133.

Benton, T. 1989. Marxism and Natural Limits: An ecological Critique and Reconstruction. *New Left Review* 178:51–86.

Buttel, F., and G. Gillespie, Jr. 1988. Agricultural Research and Development and the Appropriation of Progressive Symbols: Some Observations on the Politics of Ecological Agriculture. Cornell University Department of Rural Sociology, Ithaca.

Carson, R. 1962. *Silent Spring.* Boston: Houghton Mifflin.

Crews, T. E., C. L. Mohler, and A. G. Power. 1991. Energetics and Ecosystem Integrity: The Defining Principles of Sustainable Agriculture. *American Journal of Alternative Agriculture* 6:146–149.

De Janvry, A., and E. P. LeVeen. 1986. Historical Forces That Have Shaped World Agriculture: A Structural Perspective. In *New Directions for Agriculture and Agricultural Research*. Kenneth A. Dahlberg (ed) Totowa, Rowman and Allanheld.

Goodman, D., and M. Redclift. 1990. *Refashioning Nature: Food, Ecology, and Culture*. London: Routledge.

Goodman, D., B. Sorj, and J. Wilkinson. 1987. *From Farming to Biotechnology: A Theory of Agro-industrial Development*. Oxford: Basil Blackwell.

Hamlin, C. 1991. Green Meanings: What Might 'Sustainable Agriculture' Sustain? *Science as Culture* 2(13):507–537.

Harwood, R. 1990. A History of Sustainable Agriculture. In *Sustainable Agricultural Systems*. Clive A. Edwards, Rattan Lal, Patrick Madden, Robert H. Miller, and Gar House, eds. Soil and Water Conservation Society, Ankeny, Iowa.

Jackson, W. 1990. Agriculture with Nature as Analogy. In *Sustainable Agriculture in Temperate Zones*. Charles A. Francis, Cornelia Butler Flora, and Larry D. King (eds.). New York: John Wiley.

Lockeretz, W. 1989. Comparative Local Economic Benefits of Conventional and Alternative Cropping Systems. *American Journal of Alternative Agriculture* 4(7): 75–83.

Mann, S. A. 1990. *Agrarian Capitalism in Theory and Practice*. Chapel Hill: The University of North Carolina Press.

Myers, N. 1989. Making the World Work for People. *International Wildlife* 19:6.

O'Neill, O. 1986. *Faces of Hunger: An Essay on Poverty, Justice, and Development*. London: Allen & Unwin.

Poincelot, R. P. 1990. Agriculture in Transition. *Journal of Sustainable Agriculture* 1(1):9–40.

Redclift, M. 1987. *Sustainable Development: Exploring the Contradictions*. London: Routledge.

Ruttan, V. W., ed. 1992. *Sustainable Agriculture and the Environment: Perspectives on Growth and Constraints*. Boulder, CO: Westview Press.

Sayer, D. 1987. *The Violence of Abstraction*. Oxford: Basil Blackwell.

Shiva, V. 1988. *Staying Alive: Women, Ecology and Development*. London: Zed Books.

Smith, N. 1984. *Uneven Development*. Oxford: Basil Blackwell.

U.N. Food and Agriculture Organization. 1989. *Sustainable Agricultural Production: Implications for International Agricultural Research*. FAO Research and Technology Paper No. 4, prepared by the Technical Advisory Committee to the Consultative Group on International Agricultural Research. Rome: UN FAO.

World Bank. 1986. *Poverty and Hunger: Issues and Options for food Security in Developing Countries*. Washington, DC: The International Bank for Reconstruction and Development.

PART I

CONDITIONS OF SUSTAINABILITY: NEW CONCEPTUAL APPROACHES

The Production of Agricultural Sustainability: Observations from the Sociology of Science and Technology

Frederick H. Buttel

As the environmental challenge to modern agriculture has increased in breadth and intensity, there is a growing community of scholars, in both the natural sciences and the social sciences and humanities, who define themselves as "sustainability researchers/advocates," and who constitute what I will refer to later as the "sustainability community." Agricultural sustainability has been propelled to prominence sufficiently rapidly so that most observers agree that sustainability has been the most significant trend of the past half-decade within the land-grant system, and second only to biotechnology as the most important trend since 1980. This, of course, is potentially a very positive development, since the emergence of a multidisciplinary community that can achieve cross-fertilization across traditional disciplinary boundaries while accomplishing needed social goals—particularly this one—is refreshing and welcome.

Food for the Future: Conditions and Contradictions of Sustainability, edited by Patricia Allen.
ISBN: 0-471-58082-1 © 1993 John Wiley & Sons, Inc.

The size of and commitment along the multidisciplinary community of sustainable agriculture researchers and advocates suggest a promising future for this area of science and technology. This said, the sustainable agriculture thrust exhibits several weaknesses and vulnerabilities that suggest the need for critical self-reflection on the course that the notion of agricultural sustainability has taken and on the current sociopolitical dynamics within the natural science and social science sustainability communities. Perhaps most important, it is arguably fair to say that despite the considerable efforts that have been devoted thus far to sustainable agriculture, there are no unambiguous biological or social–institutional breakthroughs or innovations that have yet emerged or are currently in the pipeline.[1] Institutionally, federal policies remain nearly as unsupportive of sustainable agriculture as they were when sustainable agriculture emerged as a popular area of research during the mid-1980s. Funding for sustainable agriculture research is modest and vulnerable, and must be fought for during each state legislative session and federal appropriations period. Many suspect that large numbers of land-grant administrators see sustainable agriculture more as a political concession to public interest groups than as a serious area of research.

In a sense, then, one purpose of this chapter will be to understand some of the factors that have caused sustainable agriculture to have been limited in its stature and accomplishments up to this point. This chapter, however, will not be devoted entirely, or even mainly, to explaining the modest gains that have been achieved, nor will it aim to be a comprehensive assessment in this regard. In particular, the chapter will not treat the national political–economic constraints to sustainability in an exhaustive manner (see, e.g., NRC, 1989, for an important treatment of this topic), though I will take up several of what I consider to be neglected considerations in the political economy of sustainability and sustainability movements. Instead, I will concern myself primarily with constraints on agricultural sustainability owing to the fact that the sustainability community increasingly faces, self-consciously or not, several dilemmas relating to its being a component of a movement that is simultaneously rooted in (natural) science and implicated in the current milieu of international environmental activism. The approach to be taken in this chapter will be to apply the basic framework of the sociology of science and scientific knowledge to this problem. The focus will be primarily on sustainable agriculture and agricultural sustainability in the North American and other Western contexts, though many of the observations are germane to related issues such as "sustainable development" in the Third World.

WHICH SOCIOLOGY OF SCIENCE AND TECHNOLOGY?

As anyone familiar with contemporary trends in the sociology of science and technology can attest to, calling on the tools and resources of this field of study begs the question of which variant one wishes to embrace. This field has exhibited enormous shifts over the past two decades. In the 1960s and as late as the early 1970s, the sociology of science in North America was virtually coterminous with the functionalist sociology of scientists and scientific careers [typified by Merton's (1973) classic text, *The Sociology of Science*]. Initially stimulated by Kuhn's (1962) *The Structure of Scientific Revolutions,* there emerged a number of provocative arguments that the Mertonian sociology of science was incapable of treating the nature of the knowledge produced by scientists as a explanandum in its own right (e.g., Mulkay, 1969). As Woolgar (1988:26) has argued, "scientific knowledge, in this [Mertonian] view, is not amenable to sociological analysis, simply because it is its own explanation: scientific knowledge is determined by the actual nature of the physical world."

By the mid-1970s the sociology of science was in the midst of a rapid shift, in which the Mertonian emphasis on the normative or occupational structure of science was being supplanted by the "social construction" of scientific knowledge as the key problematic (e.g., Bloor, 1976; Mulkay, 1979; Latour and Woolgar, 1979). Science was increasingly conceptualized as being a relatively ordinary sphere of ideation, symbol construction, persuasion, "representation," and social knowledge, to be analyzed in a sociology of knowledge framework in much the same manner that one would analyze "nonscientific" beliefs, ideologies, and symbols. As the "strong program"—of seeing the sociology of science as being a mere branch (rather than a related and coequal intellectual cousin) of the sociology of knowledge (Bloor, 1976)—and other related projects became more popular, sociologists increasingly approached science by way of *relativizing* its knowledge claims, through the application of frameworks of social constructionism and interpretive sociology (see, e.g., Collins, 1983; Gilbert and Mulkay, 1984; Knorr-Cetina, 1981; Latour and Woolgar, 1979; Mulkay, 1979; Pinch, 1986; Collins, 1985; Woolgar, 1988). In particular, the new sociology of science prided itself on having demonstrated empirically that logic, reason, and rules of scientific method are as much or more post hoc rationalizations of scientific practices and knowledge claims than they are the determining force behind scientific knowledge creation (see the overview by Woolgar, 1988:Chapter 3).

The relativist/constructivist turn of the sociology of science and technology over the past two decades has been a step forward, since it has provided the tools for demystifying science, and for forging recognition of the fact that scientific knowledge claims being "interested" or being shaped by extrascientific social forces is normal rather than exceptional. It is important to note, however, that there are several interrelated lacunae in both Mertonian functionalism and the relativist or social constructionist traditions: both have tended to stress "science," and accordingly to study mainly the paradigmatically "basic" (and most prestigious) physical and related biological sciences (such as molecular biology); both have tended to deemphasize "technology"; both have tended to see science and technology mainly as social practices or as ideational phenomena, rather than as material-productive forces;[2] both have tended to see science as a relatively self-contained institutional ensemble; and both have been limited in exploring science–society relationships (Brante, 1986).[3] This is particularly the case with respect to (publicly funded) science as a form of state intervention. In both the Mertonian and relativist/constructivist formulations the concept of the state seldom appears.

At the same time, outside of the traditional mainstream of the sociology of science there is a tendency to reduce science to being largely, if not entirely, a material or productive force. This, as I will note later, is particularly common in sustainability circles, but is a far broader tendency. The two most common modes of approaching science from this standpoint—induced innovation in economics,[4] and the political economy of science and technology—have traditionally left little room for conceiving of science and scientific groups as institutions with their own dynamics, and have tended to presume that science can be examined by assuming the instrumental rationalism of individual scientists and their scientific institutions.[5]

These patterns are explored further in Table 1, which provides a typology of what can be termed "metaperspectives" in the sociology of science and technology. The basis of the typology is two dimensions of science and technology, both of which reflect dualities that are *a priori* reasonable views of the nature of science and technology and the role of science and technology in society. The first dualism is that which I call "deference" toward vs. "demystification" of science and technology. A deferential view toward science is based on science being viewed as either intrinsically good, on account of its distinctive decision rules—being rational, "scientific," universalistic, disinterested, etc.—or as being, in principle, socially desirable if organized appropriately or rationally. Demystification of science, by contrast, involves *relativizing* scientific knowledge claims or scientific accomplishments—as being social constructions, or as being of social scientific interest mainly because they are derivative of interests, political–economic relations, ideology, class structure, gender relations, and so on. The second dualism is that of science and technology as a normatively shaped social "practice" or

TABLE 1.   A Typology of Metaperspectives in the Sociology of Science and Technology and Major Exemplars

Science and Technology Are Understood Mainly in Terms of	Underlying Orientation toward Conceptualizing Science and Technology as Social Activities	
	"Deference"[a]	"Demystification"[b]
Science and technology as "practices," or as normatively specified or ideational phenomena	Mertonian Sociology of Science	The "strong program" (relativism, social constructionism, "interpretive") and not-so-strong versions of relativism; Latour, Woolgar et al.
	Mature Sociology of Science and Technology	
Science and technology as material-productive forces	Induced innovation; most science and technology policy studies; political economy of technological change; technology assessment	Political economy of science (J. D. Bernal, Lewontin and Levins, Noble); (most) Nelkin-style studies of technical controversies

[a] Source is taken at face value, as a unique social sphere to be distinguished from the social relations and practices of "nonscience." Science is said to have unique, desirable decision rules (universalism, disinterestedness, rationality, and so on), and as such is to be seen as being demarcated from nonscience, and/or science is treated deferentially, as being intrinsically good or as being, in principle, socially desirable if organized appropriately.
[b] Science is seen as being an essentially "ordinary" sphere of social life, to be analyzed in terms of the sociological concepts employed to depict other realms of social relations, power relations, ideology, etc. The lack of demarcation of science from nonscience is emphasized. The focus is on demystifying science, as being fundamentally ordinary or as being one of many social modalities by which interests are expressed or served.

ideational sphere on one hand, and science and technology as a material-productive force on the other.

Table 1 points to two conclusions about these essential dualisms of science and technology in society. The first is that the major prevailing perspectives in the field Mertonian-style functionalism, relativism/constructivism, induced innovation, political economy of science and technology, and so on are incomplete, since each is based in only one quadrant of the typology. The second is that a mature sociology of science and technology must transcend each of these dualisms, so as to straddle each of the quadrants of the typology and incorporate their insights.

While this is not the time and the place to stake out the major postulates of what this mature sociology of science and technology will look like,[6] the discussion that follows will be mindful of the need to premise a sound sociology of science and technology on a more diversified approach than has typically been the case. There are two basic messages of this chapter. One is that we must look beyond, while not disregarding, the reality of sustainability and sustainability research as a (potential) material-productive force. I will want to explore this dimension in the portion of the chapter that follows by taking sustainability to be not merely a social movement concern and sociopolitical demand, but also a scientific knowledge claim. The second dimension, though less well explored here, is that the notion of agricultural sustainability has been shaped in several respects by the interactions between dominant global and national social structures and the social structure of science and technological change.

AGRICULTURAL SUSTAINABILITY AS A MOVEMENT SYMBOL: ITS RISE AND CONSTITUTION

It is useful to begin with some observations about the formation of the (academic and quasiacademic) agricultural sustainability community over the past decade or so. The establishment of this scholarly community has been strongly focused on the elaboration and diffusion of the "sustainable agriculture" construct or symbol, which for purposes of this chapter can be considered to be simultaneously a social movement symbol and demand and a scientific "knowledge claim."[7] There have been a number of notable attempts to chronicle the rise of the notion of sustainability, such as W. M. Adams' *Green Development* (1990). Most such efforts, however, have been concerned with "sustainable development," the international-development variant of sustainability, and have given little attention to the relationships between what I will refer to here as the "domestic" or Western sustainability community and the formation of the sustainability symbol itself. Also, these analyses of the rise of sustainability tend to view sustainability relatively uncritically, equating its ascendancy with a sui generis phase of enlightenment in the context of increasingly severe environmental problems and of increasingly more strongly held green sentiments in society at large.

It is my argument that this received rationalist/evolutionary view of the rise of the sustainability symbol has a number of major limitations. In particular, this view ignores a number of crucial political–economic antecedents, concomitants, and implications of the sustainability movement and "sustainability science" in general, and their agricultural sustainability variant in particular. Chief among them is the fact that agricultural sustainability is simultaneously (1) a social movement ideology, of the agricultural sustainability movement that (2) has its fortunes closely tied to the international environmental movement and (3) premises its appeals and program on the legitimacy afforded by (sustainability) science and scientific research.

The sustainability movement can be seen as the most recent stage of modern agrarian struggle or protest. Accordingly, sustainability shares some important characteristics with earlier phases of agrarian struggle (e.g., "family farmism" in the United States and "basic needs" in Third World contexts). Both were mainly intellectually generated and/or framed movements. That is, while they expressed—or represented and articulated—concerns of ordinary people, they were, in the final analysis, intellectual formulations. This tendency is by no means unique to sustainability; modern social movements are increasingly characterized by the tendency of movement officials and supporters to be a career intelligentsia that represents those persons who experience the particular problems or who have aspirations that are the focus of the movement. These movement organizations also generally garner their resources from similar persons or from institutions such as philanthropic foundations (rather than the movement generating its resources mainly from the persons whom the movement speaks for or represents). Thus, in most modern social movements, those persons who experience the problems of concern to these movements tend to be uninvolved in either directing the movement or providing the resources employed by the movement. Thus, the fact that the organizational backbone of the sustainability movement consists largely of an upper-middle class of movement officials and supporters, and of a cadre of natural and social scientists in universities and related educational and research institutions, is quite unexceptional.

Both the ("classic") 1960s/1970s agrarian protest movements and the modern agricultural sustainability movement have also been subject to what I have referred to elsewhere as establishment appropriation of progressive symbols (Buttel and Gillespie, 1988). These symbols tend to proceed through a "life cycle" of establishment appropriation as follows. Initially, the progressive symbol is crystallized and framed by dissident or oppositional groups as a challenge to the legitimacy of dominant institutions and their practices.[8] If the groups and symbols involved are sufficiently efficacious in mobilizing supporters and threatening to dominant institutions, these institutions will attempt to respond to these challenges by "appropriating" or embracing the symbol themselves. In so doing these dominant institutions—such as the World Bank and other international development assistance agencies, the land-grant system, federal agencies, and so on—are typically able to demobilize the movement. This is normally done either through trivializing the symbol (e.g., the World Bank's having translated demands to deal with unequal development into a technocratic formula of "growth with equity" during the 1970s; Wood, 1986), smothering it through bureaucracy (e.g., USAID's deft refraction of demands for greater women's participation in development in the late 1970s and early 1980s, by creating a small, impotent women and development office, with inadequate resources to accomplish its goals), or by otherwise creating the impression that the problem has been solved (as in the establishment of the Environment Division at the World Bank in the late 1980s). Subsequent to establishment appropriation, activist groups opposing the priorities of agricultural development institutions find themselves facing a difficult dilemma: should they continue to hammer away at the issue, despite being hampered by the fact that the dominant institutions against which they have struggled have superficially embraced and appropriated their movement symbol? Or should they move on to other issues, movement symbols, and demands in order to sustain their ability to keep the pressure on these institutions?

Sustainability, however, differs from previous phases of agrarian struggle in several ways. Earlier phases of agrarian struggle were crystallized and mostly promulgated by social scientists and by those whose strategies and visions were at least implicitly informed by social science or humanistic perspectives or reasoning. Moreover, the persons who catalyzed earlier phases of agrarian protest often were virtually antiscience in their orientations, typically viewing modern (natural) science as a handmaiden to dominant classes and their interests (see Yearley, 1991a,b, for a general discussion of the ambivalent relationship of environmentalism and science). There generally was, in fact, considerable tension within the professorate, between natural scientists and social scientists. Now, however, the impetus for and the legitimacy of sustainability come from natural science, natural scientists, and natural science data. Social science and scientists have mostly followed (or, in some cases, their voices have not been heard). The struggle over sustainability, as currently constituted, is largely a struggle *within* the natural sciences, with members of the social sciences aligning themselves with one or the other faction within the natural sciences, usually along ideological lines (with the bulk of the sustainability community consisting of persons from the left). Nonetheless, for both sides, the appeals and claims made and mainly buttressed by natural science data, along with claims about the scientific superiority of their natural-scientific theories and approaches (e.g., sustainable approaches being nonreductionist, more capable of "systems reasoning").

Another departure of the sustainability movement from those of the 1970s and the early 1980s is that the rise of sustainability represents the first instance in which post-1960s agrarian struggle has not been focused around social justice per se (in relation to, for example, the needs and interests of "small farmers," the "family farm," or farm workers, or as criticism of industrial or corporate agriculture). This does not, of course, imply that all agricultural sustainability activists and supporters are uninterested in or unconcerned about the distributional, equity, and class dimensions of agriculture. For example, some in the sustainability community, particularly those who are most radical or activist, were actively involved in presustainability agrarian protest, and now carry on their research, activism, and educational activities in sustainable agriculture out of perceived continuity with their earlier work. Others presume that achieving environmental sustainability will lead to other desirable outcomes for agriculture—such as curbing industrial agriculture, attenuating the exodus of family farmers from the land, improving working conditions for farm workers, and creating a more equitable and effective food distribution system. Distributional, equity, and class issues, however, are typically in the background of agricultural sustainability activities, and for many enhancing the environmental sustainability of agriculture (or of "development") is an end in itself.

The "scientization"of protest or opposition within the sustainability movement has strong parallels in other environmentally related realms, particularly "global change" (e.g., Yearley, 1991a). In fact, while agricultural sustainability had a certain political and scientific momentum in the mid-1980s, this notion and program of research were propelled to particular prominence *because of, and through,* the spectacular growth of scientific knowledge and public concern about "global change" and "global warming." The rise of sustainable agriculture would not likely have occurred in the absence of mid- and late-1980s growth in environmental sentiment, which was largely anchored in rising popularization of and public concern about global environmental problems, particularly global warming, stratospheric ozone depletion, and loss of biodiversity (Yearley, 1991b). Sustainable agriculture also benefited from the fact that the international environmental movement, recognizing that Third World cooperation on international environmental problems would require concessions to their growth and development aspirations, embraced the notion of "sustainable development" (in preference to notions, such as the "limits to growth," that were roundly criticized by Third World officials, activists, and scholars in the previous decade). WCED (1987), and the many privately circulated draft documents that preceded its publication by several years, played an extremely important role in legitimating the notion of "sustainability" within international environmental circles. Though major (particularly U.S.) environmental groups had shown little interest in production agriculture (other than with respect to the health effects of agricultural chemicals) until the 1980s, the growing fascination with the notion of sustainability led these groups to focus growing attention to production agriculture technology and federal farm policy. U.S. environmental groups, for example, were integral to the LISA appropriation and passage of the Conservation Reserve Program in the 1987 farm bill.

Environmental groups had been aware for several years about the growing speculation within some quarters of the atmospheric science community that the world was in the early stage of a potentially profound global warming trend—one widely discussed concomitant of which would be rising temperatures and instability of rainfall in the temperate agricultural breadbaskets. The drought of the summer of 1988 prompted environmental groups to "go public" about global warming and greenhouse gases apparently despite some misgivings (such as being aware of the fact that the need for strict conservation would be difficult to sell in the consumerist 1980s and that reduction of greenhouse gases might virtually compel greater reliance on nuclear power). Nonetheless, there ensued a global warming bandwagon effect, which up to this time has been a major boon to environmental mobilization in general, and to agricultural sustainability efforts in particular.

A global warming- and global environmental-change-centered environmentalist program has several advantages from the standpoint of environmental activism. The global warming and global environmental change message can be very influential with the public and with policymakers, since it can be buttressed with "dread claims," e.g., about the massive biospheric disruption, coastal inundation, destruction of the agricultural productivity of the temperate agricultural breadbaskets, etc., that will occur if we fail to act now to reduce pollution of the atmosphere. Global warming and the associated concerns (e.g., rising skin cancer rates due to depletion of stratospheric ozone, biospheric disruption due to loss of tropical biodiversity) can serve as a comprehensive, *overarching* justification for a very lengthy agenda of environmental goals. The need to reduce industrial pollution, achieve greater energy conservation, conserve tropical rainforests, utilize more environmentally sound agricultural production practices, and so on can all be justified through one overarching imperative—to stem global warming rather than having to justify each on its own particular merits. This "global packaging" is thus attractive, since it obviates the need for environmentalists to struggle to achieve multiple goals in many places simultaneously. Persuasive claims about global warming and global change can also become a strong rationale to override politics-as-usual at the international and national (and often the subnational) levels. The extraordinary threat of global warming and global biospheric disruption can be said to require extraordinary measures. This can increase the likelihood of convincing many countries and groups that they must set aside their immediate, narrow political and economic interests if the needed measures—such as international conventions and protocols on CFC emissions, greenhouse gas emissions, forest policy, and biodiversity conservation—are to be achieved.

Nonetheless, for all practical purposes, "sustainable" and "sustainability" were given particular legitimacy, and were imprinted most poignantly in the minds and vocabularies of scientists, politicians, and so on, because of the urgency that grew up about responding to the threats that global environmental change posed to human living standards and well-being. Accordingly, the sustainability movement may in the long term distinguish itself by being more durable and enduring than previous (social justice) foci of agrarian struggle. In other words, sustainability may be able to avoid the cycle of establishment appropriation of progressive symbols referred to earlier, particularly if the larger "greenhouse legitimacy" of sustainability holds. It will also likely prove to be the case that the agricultural intelligentsia is more effective in mobilizing concerns rooted in the authority of natural science than it is at mobilizing and representing social justice concerns. In part, this is because the intelligentsia is relatively affluent; because it is a relatively privileged group, its long-term interests will not lie in subordinating its discourses to the interests of the poor and disenfranchised. In addition, and most importantly, the socioeconomic and fiscal bases of social justice politics as we have known it in the post-World War II period is slowly unraveling, due to the declining position of the industrial working class, and to longstanding economic stagnation that has eroded the fiscal basis of the welfare-state. One on the central characteristics of the politics of the emerging era of post-Fordism or neoconservatism [or what is alternatively called "postindustrialism" (Block, 1990) or "disorganized capitalism" (Lash and Urry, 1987)] is that social justice issues have lost their force, at least at the national political level during the current phase. One of the concomitants of post-Fordist politics and the demise of social democracy is that we typically see environmental and sustainability discourse being substituted for social justice discourse—or at best sustainability claims being a disguised means of keeping social justice on the agenda. As I will stress later, however, movement recourse to the authority of natural science has its limitations on a number of grounds.

THE DYNAMICS OF SUSTAINABLE AGRICULTURE AS A KNOWLEDGE CLAIM

Integral to the relativist/contructivist position in the sociology of science and technology is the notion that scientific practice necessarily involves "representation"—a process by which scientists seek to convincingly and authoritatively persuade others that their particular interpretations of reality (and of the implications of these interpretations for policy and practice) are valid (e.g., Lynch and Woolgar, 1990). Scientific knowledge is thus to be treated as "scientific beliefs" (in a manner much like other social beliefs, without judgment as to the truth or falsity that might eventually be assigned to such beliefs) or as "claims to knowledge." Scientific beliefs and claims to scientific knowledge are seen to be shaped by factors such as scientists' own social interests (e.g., in funding, career advance), their ideological commitments, their networks, their rhetorical skills, selectivity in the generation and transmission of data in the processes of abstraction and the symbolization of results (in tables, graphs, and pictures), and the negotiation of formal and informal agreements (or disagreements) over the meaning of the laboratory data. In addition to students and other scientists, the groups at which persuasion, representation, and claims to knowledge are aimed include powerful political and economic officialdoms, the media, the "public," and so on.

The relativist/constructivist posture does have methodological shortcomings if its critique of rationalism is pushed to the point of reducing science to mere rhetoric and discourse. Radical relativism/constructivism may also lend itself to a moral neutrality that can yield paralysis in conceiving of ways in which scientific institutions can be restructured for human benefit (Busch et al., 1991; Woodhouse, 1991). Nonetheless, if employed judiciously, this perspective can enable one to develop certain insights that are not possible if science is approached from a rationalist (or, as in Table 1, a "deferential") viewpoint in which truth or falsity of knowledge is the overriding criterion or explanandum. Here I will make a few such observations that can be derived from a relativist/constructivist view of agricultural sustainability, which is at the same time mindful of sustainability science and technology as being material-productive forces.

Sustainable agriculture, rooted as it is in the special authority of natural science as form of argument, can be treated as a scientific knowledge claim. It is a very general, and at the same time a complex, knowledge claim for several reasons. First, agricultural sustainability is rooted in a large number of scientific disciplines of the sustainability community—particularly ecology, conservation biology, the environmental sciences, and "agronomy" (broadly construed) as well as the social sciences and humanities. Second, as noted earlier, agricultural sustainability has over the past 4 or so years become a variant of greenhouse-related "motherhood environmentalism," within which "sustainability" enjoys near-icon status.

Third, while scientific knowledge claims by their nature involve implicit and explicit "nonscientific," moral, or ideological components, this is particularly the case with sustainability [see, e.g., a parallel assessment with respect to the "Brundtland Report" (IWCED, 1987) by Timberlake, 1989]. These "nonscientific" or moral components include not only those of modern internationalized environmentalism (as depicted, for example, by Yearley, 1991a), but also strong doses of agrarianism and legacies of presustainability agrarian protest and of the farmer-client- and productivist-oriented *Weltanschauung* of the land-grant universities. That fact that nonscientific components are woven into the fabric of agricultural sustainability does not, however, diminish the fact that the sustainability community buttresses its arguments primarily via the special authority of science. Thus, while agricultural sustainability was propelled to prominence by growth of concern about global change in the mid-1980s, it received an enormous "jump-start" following the publication of the *Alternative Agriculture* report of the Board on Agriculture of the NRC, which as part of the National Academy of Sciences enjoys particular prestige and influence.

Perhaps the most obvious concomitant of sustainable agriculture being rooted in science is the stress given within both the movement proper and in the sustainable agriculture research community to *scientific research, scientific research institutions, and agricultural research policy*. This emphasis on influencing science and research policy ironically reinforces the privileged role of science in shaping agriculture that was criticized in agrarian opposition movements of the social justice variety during the 1970s and early 1980s. This tendency also reinforces the shorthand postulate that technology shapes social structure, rather than vice versa, which in turn buttresses a largely technocratic conception of sustainable agriculture. In addition, the natural science and social science wings of the sustainable agriculture community approach agricultural sustainability mainly from the perspective of science and technology as material-productive forces, as presented in the typology in Table 1. This observation also pertains to the community's view of "conventional technology," as having been a relatively autonomous force that led American agriculture in undesirable directions during the post-World War II period. Put somewhat differently, insofar as the sustainability community's work remains so focused on research institutions and research politics, it has essentially come to be based on a view that desirable ("sustainability") science can and must squeeze out the bad.

As noted earlier, in natural science sustainability quarters there has been a tendency toward promulgating a broad—and thus diffuse and, I believe, largely ineffectual—umbrella over an essentially conflictual and contradictory reality: Scientists who do, or wish to do, agricultural sustainability research are quite diverse, and have quite different professional interests, worldviews, research styles, and relationships with farmers. On the one hand, the largest group in the natural science sustainability community is that of scientists who are trained in the traditional agricultural disciplines, employed in land-grant university colleges of agriculture, are interested in reorienting their departments and disciplines, who mostly do applied, problem-solving research, who typically have strong commitments to direct service to ("family") farmers, and who tend most to embrace a notion of science as contributing to an instrumentally rational productivity-oriented framework. This group also includes a good number of molecular-level scientists—e.g., the growing role of molecular biology in IPM research (Palladino, 1989). On the other hand, there is a significant cadre of scientists, many of them ecologists, who are mostly trained outside of land-grant colleges of agriculture, are more concerned bout staking out "space" for applied ecology and theoretical work in agroecology than they are preoccupied with enlightening agronomists and entomologists, who are mainly oriented to relatively basic or fundamental research, who do not necessarily see it as their role to do land-grant-style client-oriented work, and who tend less to see their work as contributing to an instrumentally rational agricultural order. While there are some instances of these two groups of scientists working together closely and there exist "mixed types," this probably occurs less often than might be imagined. Each group, given their interests and ideological worldviews, and given their professional–political commitments and the broader political–economic context, appears to have chosen to submerge their potential disagreements and to unite in a broad, but diffuse and mushy, coalition.

One consequence of this coexistence is that umbrella-type conceptions of agricultural sustainability continue to be promulgated. In part, this no doubt reflects mutual interest in prompting funders to provide broader support for sustainability research, as well as because the recent political–economic context has not been conducive to more sharply focused, let alone more uncompromising, definitions of sustainability. Another consequence has been that sustainable agriculture has been largely defined as a LISA-type approach, in which reducing and rationalizing the use of chemicals have been grafted onto the knowledge base of otherwise conventional agronomy.

How and why has the LISA conception of sustainable agriculture become the predominant one? There were a number of factors that contributed to this development. The LISA initiative (of the 1985 farm bill), for example, was largely a product of mid-1980s farm crisis, and thus of the growing pressure on the land-grant system and USDA to respond to the immediate economic needs of the victims of the crisis. Thus, the LISA initiative not only tended to direct the bulk of these funds to land-grant researchers, but also involved a very strong emphasis on *applied research* and on use of the traditional cooperative extension technology transfer network. Also, the cadre of land-grant agronomists, horticulturalists, and related agricultural scientists is much larger than that of agroecologists, tilting the center of gravity of sustainability science in the direction of the former. Finally, the LISA approach also lent itself to political compromise between the sustainability community and the agricultural establishment. This was particularly the case insofar as sustainability has come to be defined in terms of bolstering productivity (through reduction of use of purchased inputs) within a larger capitalistically or instrumentally rational framework.

Empirical evidence on the structuring of sustainability within the land-grant system suggests the crucial role that the land-grant research community has played in the formation of the sustainability community. Survey data collected among a random sample of land-grant biological scientists in 1989, the apogee of popularity of sustainability engendered by publication of the NRC (1989) report, show that scientists who support or do research on sustainability are drawn disproportionately from the traditional applied, client-oriented, productivist sector of the system (Buttel and Curry, 1992). It is particularly noteworthy that agronomists, who one decade earlier tended to be ambivalent about or critical of alternative agriculture (Busch and Lacy, 1983), were found in this 1989 study to be the most prosustainability discipline in the land-grant system.

THE VULNERABILITY OF SCIENTIZED ENVIRONMENTALISM TO "DECONSTRUCTION"

Agricultural sustainability is a good example of a symbol that serves simultaneously as a scientific concept (and knowledge claim) and as a movement ideology (of environmentally related movements). The modern environmental sciences and environmental movements, both broadly construed, exist in a state of mutual dependency and contradiction. The roots of mutual dependency are relatively obvious: The environmental movement depends on persuasive environmental science knowledge claims, and the environmental sciences depend on a politically persuasive environmental movement.

As Yearley (1991a,b) stressed, a great many modern social movements, including but not limited to environmentally related movements, find that the discourse of science is a powerful means of promoting their agendas. At the same time, scientific knowledge in general, and that deployed by movements in particular, is vulnerable to "deconstruction." Those who may seek to deconstruct scientific knowledge include not only upwardly mobile, opportunistic scientific researchers who see that debunking conventional wisdom can earn them a reputation of being innovative, but also the political opponents of the movements that deploy this knowledge (Jasanoff, 1992).

Sustainable agriculture knowledge claims are vulnerable to deconstruction on a number of counts. For example, many agricultural scientists remain vocal opponents of decreased use of agrochemicals, feel that the integrity of their earlier scientific work is under assault by those who advance sustainability claims, and accordingly are motivated to discredit sustainability science. The discourse of neoclassical economics (e.g., Knutson et al., 1990) can also be a powerful weapon of deconstruction, by generating counterclaims that the costs of implementing agricultural sustainability will be tangible and substantial and the benefits largely intangible, modest, and uncertain. Further, given the strong federal policy disincentives to sustainable agriculture practices, these practices may not deliver the goods in farmers' fields, opening the sustainability research community to criticism that their technologies are not practical or profitable.

There are also growing signs of restlessness within the scientific community about the empirical basis of much of the received knowledge about global warming and global change, to which agricultural sustainability has been symbolically associated for nearly a half decade. In particular, there is concern about whether global circulation models are sufficiently robust to make confident predictions about the global warming trend. Much the same criticism has been raised about parallel claims concerning the biosphere-threatening impacts of the loss of biodiversity (Mann, 1991). While some of this contradictory scientific evidence is being generated by persons who are politically motivated to diminish the influence of environmental groups (e.g., Marshall Institute, 1989), a good share of this concern comes from persons who are not so motivated. Bryson (1990), for example, one of the pioneers of atmospheric modeling and early prophets of CO_2-induced global warming, nonetheless maintains that evidence from computer models of atmospheric pollution and warming at this point is not sufficient to justify claims of the inevitability of a major biosphere-threatening "greenhouse effect." Thus, it is unclear whether and how long the environmental movement will be able to sustain the major stylized facts of global environmental change and global warming. The questioning, and possible undermining, of the case for global warming would very likely cause the rest of the greenhouse-linked agenda—sustainable development, biodiversity and rainforest conservation, and perhaps sustainable agriculture—to be diminished in tandem (Yearley, 1991a).[9]

SUSTAINABILITY CLAIMS AND MOVEMENTS IN POLITICAL-ECONOMIC CONTEXT: SOME SPECULATIVE OBSERVATIONS

As noted earlier, there is a general view in much, if not most, of the sustainability community that greening and sustainability have emerged relatively autonomously as a result of public recognition of the environmental excesses of Western development during the post-World War II period. This was an era that many social scientists refer to as "Fordism" (following Aglietta, 1977), due to the prominence of large-scale, mass production industry and trade unionism in its social structures, as well as social Keynesianism, the welfare-state, and social democracy in its politics. It is unmistakable, however, that the rise of post-Vietnam environmentalism ("greening") and the neoconservative cast of political arrangements in the West have emerged in tandem. It is interesting to note that there has been little attention to, much less research on, the nature of the relationship between the two. Why has this association been the case, and with what implications for green movements and sustainability? I believe we do not yet have any clear idea as to the reasons this relationship exists, and even less about the implications for the future of greening and sustainability.

There are two basic ways in which we can answer these questions. The first is that "greening" has been related to neoconservative politics by being an expression of a profound, enduring, and potentially transformative "new social movements"-like opposition to neoconservatism's fundamental tendencies.[10] Traditional left politics, focused on enhancing the status of the working class and the poor through the welfare state, protective labor legislation, regulation of capital, and so on, has been exhausted due to international capital mobility (which undermines the integrity of national economies), economic and fiscal crisis, and the decline of the industrial working class and trade unionism. In its place has arisen a new form of left politics based on new social movements concerns [the list of which, as Scott (1990) stresses, usually includes ecology, feminism, and peace]. Given the declining position of the industrial working class in the overall class structure, and thus of the key social base of left politics in Fordism, new social movements mobilization is better suited to presenting a credible opposition to Thatcherism and Reaganism than are redistributive-oriented labor/working class parties and trade unions. Most optimistically, it might be contended that neoconservative politics' Achilles heel is that it generates, in a significant structural way, green and related new social movements opposition, and that this opposition is formidable because it can mobilize large numbers from many classes and social groups.

A second, more sobering, perspective is that greening is a hegemonic—a "kinder, gentler"—symbol of these conservative regimes themselves, or a form of opposition that can readily be accommodated within neoconservatism. Some evidence in support of this contention can be found in the fact that neoconservative Western regimes (save for the U.S. government) have not overtly resisted demands to respond to the global environmental change and global warming claims of sustainability's first-cousin movement, the international environmental movement (Buttel et al., 1990). The coexistence of international environmental activism with the "structural adjustment" emphasis of the international development agency complex is further evidence in this regard (Hawkins and Buttel, 1991). In particular, international environmentalism has accommodated itself in a surprisingly thoroughgoing way to the free-market resurgence of the past decade. The movement has relied heavily on working with the World Bank/IMF establishment, such that the lever of developing-country vulnerability to termination of bridging, adjustment, and project loans is employed to implement conservation policies there. With very few exceptions, the international environmental community has pulled its punches in criticizing the "debt regime" and in actively advocating debt forgiveness, which would no doubt be the single most efficacious means of achieving simultaneously development and conservation. This shortcoming is particularly revealing insofar as oppressive debt loads virtually compel unsustainable, extractive accumulation in the Third World, and reinforce mass poverty and the share of developing-country environmental degradation that result from the desperate measures of the poor. Nonetheless, the important question remains: Will sustainability, as one of the core symbols of new social movements-type green sentiments, prove to be a mere legitimating counterideology of modern Thatcherism and Reaganism? We do not yet know, and must endeavor to find out.

A related question concerns why there has not been a demonstrable tendency toward effective corporate veto of agricultural greening and sustainability efforts. To be sure, chemical companies would like to do so, and have tried to do so (often in concert with commodity groups), with respect to the agricultural sustainability movement. But in the main they have not been very effective, and, as in the case of the National Agricultural Chemicals Association, have even attempted to demonstrate that they accept the goal of sustainability and wish to contribute to it. Why? Is it because sustainability has had enough of a bandwagon character—and has been sufficiently powerful—so that chemical companies have been neutered? Or is it that sustainability is a relatively "harmless" form of opposition to this and related industries? Or is it that some farsighted chemical companies recognize that bulk agricultural chemicals are no longer very profitable, that biotechnology products will ultimately supplant the synthetic chemical pesticides that are currently used, and that struggling to preserve chemicals markets is not a good business tactic? Again, I do not think we are yet able to answer these questions.

A related matter is that of the social-mobilizational capacity of sustainability over the long term. Sustainability is essentially an environmental ideology. Accordingly, it faces the dilemma of most extralocal (non-NIMBY) environmentalism, in that it has no "natural constituency" or self-evident "bearers." By contrast with most other left movements, such as the civil rights movement and feminism, in which each has a clearly identifiable core constituency, extralocal environmentalism does not. Environmentalism is in the interest of everyone, and thus of no one.[11]

In particular, agricultural sustainability has been advanced primarily as "consumption politics," rather than a production politics, which has tended to undergird the agendas of most conventional and historically successful national-scale social movements (Lash and Urry, 1987). What seems to unify and sustain the agricultural sustainability movement more than any other factor is its implications for "consumption:" food quality, health, landscape, and so on. While some of these consumption concerns involve privatized consumption, most relate to collective consumption (i.e., "public goods," which would accrue to all persons regardless of whether they favor or resist sustainability). That is, agricultural sustainability is advanced as a sociopolitical demand and knowledge claim mainly by persons for whom the benefits of movement success, or losses caused by movement failure, are not substantially greater than those of their opponents (save for chemical companies). This is particularly the case among the privileged groups within which the bulk of pro- and antienvironmental combat occurs. But how sustainable can sustainability mobilization be when it rests largely on consumptionism?[12]

CONCLUSIONS

The formation and rise to prominence of the "sustainability community," located primarily in universities and other research and educational institutions and consisting of a multidisciplinary intelligentsia, have perhaps constituted the principal structural change in the U.S. public agricultural research system over the past half decade.[13] But while agricultural sustainability enjoys obvious popularity, it remains a subordinate focus—i.e., an appropriated progressive symbol—within agricultural science and the land-grant system. The rise of sustainability, along with other challenges to the environmental track record of agriculture in the political sphere, has enabled the proenvironment wing within the research community to move toward parity with the forces that defend or wish to extend the chemical-intensive trajectory of U.S. agricultural development.

Thus, there has been an ongoing restructuring of political boundaries within the agricultural research community (broadly construed, to include a range of disciplines, as well as scholars in land-grant universities, non-land-grant universities, and nonacademic research and advocacy institutions). The coordinates of cleavage and contention have increasingly shifted from the traditional ones of natural scientists vs. social scientists, and of production scientists vs. "impact scientists."[14] There has emerged a new configuration in which social and natural scientists who share common (proenvironmental) worldviews and interests see themselves in opposition to their social science and natural science counterparts with contrary ideological commitments and professional interests. Struggles and conflicts over sustainability, ecological agriculture, the environmental track record of postwar agricultural research, and the environmental responsibilities of agriculture are the most important focal points of contention within agricultural research institutions and agricultural science disciplines right now. This new configuration is typified by the recent tussle between those who support the recent NRC (1989) *Alternative Agriculture* report, and those who are persuaded by CAST's (1990) parallel volume, *Alternative Agriculture: Scientists' Review,* which in the main is an uncompromising critique of the NRC report and of the notion of sustainable agriculture.

In addition to whatever analytical insights the approach of this paper might offer this analysis might also be useful in enabling the sustainability community to be more reflexive. The sustainability community should be willing to reexamine its assumptions, beginning by recognizing that all science, including that which we like and that which we do not, is affected by social and political–economic parameters. These include factors such as scientists' and professional groups' interests, the role of scientists qua scientists and as representatives of "the scientific community" to "the public" and in the halls of power, and the interplay of ideology and knowledge claims. The dynamic of vulnerability of scientific knowledge claims to deconstruction is particularly important to understand.

One of the key issues with which the sustainability community must grapple is that of the implications of the dominance of the LISA approach. There is, to be sure, much to be said for the typical LISA approach. It is generally based on relatively well-established lines of research and knowledge, often lends itself to on-farm trials, and accordingly has the potential to be delivered relatively quickly to farmers. Since organic and alternative farmer groups have long been among the more vocal clienteles pushing for this type of research and are understandably impatient for the land-grant system to deliver, this approach is best suited to satisfying those who want results yesterday—or tomorrow at the latest.[15]

This said, the fact that the land-grant sustainable agriculture program is quite strongly focused on LISA-type agronomic research involves some weaknesses. I see two particularly fundamental problems with the current approach. One is that the strengths of the LISA approach—that it is based heavily on employing existing knowledge traditions about crop rotations, integration of crop and animal enterprises, and so on—may also be precisely its longer term weakness: The gains that can be achieved may be exhausted fairly rapidly, and there will not be sufficient basic research and system-redesign knowledge in the pipeline to serve as a basis of sustainable agriculture after the turn of the century. Even over the next decade, LISA-type production systems may well have limited appeal if our commodity policies essentially continue to amount to subsidies of monoculture and continuous cropping (Faeth et al., 1991; NRC, 1989). Despite the applied character of the research, sustainable systems based on LISA approaches still may not deliver the goods in farmers' fields if the public policy environment for sustainable agriculture remains adverse, which can only contribute to the vulnerability of sustainable agriculture knowledge claims to deconstruction.

A second fundamental weakness concerns the fact that there are a number of other promising approaches to agricultural sustainability [especially agroecology and policy-led (rather than science- and technology-led) sustainability, but possibly also biotechnology; Buttel, 1993] that will have a significant role to play, meaning that some important opportunities are being foregone. One might also note that applied LISA-type agronomic research is also the type of research that is playing a declining role in the land-grant and ARS systems as they continue to shift toward basic research. This suggests the significant threat that this work would be among the first to be cut if financial support would have to revert mainly to formula or other sources of institutional funding.

The utility of critical self-reflection is by no means confined to the natural science wing of the sustainability community. In the social sciences, we see a strong tendency for the research claims of the antisustainability camp to be relativized (e.g., in terms of reflecting corporate influences, subservience to corporate agribusiness or to industrial farmers, a lack of appreciation of new scientific trends or of recognition that societal trends render chemical-intensive agriculture increasingly problematic, and so on). This type of sociology-of-science/knowledge-type account of the research claims of the natural science-based sustainability community is, however, essentially absent. Further, the social science sustainability community has been slow to develop a critique of sustainability formulations from the natural sciences (even though they have admittedly performed this task well with respect to sustainability's international cousin, "sustainable development"). Save for some of the penetrating critiques of neoclassical resource economics and welfare economics that have come from the social science sustainability community (e.g., Norgaard, 1985; Martinez-Alier, 1987), there has also been little critical self-reflection by social scientists on the literatures they have generated.[16]

The purpose of this chapter has been to explore agricultural sustainability in a novel way, which although not denying its material importance considers sustainability to be a scientific knowledge claim and a symbol or ideology of a social movement. These realities of sustainability, in my view, have major implications for its future as a movement and object of research. While the observations in this chapter are not intended to be a call for any particular actions, I believe the sustainability community can benefit from a frank encounter with issues such as those raised in this chapter.

NOTES

1. For example, there have yet to be any major technical innovations of the stature of transgenic crop varieties or genetically engineered animal growth hormones. There has been a significant national decline in use of agricultural chemicals. But this trend began before the consolidation of the sustainable agriculture research community in the late 1980s; declining chemical usage began during the mid-1980s farm crisis and seems to be accounted for more by distress in the farm economy rather than by the adoption of sustainable practices that have been developed in public research institutions.

2. Science and technology s a material-productive force would involve stress on phenomena such as how technical knowledge contributes to the restructuring of production processes, the differential impacts of technical change on various social classes and groups, and the processes by which these classes and groups develop and act on political and economic interests in scientific research priorities or in technology policy.

3. It should be noted, however, that many persons (Barnes, 1982; Bijker et al., 1987; Cozzens and Gieryn, 1990; Latour, 1987; Woolgar, 1988; Yearley, 1988) whose work has conventionally been thought to lie within the constructivist-relativism traditions have played a major role in urging that the sociology of science take up questions relating to technology and science–society relationships.

4. Note, however, that Busch et al. (1991: Chapter 2) have sought to rescue induced innovation from its singular emphasis on technology as a material force, by uniting it with Latour's (1987) "science in action" framework to yield a "supply and demand in science" perspective.

5. The distinction referred to here is that between science as an "induced institution," versus science as a distinctive, autonomous institution. A "science as induced institution" framework suggests that scientific knowledge production can be "read off" from larger political–economic or cultural forces of society at large. The latter (more defensible) perspective is based on the notion that science's very character—that it produces, certifies, and regulates knowledge—leads it to have distinctive practices, hierarchies, and norms, and thus a significant degree of institutional autonomy.

6. Some of the most encouraging progress in creating this comprehensive sociology of science and technology has been made is the sociology of agricultural science. The synthesis of the Latour actor-network framework and the induced innovation tradition by Busch et al. (1991) is particularly noteworthy in this regard. Feminist perspectives on science, the best of which straddle the two (social-constructivist and political economy of science) "demystification quadrants" of the typology in Table 1, have also made major contributions. Thus, it is no accident that Kloppenburg's (1991) paper on de/reconstructing agricultural knowledge, which is perhaps the most important recent article in the sociology of agricultural science, is built on feminist perspectives on science as well as on the frameworks of indigenous knowledge and political economy.

7. The notion of scientific knowledge claim is based on the observation that scientific ideas and data do not speak for themselves. Scientists not only generate data; they must also strive to package their ideas and data (or, in other words, create "knowledge claims") in an authoritative, persuasive way so as to convince other scientists and policymakers that their ideas are a valid reflection of natural reality, and implicitly or explicitly a claim that other frameworks of knowledge are invalid or less valid. Thus, science is, in part, a process of rhetoric and persuasion, in which scientists are active players in a game to achieve acceptance and recognition of their ideas, consistent with their interests as scien-

tists. Accordingly, the dynamics of power relations within the scientific establishment may often lead to ideas being accepted as valid, independent of the impressiveness of the evidence behind them.

8. Effective opposition requires that groups outside of the intelligentsia earnestly take up the criticism. It is, however, virtually precluded that a (nonviolent) opposition program or demand will be taken seriously in official institutions if it is not presented, at least in part, within a framework of research-based evidence. Thus, opposition invariably involves some coalition of intellectuals and nonintellectuals, with intellectuals usually becoming more dominant as an oppositional symbol becomes more persuasive and threatening.

9. It should also be noted from a Third World perspective that environmental activism in this era of greenhouse concern has become increasingly focused on rainforests and rainforest biodiversity conservation efforts. One of the potential concomitants of the growing fascination with rainforests—and of the declining interest in the ecological zones, such as most agricultural ones where the majority of the Third World rural poor live—is that attention and resources may be diverted from agricultural research and development programs. There is also growing Third World hesitation about the global change-based environmental agenda on national sovereignty and equity grounds. Many developing country political leaders—and even some Third World environmental activists—are raising objections on these grounds to the global conventions being prepared for the 1992 Earth Summit in Rio de Janeiro (Pearce, 1991). There is also now a growing, and often heated, debate over the political arithmetic of national contributions to greenhouse gases. First World environmental groups and think-tanks tend to produce high estimates of Third World sources of greenhouse gases, while many Third World scientists and researchers claim these numbers are considerably exaggerated (see Agarwal and Narain, 1990).

10. See Scott (1990) for a particularly useful overview of the theoretical and empirical literatures on "new social movements." Most importantly, Scott takes up one of the most common misunderstandings of new social movements, that they are largely left-oriented. Scott devotes an entire chapter (1990:Chapter 4) to "varieties of ideology within the ecology movement," noting that "eco-conservatism" coexists with left environmentalism. This division, I might note, is clearly manifest within the U.S. sustainability community.

11. The definition of particular issues as environmental issues, however,is normally superimposed on existing sociopolitical cleavages and phenomena. Support for or opposition to proenvironmental policies will thus tend to be shaped by the nonenvironmental material and symbolic dimensions of these issues. This accounts, in part, for the fact that opposition to environmental policies can typically be more readily accounted for than support for environmental policies.

12. There is also the broader issue, which has been taken up by O'Connor in this volume, as to whether it is plausible to conceive of environmentally sustainable capitalism. Like O'Connor, I have my doubts. It should, however, be recognized that agriculture and food in the West are but one arena of sustainability. Agriculture and food are important, but probably less so than resource use in industry, transport, and the residential sector. It is thus possible to achieve significant movement toward sustainability in agriculture and food while failing to do so in the broader sense referred to by O'Connor. Among the reasons that in the West agricultural sustainability has greater possibilities than the larger ecological agenda is the fact that environmentally sound agriculture is consistent with—and, in fact, is one of the least expensive and hence most attractive means of—agricultural supply control. Much of the environmental progress achieved in the past two

U.S. farm bills has been because their environmental provisions have passed muster with the Reagan and Busch administrations on account of their being able to limit output without being a mandatory program or involving substantial budget outlays.

13. Agricultural sustainability, and the broader matter of agriculture–environment relationships, has been crucial in this restructuring process, though issues such as corporate involvement in agricultural research, industry–university relationships, the desirability of biotechnology, and the like have been important as well.

14. The production-impact science distinction (between sciences whose principal roles are, respectively, to improve production efficacy vs. assessing the environmental and human impacts of new technologies) was originally developed by Schnaiberg (1980). I have employed this distinction (Buttel, 1985) to characterize the land-grant system's internal cleavages during what can now be seen as the "presustainability" era.

15. Indeed, sustainability activists have played a very influential role in leading to the predominance of highly applied research within the federal LISA initiative, and of the LISA approach within sustainable agriculture research in general. This community has tended to be suspicious about "basic research" in the land-grant system, typically seeing this type of research as being irrelevant to farmers' needs at best and potentially inimical to their interests at worst (particularly if it involves "biotechnology" or connections with multinational agroinput firms). In some state/land-grant university contexts, basic research is considered by activists to essentially be a mantle taken up by scientists and administrators out of narrow careerism and to insulate research problem choices from public accountability. Thus, ("basic") agroecology, due at least partly to its equation with ostensibly nonaccountable basic research in general, has had little support within the agricultural sustainability movement. As will be noted shortly, this may prove to be problematic for the long-term potential of the movement.

16. Perhaps the chief contribution of the social sciences has been for a goodly number of social scientists to join with some representatives from the natural sciences in arguing that sustainability should be seen in social justice, and not merely biological or physical, terms (e.g., Altieri et al., 1987). This is mostly a welcome challenge to the more common technocratic conception of sustainability. Unfortunately, when these additional concerns and goals that are brought forward by social scientists are superimposed on the current "all things to all people" conception of sustainability, the result is to make sustainability discourse even more general and vague.

REFERENCES

Adams, W. M. 1990. *Green Development*. London: Routledge.

Agarwal, A., and S. Narain. 1991. Global Warming in an Unequal World: A Case of Environmental Colonialism. *Earth Island Journal* (Spring):39–40.

Aglietta, M. 1977. *A Theory of Capitalist Regulation*. London: Verso.

Altieri, M. A., with contributions by R. B. Norgaard, S. B. Hecht, J. G. Farrell, and M. Liebman. 1987. *Agroecology*. Boulder, CO: Westview Press.

Barnes, B. 1982. The Science-Technology Relationship: A Model and a Query. *Social Studies of Science* 12:166–172.

Bijker, W. E., T. P. Hughes, and T. Pinch (eds.). 1987. *The Social Construction of Technical Systems*. Cambridge, MA: MIT Press.

Block, F. 1990. *Postindustrial Possibilities*. Berkeley: University of California Press.

Bloor, D. 1976. *Knowledge and Social Imagery*. London: Routledge & Kegan Paul.

Brante, T. 1986. Changing Perspectives in the Sociology of Science. In *The Sociology of Structure and Action*. U. Himmselstrand (ed.), pp. 190–215. London: Sage.

Bryson, R. 1990. Will There Be a Global 'Greenhouse' Warming? *Environmental Conservation* 17:97–99.

Busch, L., and W. B. Lacy. 1983. *Science, Agriculture, and the Politics of Research*. Boulder, CO: Westview Press.

Busch, L., W. B. Lacy, J. Burkhardt, and L. E. Lacy. 1991. *Plants, Power, and Profit*. Oxford: Basil Blackwell.

Buttel, F. H. 1985. The Land-Grant System: A Sociological Perspective on Value Conflicts and Ethical Issues. *Agriculture and Human Values* 2 (Spring):78–95.

Buttel, F. H. 1993. Sociology in and the Sociology of Agricultural Sustainability: Some Observations on the Future of Sustainable Agriculture. *Agriculture, Ecosystems, and Environment* 25:in press.

Buttel, F. H., and J. Curry. 1992. The Structuring of Sustainable Agriculture in Public Research Institutions: Results from a National Survey of Land-Grant Agricultural Scientists. *Impact Assessment Bulletin* 10:7–26.

Buttel, F. H., and G. W. Gillespie, Jr. 1988. Agricultural Research and Development and the Appropriation of Progressive Symbols: Some Observations on the Politics of Ecological Agriculture. *Cornell Rural Sociology Bulletin* No. 151, July.

Buttel, F. H., A. Hawkins, and A. G. Power. 1990. From Limits to Growth to Global Change: Constraints and Contradictions in the Evolution of Environmental Science and Ideology. *Global Environmental Change* 1:57–66.

Collins, H. M. 1983. The Sociology of Scientific Knowledge: Studies of Contemporary Science. *Annual Review of Sociology* 9:265–285.

Collins, H. M. 1985. *Changing Order*. London: Sage.

Council for Agricultural Science and Technology (CAST). 1990. *Alternative Agriculture: Scientists' Review*. Ames, IA: CAST.

Cozzens, S. E., and T. F. Gieryn (eds.). 1990. *Theories of Science in Society*. Bloomington, IN: Indiana University Press.

Faeth, P., R. Repetto, K. Kroll, Q. Dai, and G. Helmers. 1991. *Paying the Farm Bill: U.S. Agricultural Policy and the Transition to Sustainable Agriculture*. Washington, D.C.: World Resources Institute.

Gilbert, G. N., and M. J. Mulkay. 1984. *Opening Pandora's Box*. Cambridge: Cambridge University Press.

Hawkins, A., and F. H. Buttel. 1991. The Political Economy of Sustainable Development. *La Questione Agraria* 33:71–94.

Jasanoff, S. 1992. Science, Politics, and the Renegotiation of Expertise at EPA. *Osiris (new series)* 7:195–217.

Kloppenburg, J., Jr. 1991. *Social Theory and the De/Reconstruction of Agricultural Science: Local Knowledge for an Alternative Agriculture*. Rural Sociology 56:519–548.

Knorr-Cetina, K. D. 1981. *The Manufacture of Knowledge*. Oxford: Pergamon.

Knutson, R. D., C. R. Taylor, J. B. Penson, and E. G. Smith. 1990. *Economic Impacts of Reduced Chemical Use*. College Station, TX: K & Associates.

Kuhn, T. 1962. *The Structure of Scientific Revolutions*. Chicago: University of Chicago Press.

Lash, S., and J. Urry. 1987. *The End of Organized Capitalism*. Madison: University of Wisconsin Press.

Latour, B. 1987. *Science in Action*. Cambridge, MA: Harvard University Press.

Latour, B., and S. Woolgar. 1979. *Laboratory Life*. Beverly Hills, CA: Sage.

Lynch, M., and S. Woolgar (eds.). 1990. *Representation in Scientific Practice*. Cambridge, MA: MIT Press.

Mann, C. C. 1991. Extinction: Are Ecologists Crying Wolf? *Science* 253:736–738.

Marshall Institute. 1989. *Scientific Perspectives on the Greenhouse Problem*. Washington, DC: Marshall Institute.

Martinez-Alier, J. 1987. *Ecological Economics*. Oxford: Basil Blackwell.

Merton, R. K. 1973 [1938]. *The Sociology of Science*. Chicago: University of Chicago Press.

Mulkay, M. 1969. Some Aspects of Cultural Growth in the Natural Sciences. *Social Research* 36:22–52.

Mulkay, M. 1979. *Science and the Sociology of Knowledge*. London: Allen & Unwin.

National Research Council (NRC). 1989. *Alternative Agriculture*. Washington, DC: National Academy Press.

Norgaard, R. N. 1985. Environmental Economics: An Evolutionary Critique and a Plea for Pluralism. *Journal of Environmental Economics and Management* 12: 382–394.

Palladino, P. S. A. 1989. Entomology and Ecology: The Ecology of Entomology. Unpublished Ph.D. dissertation, University of Minnesota.

Pearce, F. 1991. North-South Rift Bars Path to Summit. *New Scientist* 132(23 November):20–21.

Pinch, T. J. 1986. *Confronting Nature*. Dordrecht: Reidel.

Schnaiberg, A. 1980. *The Environment*. New York: Oxford University Press.

Scott, A. 1990. *Ideology and the New Social Movements*. London: Unwin Hyman.

Timberlake, L. 1989. The Role of Scientific Knowledge in Drawing up the Brundtland Report. In *International Resource Management*. S. Andresen and W. Ostreng (eds.), pp. 117–123. London: Bellhaven Press.

Wood, R. E. 1986. *From Marshall Plan to Debt Crisis*. Berkeley: University of California Press.

Woodhouse, E. J. 1991. The Turn to Society? Social Reconstruction of Science. *Science, Technology, and Human Values* 16:390–404.

Woolgar, S. 1988. *Science: The Very Idea*. London: Tavistock.

World Commission on Environment and Development (WCED). 1987. *Our Common Future*. New York: Oxford University Press.

Yearley, S. 1988. *Science, Technology, and Social Change*. London: Unwin Hyman.

Yearley, S. 1991a. Greens and Science: A Doomed Affair? *New Scientist* (13 July): 37–40.

Yearley, S. 1991b. *The Green Case*. London: Harper/Collins.

CHAPTER 2

Political Ecology of Sustainable Rural Development: Dynamics of Social and Natural Resource Degradation

Lori Ann Thrupp

I n spite of the recent proliferation of reports about environmental degradation and "sustainable development," pervasive social–ecological predicaments continue and are poorly understood in most countries of the world.
The pernicious erosion of both human livelihoods and natural resource conditions are all too familiar. Moreover, powerful impediments still obstruct
efforts to solve the problems. Superficial rhetoric is no longer sufficient.
Serious efforts are needed to improve comprehension and to identify and
implement effective changes.

 An illuminating approach emerging in sustainable development discourse
is "political ecology" (e.g., Blaikie and Brookfield, 1987; EPOCA, 1989; Milliken, 1992; Schmink and Wood, 1987). This approach links theories of ecology and political economy. This chapter presents a political ecology approach
to analyze "unsustainable" patterns of development and ways to ameliorate
degradation of natural and human resources in the South.[1] It begins with a
definition of the conceptual approach, its central premises, and a summary

Food for the Future: Conditions and Contradictions of Sustainability, edited by Patricia Allen.
ISBN: 0-471-58082-1 © 1993 John Wiley & Sons, Inc.

of the theoretical underpinnings and the main arguments of political ecology. The second part of the chapter illustrates the use of this analytical framework to identify the socioeconomic repercussions and root causes of resource degradation (i.e., deforestation and soil erosion) and concomitant patterns of agricultural development, focusing on Central America. The final section briefly discusses policy implications.

In essence, the central propositions of the chapter are as follows:

- The causes of natural and human resource degradation consist not only of ecological factors and technological "errors"/problems, but most centrally socioeconomic and political factors that determine why/how people use land, resources, and technology.
- Overcoming social and ecological problems requires not only technical and ecological changes, but more crucially, changes and actions that confront social and political economic causes.

THE ANALYTICAL APPROACH OF POLITICAL ECOLOGY

The explanation of environmental unsustainability requires an approach that accounts for complex dynamic interactions and conflicting interests—which are inherent to ecological and social realities. This implies a need for an innovative, integrated, and interdisciplinary paradigm, as opposed to a conventional approach, such as physics or microeconomics, which is reductionist and mechanistic. The approach to be used here to fulfill these requirements is *political ecology,* which "combines the concerns of ecology and a broadly defined political economy" (Blaikie and Brookfield, 1987). Variations of the perspective have been used for studies of soil erosion, deforestation, agricultural technology, and land-use change and degradation, including work by Redclift (1987), Cockburn and Hecht (1989), Leff (1986), Yapa (1979), Schmink and Wood (1987), EPOCA (1990), Thrupp (1988, 1990), Campbell and Olson (1991), and CNS (1989),[2] although many of these authors do not always use the term "political ecology." This interdisciplinary paradigm is intended primarily for elucidating and analyzing causes—particularly the structural roots—of social/ecological problems (Blaikie and Brookfield, 1987; Buttel and Sunderlin, 1988); yet it can also help to supplement comprehension and assessments of effects. Political ecology is a flexible and evolving approach that can be adapted for explaining various kinds of resource exploitation, social and ecological degradation, and impacts of agricultural technology and other "sustainability" issues.

Central Characteristics and Premises

Several central themes or premises of political ecology can be identified, based on an analysis of literature in this field. Although they are articulated in diverse ways, some central generalized premises include the following:

- Environmental problems are not only "green" issues, such as deforestation, loss of biodiversity, and habitat destruction, but also what may be called "brown" issues, e.g., sewage contamination, pollution, and occupational hazards that harm human health. Both green and brown predicaments are generally linked to human degradation: poverty, social inequity, injustice or oppression, alienation, sickness, and/or violation of fundamental human rights. They directly affect present societies, not just future generations.

- In evaluating the significance of any specific natural/human resource predicaments, it is essential to assess the qualitative as well as the quantitative impacts, to identify the distribution of the effects among social groups, and to recognize that the poor and minorities (e.g., agricultural workers, indigenous people, or racial minorities) are often the main victims of degradation (Bullard, 1987; Bryant and Mohai, 1990; Perfecto, 1990).

- Nature and society interact dynamically and dialectically; and similarly, social conflict, uncertainty, and nonequilibrium change are inherent in development processes. (This premise contrasts with the assumptions of some approaches or models, such as certain computer simulation models, which presume that certainty and equilibrium prevail in society.)

- It is essential to understand the historical socioeconomic context and political–power relations (including the role of the state, structure and organization of production, economic "imperatives," and prevailing patterns of development), which circumscribe each environmental dilemma.

- Any given environmental problem has multiple causes, rather than a singular cause. Both ecological and technical factors influence and constrain production and development, and technologies are not "neutral;" but these factors alone rarely determine the outcomes, and therefore, a political ecologist avoids ecological determinism (i.e., the view that biophysical factors are the causal determinants) and technological determinism (i.e., the view that technologies are the determinants).

- Poverty both contributes to and results from environmental degradation; but poverty is not a root cause of degradation, and poor people should not be blamed for the problems, because poverty is derived from broader socioeconomic inequities and structures tied to uneven development processes. On the other hand, certain forms of affluence—e.g., exploitation by wealthy enterprises and sectors of society, profligate consumption patterns, and wasteful production systems for the purpose of accumulation—are frequently the central root causes of degradation of natural and human resources.

- Similarly, "population pressures" may contribute to and result from resource degradation; but population growth itself is seldom the determining cause (e.g., Blaikie and Brookfield, 1987; Campbell and Olson, 1991). Population determinism is thus avoided as an explanation in this view. Rather, it is recognized that problems arising from population pressures generally are derived from broader patterns of development, such as inequities in distribution of population and resources, lack of economic opportunities and education for poor women, control over resources by social groups, the nature of peoples' productive activities, and culture-specific gender relations as well. In other words, demographic interactions with resource degradation must be understood as highly complex and dynamic factors that are shaped by the characteristics of production/technology and social–economic forces.

- In the political ecology perspective, goals of "sustainable development" emphasize or require social equity and respect of human rights, as well as ecological soundness and economic productivity. This does not refer only to rights of *future* generations (as stressed by the report "Our Common Future" by the Brundtland Commission); rather, these aims are urgent for *present* societies—i.e., *intra*generational rights. This suggests a divergence from purely protectionist environmental areas (e.g., focusing on parks and protected areas), and criticism of opportunistic "green" investments of some large enterprises such as power companies, which are making superficial technical changes or donations to environmental issues, yet continue exploitative practices for their own accumulation interests. Instead, the implications from a political–ecology perspective uphold social justice and human rights as essential factors to overcome resource-related dilemmas. Means to these aims may include major political changes and challenging power relations, empowerment of marginalized people, including women, poor classes, and ethnic minorities, upholding cultural and epistemological diversity, and legitimization of local knowledge that has been displaced with the hegemony of Western scientific reductionism. Thus, this alternative vision embodies a challenge to the prevailing paradigm of development that has often perpetuated socially and environmentally unsustainable conditions.

All analysts with a political–ecological orientation may not agree with all of these premises as they are articulated above, but are likely to agree with the general perspectives. The nuances will obviously vary, depending on the subject or area under consideration.[3]

Analytical Methods

Methodologically, to identify the causes and significance of problems, a political ecologist systematically analyzes interactions between resource use and social dynamics at a local level and responses and influences of political–economic processes and policies at a macro level. This requires the acquisition of information about the perceptions of people—i.e., learning from individuals involved in decision making, generally using anthropological or sociological methods. This has been called a "local level" or "bottom-up" orientation (Blaikie and Brookfield, 1987; Little and Horowitz, 1987). The analyst generally undertakes systematic interviewing and/or participatory action research methods with local-level informants, such as farmers, heads of rural households, or state agricultural officials, who are actors affecting the particular resources or technology. For example, in analyzing deforestation and soil erosion in developing countries, the analyst focuses on the farmers' use of land and forest resources, interactions between the physical processes, the changes in livelihoods of different social groups, and peoples' reactions to alternatives being proposed by different institutions.

The approach requires more than local-level research, however. The social dynamics of environmental degradation and the possibilities of overcoming the problems need to be understood in the national, regional, and world contexts (Barraclough and Ghimire, 1990). For example, to understand causes of soil erosion in a particular farming community, it is not enough to examine only the soil quality and the location-specific tillage methods, but to systematically analyze other societal factors that influence the farmers' practices. Thus, an important part of the research method is to analyze the historical and socioeconomic (or structural) context in which the local problem is situated, and similarly, to trace the links of causation to factors in the wider political economy. This requires the examination and documentation of historical and economic development processes, power relations, land tenure systems, and institutions surrounding the use of resources and technology. The analyst elucidates the interplay of power among different social groups, the control of information and technology, and the role of the state and interests of private enterprise that influence agricultural development, land-use, human, and natural resources.

The scope of the research may also encompass the identification of ways to ameliorate the problems or to achieve sustainable forms of development. Such findings are derived partly from the analysis of root causes, and are thus based on understanding the interplay of social and ecological conditions at both the local and macro levels. A central challenge is to identify changes in perceptions, actions, and policies to promote more equitable and environmentally sound development—for social and economic goals. Important questions in addressing agroecological dilemmas are how to resolve the competing or conflicting resource-related interests of different social groups and of unborn generations, and how to uphold the justified claims and rights of the poor and dispossessed.

Theoretical Foundations

The theoretical foundations of political ecology are derived from ecology and political economy. Ecology entails "the study of interactions between organisms, populations, communities, and their physical environment" (adapted from Odum, 1983). Ecology is "holistic" or "systemic" as opposed to reductionist, meaning that it emphasizes the understanding of parts of a systems in their functional relationships with the whole; and it upholds an evolutionary perspective of development, which suggests that systems always change and evolve incrementally (Spooner, 1986). Although ecology consists of various paradigms, it is usually considered a natural science, within biology. However, many social scientists have developed ecological approaches—such as human ecology, cultural ecology, or political ecology; and they study interactions of human activities in their environments, generally borrowing concepts and analogies from biological ecology. For example, ecological concepts and principles that are relevant to this analysis include ecosystem (particularly "agroecosystem"), i.e., a unit that encompasses biotic relationships, characterized by nutrient and energy flows (Conway, 1987; Altieri, 1983), and natural processes governing biorhythms and growth, including respiration, photosynthesis, and energy and nutrient cycling (Dasmann et al., 1973). Other concepts adapted and acknowledged by social-ecologists include diversity, stability, sustainability, adaptation, and coevolution. In a political ecology view, understanding ecological principles and the inherent capacities and constraints of nature is considered essential in order to explain many aspects of environmental degradation.

However, in a political–social ecology approach, many analogies and concepts from biological systems are inappropriate and insufficient. Ecological principles are limited for understanding many kinds of socially related environmental problems, mainly because (1) they do not account for the purposive nature of human behavior and neglect politics, power, ethics, sociocultural values, and conflicts that influence the use of technology and resources; (2) relying on animal analogies for human behavior, as done by early sociobiologists, can be misleading or erroneous; and (3) many biological-based ecologists view humans as necessarily "intrusive" to nature, while, in contrast, a social-oriented ecologist sees people as part of the wider environment. Moreover, the uncritical adoption of biological principles in previous sociobiology theories has had negative or regressive sociopolitical implications.[4]

Realizing such limitations of ecology leads to the need for the second (and more important) conceptual component of the political ecology approach, which is *political economy*. This illuminates key social factors that shape peoples' actions and that cannot be explained by ecology alone. Political economy also has a long history and numerous interpretations, which do not require elaboration in this chapter. As used here, corresponding to the perspective used by Blaikie (1985) and others in this school, political economy is mainly concerned with social–structural formations, economic forces, and political–power relations that generally underlie and determine resource exploitation and development processes (e.g., Blaikie and Brookfield, 1987; Milliken, 1992; Watts, 1983; Redclift, 1987; Cockburn and Hecht, 1989). The specific determinants obviously vary, depending on the particular problem being analyzed; but they have included, for example, inequities or disparities in control over resources and land and technology, economic imperatives shaped by global market conditions, state-supported incentives for resource exploitation, distribution of benefits and costs of development among social groups, North–South political–economic relations, historically rooted social–power conflicts, and capital's appropriation of the rural process (e.g., Blaikie, 1985; Cockburn and Hecht, 1989; Schmink and Wood, 1987; Goodman et al., 1989; Thrupp, 1988). Insights are drawn from social theory, political geography, neodependency theorists, ecodevelopment, and other innovative political or social approaches (e.g., Leff, 1986; Buttel and Sunderlin, 1988; Cockburn and Hecht, 1989; Campbell and Olson, 1991; CNS, 1989; Goodman et al., 1989; Williams, 1986; Barraclough and Ghimire, 1990).

Framework for Analysis of Social–Ecological Problems

A central task of most political ecology analyses is to explain the causes of a given social–environmental problem. The general framework of explanation is illustrated simply in Figure 1. A central proposition is that there are multiple interlinked causes: (1) ecological phenomena such as intrinsic qualities of the natural resource base, (2) technical factors, such as types of land use practices or technologies, and (3) social and structural factors, situated in the realm of political economy, that ultimately determine why and how people use or undertake particular practices and technologies. In general, when referring to land use dilemmas and agroecological degradation,

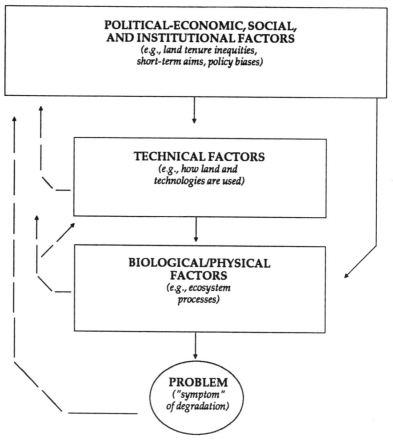

Figure 1. Causes of natural and human resource degradation in a political ecology approach.

this approach suggests that the most important causes include the inequitable control of resources, short-term economic interests and resource exploitation, and skewed policy incentives (i.e., state influences)—embedded in the prevailing patterns of uneven development (see Fig. 1).

This general framework shown in Figure 1 can be adapted to a variety of problems, such as land or forest degradation, erosion, water depletion, and effects of agricultural technologies. The specific causal factors vary from case to case; but the framework can be used to conceptualize a "chain of explanation" (Blaikie and Brookfield, 1987) to identify the most important determinants. This framework for understanding the nature and causes of the crises and the failures of remedial measures can also assist in identifying effective changes.

AGROECOLOGICAL DEGRADATION IN CENTRAL AMERICA: SOCIAL DYNAMICS, IMPACTS, AND CAUSES[5]

Background

The analysis of agroecological conditions in Central America illustrates how a political–ecology approach can improve understanding of social dynamics of degradation. A case study of deforestation, agricultural development, and conservation efforts was undertaken by the author in Costa Rica, with aims to identify the socioeconomic impacts and root causes of deforestation and land degradation, and to determine implications for effective change. The study went beyond previous scientific studies that described only biophysical and ecological repercussions. This analysis concerned mainly Costa Rica, but the situation here exemplifies typical conditions in much of the country, and in many areas of Central America. This analysis started with a "people-centered" local-level methodology, focusing on perceptions of farmers and representatives of government agencies involved in and affected by resource use and degradation. It therefore entailed systematic interviewing of these people and observation of the physical conditions and practices at the local level.[6] Simultaneously, the study included the analysis of historical political–economic forces that shape the processes of land-use change. The main findings, arguments, and lessons of the analysis are highlighted below.

Social–Ecological Context and Linkages

In Central America, as in many regions of the world, deforestation and concomitant agricultural expansion and land-use changes in Central America do *not* necessarily result in deleterious effects, although some environmental analysts (concerned about biodiversity loss) have created foreboding images of this process. The changes have generated short-term benefits: That is, in many areas, fertile soils have enabled production of a variety of crops, and agriculture has opened new opportunities and economic benefits, accruing mainly to certain people who control land. The influential connection to export markets has provided sources of foreign exchange and has benefited North American consumers who enjoy low-priced beef from Central America.

However, there are also contradictions and problems associated with these changes, manifested as both negative ecological impacts, and adverse socioeconomic impacts and inequities. Central America is characterized by increasingly serious degradation of natural resources, which has created high social costs (e.g., Leonard, 1987; Williams, 1986). The estimated rates of deforestation reached 40,000–50,000 hectares/year in the region during the

TABLE 1.   Shifts in Land Use in Latin America: Total and Per Capita[a]

	1968 (m ha)	1986 (m ha)	1968 (ha/person)	1986 (ha/person)
Agricultural	655	723	2.8	1.8
of which arable	140	175		
Forest	1,058	974	4.3	2.5
Pasture	515	548	2.1	1.4

Source: Goodman and Redclift (1991).
[a] At least 2.5% average population growth per annum.

1970s and 1980s (see Table 1), mostly for agriculture—yet much of this took place on land unsuited for sustained agricultural production. Approximately 40% of the land in the region is affected by severe levels of erosion, which, in turn, thwarts agricultural production, disrupts watersheds and hydrological cycles, and leads to devastating floods. Despite recent efforts to alleviate the problems and to pursue sustainable development, the resource degradation has worsened.

For example, the hilly landscape throughout much of Costa Rica is predominantly pastureland, which is mostly denuded of vegetation, highly eroded, impacted, parched, and cracked, and marked by deeply etched cattle paths. During the dry season (October through April), thin cows are scattered sparsely across parched dry brown grasses, seeming to cling precariously to the steep slopes. In the wet season (May through September), landslides, astounding rainfall runoff, and floods are frequent reflections of the deterioration. Other characteristics in the area include silting of riverbeds, and depletion of groundwater reserves. The grassy eroded landscape is also broken by patches of cultivated land with a diversity of crops, especially coffee, grains, and occasionally vegetables and fruits, and spots of natural vegetation, as well as unpaved access roads and small houses. These physical-ecological characteristics are common throughout many areas of Central America.

Farmer interviews undertaken by the author, in the region of Puriscal, reveal that land degradation posed a significant problem for agricultural production. Numerous farmers declared that their land was "tired" and/or "sterile." Yields were declining over time for all farmers, and the majority suffer problems from erosion and landslides on steep slopes. They also realized consistently that their own cultivation and grazing of the land contributed to this degradation; but they emphasized the imperative to continue producing for livelihoods, and they had no choice other than to continue using the land. A common response to the declining fertility is to apply chemical fertilizers, but this becomes a financial burden and is unaffordable for many; most said that they needed to buy more and more fertilizer each year to get effective results. Declining availability of fuel wood was also seen as a problem by about half of those interviewed. A majority of the people say that water scarcity is a serious problem for families during the dry summer months. The lack of water is seen to be worsening over time. Many farmers explained that the growing aridity was associated with the increasing scarcity of trees.

The numerous families affected by soil degradation have consequently suffered significant socioeconomic hardships. Although the families included in the Costa Rica study are not starving and are relatively healthy and well-off compared to many small farmers in Central America, they felt that the quality of their lives had deteriorated over the past several years because of the decline in productivity. In efforts to increase incomes, the majority have diversified away from production of the traditional corn and beans, and have tried growing different crops. Some products, such as coffee, have offered temporary improvements, but, given increasing soil erosion, lack of land and markets, and fluctuating prices, the small farmers rarely sustain prolonged benefits. Some of the farmers have converted land to pastures, yet cattle in small-scale farms are rarely lucrative, and generally displace land for food crops. Many families have sought alternative sources of income besides farming income from their own land, including wage labor on large farmers' land. In other parts of Central America besides Costa Rica, the documented socioeconomic impacts are more serious: In cases such as Guatemala and El Salvador, the increasing degradation of marginal lands and resource scarcities have exacerbated poverty, contributing to lack of food and malnutrition among the poor (e.g., Williams, 1986; Durham, 1979; EPOCA, 1989, 1990).

Land degradation combined with the appropriation of land by large cattle ranchers has also contributed to a high rate of migration from rural areas. Simultaneously, the conversion to pastures has contributed to high levels of unemployment, since cattle ranching uses about one-fifth of the labor requirements as most perennial crops. Many of the landless migrants have joined the growing numbers of unemployed or low wage labor force, usually living in slums of the capital city (Williams, 1986; Parsons, 1976). In some areas, such as in El Salvador, outcomes of the degradation and displacement processes include social upheaval, peasant rebellions, and violent conflict (e.g., EPOCA, 1989, 1990; Durham, 1979). Another outcome of the process has been called the "protein flight" (Myers, 1982). That is, the Central American countries export over half of its beef, mostly to the United States. Per capita beef consumption decreased during the 1970s, despite the increase in cattle production (Williams, 1986). Moreover, cattle pastures and export crops have replaced basic grains and food crops in many areas (Parsons, 1976). Although the exports earn foreign exchange used for purchasing food imports, this substitution has contributed to food deficits for the poor in some situations. In many rural areas, protein–calorie deficiency is apparent among the poorest; this is a result of a not only pasture expansion, but also land concentration and inequities that exacerbate poverty (Williams, 1986; Leonard, 1987; Barry, 1987). Dependency on export beef markets also creates vulnerability to fluctuating international market prices.

In sum, the land-use change process described above has often contributed to human or social degradation—i.e., impoverishment and marginalization—of the rural poor, as well as ecological degradation.

CAUSES OF DEFORESTATION AND RESOURCE DEGRADATION

A central part of the political ecology analysis is to determine the root causes of the social–ecological problems. As a preface, it should be realized that many previous reports have generated confusion about the causes of deforestation and erosion, and often they mistake the mechanisms, such as the particular practices or technologies, for the underlying root causes. Moreover, many authors have blamed poor migrants or "slash-and-burn peasants"—a term used pejoratively (and wrongly) to evoke images of "destructive" or "ignorant" people. They fail to realize that the people are often forced into such actions as a result from larger socioeconomic processes. Blaming these people is thus accusing the victims. On the other hand, macrolevel explanations, such as the accumulation of capital, and sometimes treated so superficially that they cannot be linked to the micromechanisms (Barraclough and Ghimire, 1990). Poverty and population growth are also commonly imputed for environmental degradation; but these also can be facile or overly simplistic explanations, since these factors are symptoms of unequal exploitative development, absence of employment alternatives, and skewed national development strategies.

The immediate mechanisms stimulating deforestation are fairly well-known and documented. In most Latin American countries, these proximate factors are expansion of commercial agriculture and cattle ranching, the growth of timber industry and industrial mining, and opening agricultural frontiers (WRI, 1990; WCED, 1987; Barraclough and Ghimire, 1990). The exploitation of forests for fuel wood is another immediate cause of deforestation, but this generally has a minor contribution in Latin America. Small peasant producers and landless laborers often are agents of land clearing, but rarely to increase their own subsistence crops; they provide low-cost labor for medium and large cattle ranchers. Other less considered immediate causes are urban and industrial wood demands.

An understanding of these immediate causal mechanisms is useful; but a political ecology approach requires delving deeper to reveal the causal roots. The explanation needs to encompass an understanding of bioecological constraints (e.g., natural characteristics and processes) and technological factors (e.g., specific methods of land/forest use); but most importantly, it requires analysis of socioeconomic and policy and political processes that underlie the mechanisms. As explained below, the "symptoms" of degradation are rooted in development patterns and socioeconomic structures in the region. These factors are illustrated schematically in Figure 2.

First, *biological and ecological factors,* which are essential conditions of production (O'Connor and Cockburn, 1989), underlie the degradation of natural resources. The particular characteristics of soils, plants, fauna, water, and other resources, along with the complex interactions of energy and nutrient cycling, photosynthesis, and evapotranspiration, need to be understood in order to appreciate how land and ecosystems deteriorate. In many areas of Central America, the soils were fertile and rich when they were first cultivated many decades ago, but their fertility has been lost through leaching and nutrient depletion over time. In other tropical rainforest areas, where soils are naturally lateritic and thin, agricultural production cannot be sustained beyond a few years, because of natural leaching processes (Dasmann et al., 1973; Altieri, 1983). In addition, studies have shown that monocultural production systems, in contrast with diverse heterogeneous production systems, have inherent ecological features that make them vulnerable to and responsible for resource degradation. These ecological principles must be understood on a case-by-case-basis, and they constitute constraints that require attention.

Second, *technical dimensions*—meaning how the resources and technologies are used—are also important contributing causes to degradation. For example, the intensification of production (e.g., continual land use without fallow), absence of soil conservation measures, replacement of diverse systems by monoculture, grazing excessive numbers of cows per unit of land, burning of crop residues each season, heavy use of chemicals, and exploitative timber harvesting methods have contributed to problems. On the other hand, logging and farming, when carefully managed and carried out, not necessarily degrade forests and land (Barraclough and Ghimire, 1990). Methods of forest management and sustainable agriculture practices (such as soil conservation measures, agroforestry, and intercropping) can be used to work toward sustainable development aims. Thus, the variable nature of technical factors must be acknowledged in the explanation of degradation.

The most important causes, however, are *socioeconomic factors* that lie within the realm of political economy (LACCDE, 1990; Redclift, 1987; Williams, 1986). In Central America, the underlying roots of degradation pertain to the uneven development patterns of land use and the nature of production,

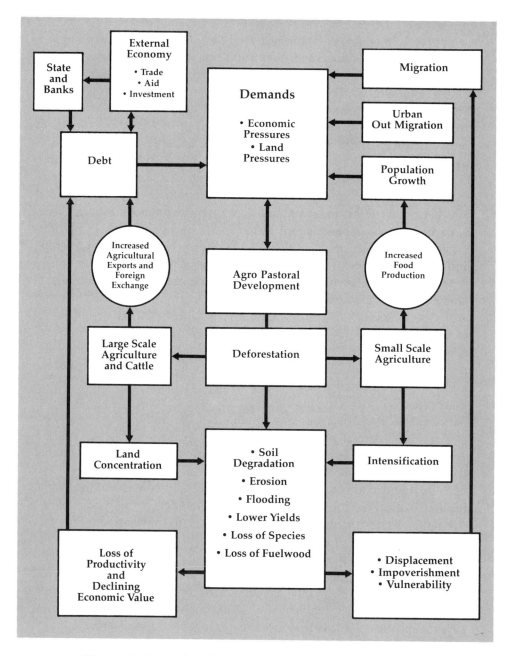

Figure 2. Agricultural land use change and repercussions.

characterized by concentration of land, export-oriented commercialization, intensification of production, socially stratified agrarian structures, marginalization, and displacement of the poor—which are also upheld by land tenure systems, powerful private interests, and state policies. Understanding these key dynamic factors requires a summary of the historical background.

Over the past four centuries, the rural forested landscape of Central America has been cleared over time primarily for agricultural purposes. Shifting agriculture, the predominant traditional form of agriculture first practices by indigenous people and later adopted by some peasant farmers, was productive and well-adapted for the ecological and social conditions in many areas. However, over time, the typical process of socioeconomic change in the region involved displacement (or disruption) of shifting agricultural practices, expansion of production, increasing permanence, intensification, and commercialization of farming, and growing rural populations, and penetration of the market economy. The late 1800s and 1900s brought the accelerated growth of agroexport production of coffee, bananas, and cotton, which have become the highest value crops, supported by well-developed marketing industries and transnational companies. Starting in the late 1960s, cattle production for export beef markets developed rapidly, resulting in extensive expansion of pastures in the region (Williams, 1986; Thrupp, 1980a, as illustrated in Table 2).

These prevailing patterns of agroexport growth and land-use change—aimed for maximizing short-term yields and growth of revenues—have been mandated by international development agencies and financial institutions. In turn, these imperatives have been supported by the

TABLE 2. Land Use Changes in Central America: 1960–1980

	Percentage of Land Devoted to					
	Forest		Cultivated		Pasture	
	1960	1980	1960	1980	1960	1980
Belize	NA	44	NA	3	NA	2
Costa Rica	56	36	9	10	19	31
El Salvador	11	7	32	35	29	29
Guatemala	77	42	14	17	10	8
Honduras	63	36	13	16	18	30
Nicaragua	54	38	10	13	14	29
Panama	59	55	7	8	12	15
Total for the region	61	40	11	13	15	22

Source: Leonard (1987:99).

State, through policy incentives and institutional and political systems (Williams, 1986; Panos, 1990). They are also promulgated by dominant business interests, such as some transnational investors and associations of large-scale producers. Since the 1970s, the Central American economy is based on an export-dependent development model and the pursuit of "comparative advantage," which hinges on the export of coffee, bananas, beef, and sugar, in exchange for manufactured goods from the United States and other industrialized countries. The state has also promoted deforestation, through the colonization/land-claim laws that reward people title for "clearing" land, and through incentives for the establishment of pastures. Land speculation by large investors has also been a strong impetus behind the expansion of the frontier (Hartshorn et al., 1982, Leonard, 1987).

Although the changes in land use and agrarian structure have undoubtedly generated benefits, as mentioned previously, the forest-to-pastures process is also linked with the increasing concentration of land ownership and of power—i.e., growing disparities between the poor and large producers, displacement of small farmers into marginal areas, and creation of a growing cadre of landless laborers. Typically, poor farmers, lacking secure tenure, are forced into steeply sloped areas, less suitable for production, where degradation ensues. This process has involved structural dualism of the agrarian economy—i.e., stratification and disparities between rich and poor—throughout the region: That is, the benefits have been distributed inequitably, and have seldom "trickled down." For example, in Costa Rica, 37% of landholders and small farmers own just 1% of all farmland, while the top 1% of large landholders own more than 25% of the farmland (Hartshorn et al., 1982). In other countries of Central America, the inequitable distribution in the agrarian structure is even more severe, as documented in numerous studies and census records (e.g., Williams, 1986; LACCDE, 1990).

In addition, the land tenure systems in the region contribute to the deforestation processes and institutionalize the agrarian structures. The inequities are consolidated by land tenure system and land titling processes that favor large estates and enterprises (de Janvry, 1981). Property laws commonly discriminate against traditional common property regimes of indigenous people in forest areas, as well poor peasant farmers (Barraclough and Ghimire, 1990). These systems manifest the power relationships of each country. Although several of the countries have attempted "land reform" policies, the land tenure laws and institutions governing property rights remain entrenched and shape the framework of incentives and constraints affecting land use.

Furthermore, policies at the international, national, and local levels underlie the degradation processes (LACCDE, 1990; Repetto and Gillis, 1988). For example, state incentives for the promotion of export crops, commercial livestock, and natural forest industries (such as credit policies) are among the many policies that contribute to both ecological and social degradation. Many poor families have been encouraged to migrate to marginal forest frontiers through land colonization policies. Structural adjustment strategies of the late 1980s have created strong pressures for expanding agroexports. This has resulted in an increase in the growth and value of nontraditional export crops (such as palm oil and tropical fruits), yet adjustment policies have exacerbated the trade deficit, contributed to increasing socioeconomic inequality, increased farmers' vulnerability to fluctuating international prices, and aggravated pressures on resources, exacerbating ecological degradation (McAfee, 1991). These national-level policies are generally conditioned and mandated by policies of international agencies in the North, and the policies contradict efforts for alternative forms of development.

In sum, the processes and patterns inherent in this conventional growth model are central causes behind the social and ecological unsustainability in the region.

RESPONSES TO DEGRADATION: STATE AND LOCAL REACTIONS

In recent years, in response to the resource-related problems, environmental scientists, development donors, government agencies, nongovernment organizations (NGOs), and some private producers have attempted to ameliorate degradation and to pursue "sustainable" development approaches in Central America, as in many parts of the world. Many of the early efforts focused on four approaches: "preservation" measures, such as formation of parks and forest reserves, sometimes financed through debt-for-nature swaps; laws intended to restrict deforestation or to promote reforestation (e.g., Costa Rica's Reforestation Law, which offers an economic incentive of a tax deduction for commercial reforestation); research on ecology and forestry; and establishment of tree nurseries for commercial forestry. In addition, some government agencies have banned the use of fire to clear agricultural land, have begun soil conservation programs in some areas, and occasionally support education campaigns to encourage reforestation or conservation. These kinds of environmental policies are often initiated and promoted by international and local environmental organizations, whose main interests include "saving" forests and biodiversity and/or off-setting carbon losses.[7]

In some cases, these remedial measures have helped to alleviate environmental problems to some extent, and sometimes to improve environmental management capacities and public awareness. However, many of the existing conservation policies and measures lack efficacy, and they often do not alleviate or prevent the multidimensional problems. Deforestation, land degrada-

tion, and soil erosion have continued at alarmingly high rates, with conse-
quent socioeconomic losses (LACCDE, 1990). For example, in Costa Rica,
the rate of deforestation continues at approximately 40,000 hectares/year,
while reforestation covers only several hundred hectares annually (WRI,
1990).

Why have many of the strategies been ineffective? One of the main reasons
is that the projects, policies, and institutions seldom address the *roots* of the
problems, and similarly, they tend to overlook or ignore the social–political
dynamics and human dimensions that are inextricably tied to natural resource degradation. Many of the existing responses are merely superficial
palliatives. More specifically, government agencies have seldom made serious
attempts to address the structural and policy factors, such as resource inequities, the dependence on the agroexport model, and short-term imperatives
imposed by international financial institutions. These macroeconomic factors
remain as entrenched impediments to truly sustainable development. Furthermore, through land use and economic policies, government agencies and
donors support the expansion of the frontier and intensification of agroexport
production with increased agrochemical inputs, as mentioned earlier. And
these contradictory policies aggravate degradation and inequities (Thrupp,
1990; Panos, 1990; Redclift, 1987; Leonard, 1987). In this study, the rural
peoples' perspectives reveal weaknesses in the main conservation policy approaches. Analyses of local perceptions have shown that farmers are often
aware of reforestation efforts and campaigns for conservation. For example,
in the Costa Rica case study, 90% of the farmers interviewed in Puriscal
have heard of reforestation, and the majority have a positive reaction to the
idea of planting trees. This high level of awareness is largely attributable to
media campaigns by the Forest Service and other NGOs. However, the farmers often had negative views and criticisms of the main kinds of reforestation
policies being pushed by the Forest Service or other agencies. In particular,
the creation of "untouchable" parks and reserves in much of the country's
land was criticized by farmers, especially by those who have very small farms
or who rent land. They disapproved mainly because they felt that the reserves
did not benefit them, and serve mainly the interests of tourists and city
people. They resent the use of land for such purposes, given their own lack
of access to land. Moreover, those living close to reserves complained that
wild animals from the forests harm their crops and chickens. The government's appeal to parks and biological diversity preservation is thus ineffective
and inappropriate to many rural people.[8]

Another problem concerning the creation of reserves is difficulty in enforcement of the closed borders, given pressures of large-scale enterprises
for wood and land, and given inequities and scarcities of local people. It is
very difficult, if not impossible, for forest guards to patrol all of the state-

protected forest and to stop encroachment. While small-scale hunting or fuel-gathering activity of small farmers seldom harms the forest ecosystem, other illegal and extractive logging practices by industries are unsustainable. As long as the policies do not confront the economic pressure behind deforestation, prohibitive measures alone are ineffective.

In Costa Rica, the state reforestation measures are not appropriate for small farmers' interests. For example, the farmers interviewed were uninterested in the large-scale forest plantations, logically because planting trees in monocultures for commercial purposes is impossible for most small farmers who must dedicate their small plots to food crops to feed their families. The reforestation tax deduction policy is biased to the advantage of wealthy landowners: It is impossible for most small farmers to take advantage of this policy, mainly because the tax law exempts poor people from paying income taxes. Farmers point out that the government agencies promote contradictory policies. For example, in Costa Rica, the Ministry of Agriculture and commerce policies encourage farmers to clear land and increase production, especially agroexports to earn dollars. In contrast, the Forest Service admonishes people for clearing land and insists on growing trees, not crops. As one farmer explained:

> The government tells us that we need more grain and more crops for export. The Agriculture Ministry encourages us to grow food to clear the land of trees. Yet, the other guys in the government's forest department tell us we can't cut down the trees. They don't even care that they give us conflicting messages.[9]

Beyond this, another policy conflict concerns land-use and settlement laws, because traditional laws in most Latin American countries require people to "clear" (meaning deforest) virgin land in order to claim title to farms. In stark contrast, the Forestry Laws prohibit or restrict clearing. Such contradictions create confusion and cynicism by local communities.

However, the peoples' antipathy to some of the government's conservation initiatives does not mean that they are unaware and uninterested in trees and in conservation. On the contrary, the people who have farmed in such areas for many years generally have remarkable concern, consciousness, and knowledge of the values of natural resource conditions, but their perspective and reasons differ from those of the government agencies and "environmentalists." With accumulated experience over many years, many small farmers become skilled resource managers, knowing how to manage and produce efficiently in complex diverse systems, learning through trial and experimentation, and adapting to changing conditions, in many cases. Although sometimes traditional knowledge becomes displaced and disrupted over time, or becomes poorly adapted under present circumstances, many studies have shown that farmers have a rich understanding of managing soils, pests, water, cropping patterns, genetic resources, and sometimes trees (Chambers et al., 1988). For example, in Costa Rica, trees and vegetative cover are valued by many small and medium farmers, mainly for their productive and agroecological purposes—i.e., fuel, fodder, fruit, shade, fence posts, retention of soil, and nutrients/manure from leaf fall. In fact, in Costa Rica, agroforestry is a long tradition practiced over decades and even centuries for numerous small-scale farmers, although they may not use the term. More than half of the farmers interviewed had regularly planted trees of some kind within the last few years. The trees planted included fruit trees, native hardwood species for fences and fuel wood, native leguminous trees, and sometimes exotic species, such as pine, cedar, and eucalyptus. The trees were usually planted around the borders of farms as "live fences," next to their homes, and mixed into the fields, as shade trees. The farmers clearly recognized and emphasized the values they gain from incorporating trees in their farms. They not only identified the multiple benefits from tree products, but also explained in more detail that cover trees in their small plots of coffee provided the "health" and protection of the farm—providing not only shade, water, soil retention, and rich natural fertilizer from the leaf-fall and from the roots of nitrogen-fixing species. Their appreciation of these benefits was not derived from "formal scientific" literature or instruction, but from years of experience, experimentation, and close knowledge of the resources, crops, and vegetation.

It therefore is clear that many farmers value planting trees to improve the productivity of their crops, income, and land, and not for the reasons that are emphasized by the Forest Service and conservation agencies. Of course, some of the farmers had not planted trees, due to lack of space, time, money, knowledge of where seedlings were available.

Other kinds of conservation measures are practiced to some extent. The use of fire to burn crop stubble and weeds has decreased in Costa Rica, mainly due to the strict prohibitions (accompanied by a fine for violations) and the belief that burning "sterilizes" the soil. Allowing the land to lie fallow was recognized as an ideal measure to "give tired soils a rest," but this is viewed as impossible for all small farmers interviewed, because of limitations in space and pressures to produce crops continually. Soil conservation measures, such as terracing, are not traditional practices in Costa Rica, as in much of Central America, and have not been part of conventional Green Revolution methods. In recent years, due to efforts by extension workers and some innovative farmers, soil conservation techniques have been adopted by some farmers, but by less than a fourth of the farmers interviewed in Costa Rica. Maintaining terraces is seen as a frustrating and overly labor-intensive job and is easily impeded by the force of rains. Many expressed interest in learning more efficient and less labor-intensive methods of soil conservation.

In sum, local perspectives on the policies and natural resources provide useful insights on the political–economic constraints of existing structures and responses. These findings also show that efforts of government agencies rarely address the main causes of the problems, and, in fact, perpetuate contradictory policies that underlie degradation. Given these weaknesses, neglect of farmers' views, and conflicts, it is not surprising that both social and ecological degradation has continued.

IMPLICATIONS AND CONCLUSION

A political ecology approach not only provides insights into the causes and social consequences of social and agroecological degradation, but it also assists in the determination of implications for change. Lessons from both the local level and wider political economic context suggest needs for changes and progressive ways to promote sustainable development, meaning equitable as well as environmentally sound and productive. It is useful to summarize several general approaches and lessons.

One of the central messages is that the *causes* of the problems must be confronted. To address the technical and ecological causes, the analysis suggests a need for technical changes, such as soil conservation, agroforestry, integrated pest management, and building of indigenous sustainable agroecological methods, and ecological changes, such as developing more *diversity* (and diversification) in cropping systems to break the dependency on monocultures, and acknowledging and respecting ecological factors such as inherent constraints of tropical soils.

However, it is not enough to address the technical and ecological dimensions, and these changes alone rarely confront the deeper roots of the problems. The more important lesson is that addressing the *social and political economic* causes is usually essential to resolve these conflicts, and is often a prerequisite to technical changes as well. Although some analysts have advocated pricing and economic measures, such as raising prices of land or environmental taxes, these measures still may be insufficient to induce preventive changes, partly because they do not deal with fundamental problems rooted in economic growth models, inequities in access to resources, and structural patterns of production. Other socioeconomic and political transformations must therefore be considered:

Major changes in the *prevailing economic policies and models* of development may help to prevent and alleviate these problems. Particularly important is the need to change economic policies, such as structural adjustment requirements, which are imposed from the United States and Northern international financial agencies. At the same time, major policy reforms are needed in the South, for example, to eliminate certain subsidies, colonization policies, and land tenure laws that induce overexploitation and degradation of resources. Moreover, concerns about "sustainability" must be incorporated into models of development in radically new ways, not just "tacked on" to existing models in isolated and minor areas. These changes may appear to be idealistic, given the pervasive international political and economic interests. Yet, the growing recognition of economic losses and risks of unsustainable patterns is provoking outcry and upheaval among the poor and among international analysts, which has also initiated discussions in international policy arenas and donor agencies, which has helped to contribute to some policy changes.

Comprehensive changes in *land tenure and titling systems* may also be required in many cases to ensure more equity and security of rights to the poor and to prevent displacement onto marginal lands. Again, given the pervasive political structures, this may appear to be an idealistic appeal. Yet, experiences show that resource and social conditions can be improved when people have security and access to land and resources. Such changes are necessary as a part of fundamental human rights. In addition, basic land use laws also require substantive reforms. For example, land-claim and colonization laws need to be revamped to avoid settlements in marginal non-productive areas. Additional resource and forestry laws also may be useful, but laws are useless unless they are sensitive to social needs, and unless they are backed by political support and legal power.

Another central implication is that local people, especially marginalized poor groups, indigenous peoples, and women, must be full participants in determining needs and responses, and in carrying out changes. *Popular participation* in the design of strategies, policy decisions, and project implementation is important in order to address the problems of social and ecological degradation. To be truly effective, participatory approaches must enable local people to take leadership roles in all stages of development processes and projects. For example, the active participation of poor communities in tree-planting and soil conservation programs (e.g., World Neighbors, Plan Sierra, Farmer-to-Farmer) has shown considerable potential and success for improving the socioeconomic and ecological conditions of the poor. Involving small farmers, women, and workers in natural resource management projects can again serve a purpose of empowerment of those who seldom benefit from such activities. In these kinds of efforts, farmers are leaders and/or equal partners with scientists and technicians. Local efforts may be more successful if they lead to the organization and strengthening of political bargaining power of local people.

Similarly, some popularly based "social" movements have emerged throughout Latin America in relation to resource access issues, and these movements in some cases have been effective and have attracted allies at the international level, enabling local people to become influential political actors affecting policies and even development agencies (e.g., Escobar and Thrupp, 1990). Spontaneous social movements of individuals and groups affected by deforestation and land degradation can sometimes result in conflicts and controversies, and have occasionally resulted in struggles and repression by powerful groups (e.g., Cockburn and Hecht, 1989). Yet, successes of local participation and mobilization strategies have continued to motivate people to struggle for their rights for resource rights, cultural integrity, a healthy environment, and/or their livelihoods.

Improving peoples' *access to information and alternative agroecological technologies*, including agroforestry systems, nonchemical methods of pest control, soil conservation methods, and use of organic manures, is also important to induce change. In many cases, this does not imply dependence on "appropriate technologies" from outside, but rather, legitimizing, building, and using the indigenous knowledge and innovativeness about farming. Although glib "romanticization" of all traditions would be misleading, it is crucial to legitimize, rather than displace, the effective methods, and to selectively incorporate new ideas and methods from foreign sources. This approach of validating and strengthening local capacities can serve as a form of empowerment for local people. Policy promotion of alternative nonchemical methods and agroforestry also can constitute effective change.

Counteracting short-term imperatives behind the overexploitation of resources is also necessary to promote land use practices that ensure long-term returns. This requires relieving pressures on farmers to produce high yields for maximum short-term returns, and incentives for adoption of long-term planning and time horizons. Developing this change poses major practical difficulties, because the prevailing short-term motives of resource extraction reflect strong competitive pressures for growth in markets, foreign exchange, and mandates of international banks and development agencies. New bargaining tools, debt negotiations, and shifts in power in the 1990s may help countries of the South to change policies, alleviate debt burdens, and enable producers to alleviate dependency on inequitable international market and trade relations.

The identification and writing of such policy implications do not easily translate into effective actions. Undoubtedly, there are deeply entrenched political impediments to these kinds of changes. Nevertheless, the use of a "political ecology" approach—combining the concerns of ecology and political economy—contributes to deeper understanding of these dilemmas and potential solutions. This kind of innovative transdisciplinary perspective may help provide insights about how to overcome both human and natural re-

source degradation, and how to integrate social justice, environmental soundness, and economic well-being for radically different "sustainable" and people-centered patterns of development.

ACKNOWLEDGMENTS

I wish to express my appreciation to the following people for comments, suggestions, and
ideas regarding drafts of this chapter: David Campbell, Diana Liverman, Patricia Allen, Richard
Norgaard, Fred Buttel, Laskman Yapa, and Ben Wisner. I also appreciate the insights, contributions, and cooperation of many small farmers and analysts I have interviewed in Central
America. The views expressed in this paper are those of the author and do not necessarily
represent positions held by the World Resources Institute.

NOTES

1. The term "South" is used in this paper instead of "Third World," and "North" is used
   instead of "First World." This terminology is increasingly seen as more acceptable than
   other terms among development analysts.
2. Additional authors who have used variations of a "political ecology" perspective to analyze
   pesticides include Elling (1977), Laurel (1979), Perfecto (1990), and Wright (1990).
3. Political ecology differs significantly from systems analysis, mainly because of its theoretical
   foundation of social/political theory and its stress on political economic parameters.
4. See, e.g., the renowned work of Wilson (1976). Some of this work and the work of other
   sociobiologists analyzed animal (e.g., monkey) behavior to explain human behavior/social
   relations.
5. The information of this case study is derived from field research undertaken by the author
   in Costa Rica. Detailed findings are in a thesis: Deforestation, Agricultural Development,
   and Cattle-Expansion: An Integrated Approach to Land Use Transformation in Costa Rica.
   Stanford University, Latin American Studies Program.
6. A case study was undertaken by the author in the region of Puriscal, in the Central Valley
   of Costa Rica, which manifests typical patterns of agricultural development and resource
   degradation. This entailed systematic interviews with 120 small and medium scale farmers
   (Thrupp, 1980). Informal interviews were also carried out throughout other parts of the
   country as a follow-up. Other studies (e.g., Williams, 1986) also are helpful.
7. Information on these policies and government responses was obtained through interviews
   with policy officials, government documents, and sources such as Hartshorn et al. (1982),
   Leonard (1987), LACCDE (1990), and Thrupp (1990). More details on policies in each
   sector is available in these sources.
8. See also Anger (1989), which confirms similar community reactions to parks/reserves in
   the Atlantic.
9. Interview of farmer in Puriscal, Costa Rica, 1986.

REFERENCES

Altieri, M. 1983. *Agroecology: The Scientific BAsis of Alternative Agriculture*. Boulder, CO: Westview Press.

Anger, D. 1989. No Queremos El Refugio: Conservation and Community in Costa Rica. *Alternatives* 16(3):18–21.

Barraclough, S., and K. Ghimire. 1990. *The Social Dynamics of Deforestation in Developing Countries: Principal Issues and Research Priorities*. Discussion Paper 16, United National Research Institute for Social Development, New York.

Barry, T. 1987. *Roots of Rebellion: Land and Hunger in Central America*. Boston: South End Press.

Blaikie, P. 1985. *The Political Economy of Soil Erosion in Developing Countries*. Essex, England: Longman Scientific & Technical (John Wiley).

Blaikie, P., and H. Brookfield. 1987. *Land Degradation and Society*. London: Methuen.

Bullard, R. 1987. Environmentalism and the Politics of Equity: Emergent Trends in the Black Community. *Mid-American Review of Sociology* (12)2:21–37.

Burbach, R., and P. Flynn. 1980. *Agribusiness in the America*. New York: Monthly Review Press, North American Congress on Latin America.

Buttel, F., and W. Sunderlin. 1988. Integrating Political Economy and Political Ecology: An Assessment of Theories of Agricultural and Extractive Industry Development in Latin America. Rural Sociology, Cornell University, New York.

Byrant, B., and P. Mohai (eds). 1990. *Proceedings of the Michigan Conference on Race and the Incidence of Environmental Hazards*. Ann Arbor, MI: University of Michigan School of Natural Resources.

Campbell, D., and J. Olson. 1991. *Framework for Environment and Development: The KITE*. Occasional Paper No. 10, Department of Geography, Michigan State University.

Chambers et al. 1988. *Farmer First: Farmer Innovation and Agricultural Research*. London: Intermediate Technology Publications.

Cockburn, A., and S. Hecht. 1989. *Fate of the Forest, Developers, Destroyers, and Defenders of the Amazon*. New York: Routlege, Chapman and Hall.

Conway, G. R. 1987. The Properties of Agroecosystems. *Agricultural Systems* 24: 95–117.

CNS. 1989. *Capitalism, Nature, and Socialism*. Santa Cruz, CA: First Edition.

Dasmann, R., J. P. Milton, and P. H. Freeman. 1973. *Ecological Principles for Economic Development*. New York: John Wiley.

de Janvry, A. 1981. *The Agrarian Question and Reformism in Latin America*. Baltimore, MD: Johns Hopkins University Press.

Durham, W. 1979. *Scarcity and Survival in Central America: Ecological Origins of the Soccer War*. Stanford, CA: Stanford University Press.

Elling, Ray. 1977. Industrialization and Occupational Health in Underdeveloped Countries. *International Journal of Health Services* 7(2):209–236.

Ellis, Frank. 1978. Las Transnacionales del Banano en Centroamerica, Editorial Universitaria Centroamericana, San Jose, Costa Rica.

EPOCA. 1990. *El Salvador: Ecology of Conflict.* Green Paper #4, Environmental Project on Central America, San Francisco, California.

EPOCA. 1989. *Central America: Roots of Environmental Destruction.* Green Paper #2, Environmental Project on Central America, San Francisco, California.

Escobar, A., and L. A. Thrupp. 1990. The Unsustainable Paradigm of Sustainable Development. Unpublished Manuscript, Department of Sociology, Smith College, Massachusetts.

Goodman, D., and M. Redclift (eds). 1991. *Environment and Development in Latin America: The Politics of Sustainability.* Manchester: Manchester University Press.

Goodman, D., B. Sorj, and J. Wilkinson. 1989. *From Farming to Biotechnology: A Theory of Agro-Industrial Development.* Oxford: Basil Blackwell.

Hartshorn, G. et al. 1982. Costa Rica: Country Environmental Profile, Tropical Science Center and U.S. Agency for International Development, San Jose, Costa Rica.

Hilje, L., L. Castillo, L. A. Thrupp, and I. Weseling. 1987. El Use de los Plaguicidas en Costa Rica, Editorial Universidad Estatal a Distancia, San Jose, Costa Rica.

IUCN. 1987. Conservation and Equity, International Union for Conservation and Nature, Geneva, Switzerland.

LACCDE. 1990. Our Own Agenda, Latin American and Caribbean Commission on Development and Environment, Inter-American Development Bank, United Nations Development Programme, New York.

Laurel, A. C. 1979. Work and Health in Mexico. *International Journal of Health Services* 9(4):543–568.

Leff, E. 1986. *Ecologia y Capital.* Mexico: Universidad Nacional.

Leonard, J. 1987. *Natural Resources and Economic Development in Central America: A Regional Environmental Profile.* Washington, DC: International Institute of Environment and Development. New Brunswick, NJ: Transaction Books.

Little, P. D., and M. M. Horowitz. 1987. *Lands at Risk in the Third World: Local-Level Perspectives.* Boulder, CO: Westview Press.

Lutz, E., and Daly. 1990. Incentives, Regulation, and Sustainable Land Use in Costa Rica. *Environment Working Paper #34.* Washington, DC: The World Bank.

McAfee, K. 1991. *Storm Signals: Structural Adjustment and Development Alternatives in the Caribbean.* Boston: South End Press.

Milliken, B. 1992. Tropical Deforestation, Land Degradation, and Society: Lessons from Rondonia Brazil. *Latin American Perspectives* 19(1):45–72.

Myers, N. 1982. The Hamburger Connection. *Ambio* 10(1):3–9.

O'Connor, J., and A. Cockburn. 1989. Socialist Ecology: What It Means, Why No Other Kind Will Do. *Zeta Magazine* February: 15–30.

Odum, H. T. 1983. *Systems Ecology: An Introduction.* New York: John Wiley.

Panos. 1990. Hacia Una Centroamerica Verde, The Panos Institute and Departamento Ecumenico de Investigación, San José, Costa Rica.

Parsons, James. J. 1976. Forest to Pastures: Development or Destruction? *Revista Biologica Tropical* 24:121–138.

Pearse, A. 1980. *Seeds of Plenty, Seed of Want: Social and Economic Implications of the Green Revolution.* Oxford: Clarendon Press.

Perfecto, Y. 1990. Pesticide Exposure of Farmers and the International Connection. In *Race and the Incidence of Environmental Hazards.* Bryant B. and P. Mohai (eds.). University of Michigan, School of Natural Resources.

Redclift, M. 1987. *Sustainable Development: Exploring the Contradictions.* London: Methuen.

Repetto, R., and M. Gillis. 1988. *Public Policies and the Misuse of Forest Resources.* Boston: Cambridge University.

Schmink, M., and C. Wood. 1987. The Political Ecology of Amazonia. In *Lands At Risk: Local Level Perspectives.* P. D. Little and M. M. Horowitz (eds.). IDA Monographs in Developmental. Anthropology, Boulder, CO: Westview Press.

Spooner, B. 1986. Insiders and Outsiders in Baluchistan: Western and Indigenous Perspectives of Ecology and Development. In *Lands At Risk: Local Level Perspectives.* P. D. Little and M. M. Horowitz (eds.). Boulder, CO: Westview Press.

Thrupp, L. A. 1980a. *Deforestation, Agricultural Development, and Cattle Expansion in Costa Rica: An Integrated Approach to Problems of Land-Use Transformation.* Human Biology Program, Latin American Studies, Stanford University, Stanford, California.

Thrupp, L. A. 1980b. The Peasant View of Conservation. *Ceres* 14(4):31–34.

Thrupp, L. A. 1988. The Political Ecology of Pesticide Use in Developing Countries: Dilemmas in the Banana Sector of Costa Rica. Ph.D. Dissertation, Institute of Development Studies, Succex, England.

Thrupp, L. A. 1989. Legitimizing Local Knowledge: From Displacement to Empowerment for Third World People. *Agriculture and Human Values* Summer:13–24.

Thrupp, L. A. 1990. Environmental Initiatives in Costa Rica: A Political Ecology Perspective. *Society and Natural Resources* 3:243–256.

Watts, M. 1983. *Silent Violence: Food, Famine, and the Peasantry in Northern Nigeria.* Berkeley, CA: University of California Press.

WCED. 1987. *Our Common Future, World Commission on Environment and Development* (Brundtland Report). Geneva: Oxford University Press.

Williams, R. 1986. *Export Agriculture and the Crisis in Central America.* Chapel Hill, NC: University of North Carolina Press.

Wilson, E. O. 1976. *Sociobiology: A New Synthesis.* Cambridge: Cambridge University Press.

Wright, Angus. 1990. *The Death of Ramon Gonzalez: The Modern Agricultural Dilemma.* Austin, TX: University of Texas Press.

WRI. 1990. *World Resources Report.* Washington, DC: World Resources Institute.

WRI. 1991. *Accounts Overdue: Natural Resource Depreciation in Costa Rica.* Washington, DC: World Resources Institute and Tropical Science Center.

Yapa, L. 1979. Ecopolitical Economy of the Green Revolution. *The Professional Geographer* 31(4):371–376.

CHAPTER 3

Regenerative Food Systems: Broadening the Scope and Agenda of Sustainability

Kenneth A. Dahlberg

This overview seeks to broaden the scope of the theoretical, research, and policy debates over sustainable agriculture. It does this in several ways. It argues that we must go beyond the typical narrow focus on production (agriculture) to a broad analysis of complete food *systems*—which include not only production, but processing, distribution, use, recycling, and waste disposal. These food systems operate at a number of different levels—ultimately from the household to the international. In terms of theory, what I term "contextual analysis" is used to help sort out these levels and how they relate one to another. A broadened time horizon is also required, one that includes the ecological and evolutionary dynamics of regenerating both natural and human systems. Broadening the spatial and temporal scope of analysis also requires a broadening of evaluative criteria. At each level, basic goal and value assumptions must be brought out and examined in terms of their relevance and adaptability.

Such a broadening offers a better understanding of the basic structures and dynamics of food systems, one that should also be of value in developing the strategies—at each level—for making such systems more regenerative.

Food for the Future: Conditions and Contradictions of Sustainability, edited by Patricia Allen.
ISBN: 0-471-58082-1 © 1993 John Wiley & Sons, Inc.

After discussing some of the more general theoretical, conceptual, and evaluative questions, the broadened policy and research agendas that flow from this approach are outlined at each level.

THE THEORETICAL FRAMEWORK

To begin, what are the theoretical underpinnings of regenerative systems? Clearly systems approaches and theories are involved. In addition, a key theoretical point—and one that is often neglected—is that regenerative (or sustainable) *food* systems need to be understood both as part of many larger systems and as made up of many smaller systems. The health and regenerative capacity of both the larger and smaller systems will have a great bearing on the regenerative capacity and sustainability of food systems (which also can be structured and operate at several levels—from the household to the neighborhood to the regional, etc.). Both these larger natural, social, and technological systems and the particular food systems need to be analyzed in terms of their regenerative processes and cycles, something that requires an ecological–evolutionary framework.

Importance of a Systems Approach

Systems approaches are fundamental to evolutionary theories and involve an ongoing challenge to linear and reductionist models. The latter employ narrow temporal and analytic foci in order to seek precision and predictability and tend to ignore all informal and nonmeasurable phenomena. Also, they assume that the system they are analyzing is closed and unaffected either by evolutionary processes or catastrophe. If one recognizes these limitations and understands the results as a single "snapshot" of a carefully defined segment of open and evolving systems, such models can have a certain utility. However, proponents often seek to define as irrelevant or unscientific any consideration of other than the narrow and measurable phenomena they study. More significantly, these linear, reductionist, and functionally specialized approaches have come to predominate in most institutions—whether in academia or in the world of bureaucracies.

The results can be seen in each economic sector—where these theories and methodologies have been combined with powerful economic interests to yield our currently dominant economic growth ideologies and production paradigms. For example, in agriculture, to pursue an evolutionary/systems approach seriously would require not only shifts in policy priorities and bureaucratic structures, but a shift in scientific methodologies: from the bivariate statistical approaches of classical experiment station research to the systems dynamics approaches of ecologists (Conway, 1985).

Contextual Analysis

Such systems dynamics approaches are based on ecological and evolutionary approaches. In recent years ecologists have developed "hierarchy theory" to refine these by employing different concepts and units of analysis for each relevant time frame and level of analysis. This approach is similar to what I have termed "contextual analysis," the main difference being that in addition to examining natural systems at each level of analysis, I also include social and technological systems (Dahlberg, 1979). Also, the term "contextual" does not carry all of the historical overtones of governance by a superior elite from the top down that "hierarchy" does. Indeed, the main thrust of much of the literature on "hierarchy" and "sustainability" is that it is the health of the "lower" level units—whether cells in organs, microorganisms in soil, or individuals in social units—that ultimately determines the health of the "higher" level systems, although certainly the latter can have a significant influence on the health of the former.[1]

At a minimum, contextual analysis needs to be used to determine what are the key processes and structures of a system at one level of analysis and how that system is influenced by the systems above and below it—which have their own distinct processes and structures (each requiring other concepts and measures). In addition, the goals and values relevant at each level for the health and regenerative capacity of the system need to be included. This approach is in striking contrast to conventional assumptions that there are universal, value-free concepts that cut across time, space, and levels.[2] Table 1 seeks to illustrate how goals and values become broader or narrower according to their level (and scope) as well as how they vary along a spectrum of dominant and alternative values (Dahlberg, 1986). Conventional "universal-generalization" models also risk falsely "projecting" what may be useful concepts, measures, and/or goals at one level of analysis to other levels or time frames. Finally, because of Western cultural and industrial biases, such models also tend to neglect and undervalue natural, rural, and informal systems (Dahlberg, 1988).

Contextual analysis must also be concerned with the structure and interrelationships of the three basic subsystems that can be identified at each level of analysis: the natural, social, and technological. In natural systems, a diverse structure (that is, an ecosystem with many distinct species and populations widely distributed in space) offers more resilience and adaptiveness. In contrast to the evolutionary tendency for natural systems to become more complex, industrial societies have developed by simplifying and exploiting social and natural systems—replacing structural diversity and complexity with mo-

TABLE 1.   Goals Held by Dominant and Alternative Groups

Group/Level Involved	Dominant Positions		Alternative Positions	
	Goals (Professed and/or Operative)	Underlying Ethics and Values	Goals (Professed and/or Operative)	Underlying Ethics and Values
Farmers	Family support	Rural conservatism	Family and community support	Family/group self-reliance
	Make money, have a high standard of living	Individualism		Rural community
	Produce more through specialization by crop/commodity		Diversified farming/ homesteading	Integrated way of life
	Stewardship of the land	Love of nature	Conservation of energy, soil, and local species	Harmony with nature
	Fighting world hunger	Moral concern	Social justice	Moral concern
Agriculture as a sector	Increased production	Corporate and market economy	Having nutritious/ healthy food	Informal and cooperative approaches
	Stable prices and markets domestically		Sustainable production	Regenerative systems
	Expanding foreign markets		More local and regional markets (formal and informal)	Local and regional self-reliance
	Profitable operation		More small farms	Voluntary simplicity
	Specialization by commodity		Farm and regional diversity	Recycling systems
National	Increased production	Economic growth	Having nutritious/ healthy food	Sustainable economic growth
	Cheap food	Science and technology linked to progress		Respect of nature and ecosystems
	Foreign exchange and aid	National power	Rural revival	Cultural and personal contentment
	Industrialization of agriculture and urbanization		Rural revival and decentralization	
International	Elimination of hunger (through trade and aid)		Elimination of hunger (through local production)	
	Agricultural development	National sovereignty/ planning	Rural and ecodevelopment	A globe of villages
	Economic development	Expanding international markets and trade	Cultural development	Greater autarchy
Global	Balance between food, population, and resources	Western	Balance between food, population, and resources	Recessive Western plus non-Western
		Anthropocentric	Conservation of genetic and biological diversity	Inclusionist

Source: Dahlberg, 1986.

nocultures arranged in very *complicated* patterns, something that is often confused with *complexity*. The symbol of early industrial society, the clock, is complicated, but simple in structure, i.e., it has only four or five species (gears, springs, bearings, hands, etc.) and no redundancy. Consequently, the loss of a single gear tooth will immobilize it. It, like the factories and bureaucracies of industrial society, is based on functional specialization. The society-wide institutionalization of this has had profound consequences for the development of industrial society (Bennett and Dahlberg, 1990).

The disruption and simplification of traditional agricultural and rural systems that have accompanied their progressive incorporation into modern industrial society were made possible by the massive use of fossil fuels, something that has also generated pollution, acid rain, and greenhouse gases. Industrial agricultural systems are increasingly being spread worldwide through the green revolution and its institutionalization (Dahlberg, 1990b). When measured by conventional economic measures—which have short time horizons and ignore the large social and environmental disruptions and costs—these largely monocultural systems appear to be quite productive. However, given their dependence on fossil fuels and their lack of resilience or adaptivity, they are not sustainable over the longer term. Also, their complicated (as distinct from complex) structure makes them vulnerable to disruptions or macrochanges. Thus, like the trends in industrial society more generally, they are taking us in dangerous and unsustainable directions.

Another key aspect of contextual analysis is the inclusion of technological *systems.* Analysis of technologies, much less technological systems, has been seriously complicated by another set of Western cultural beliefs. While individual technologies are believed to be neutral (what may be termed the "myth of technological neutrality"), their general societal spread is (curiously) understood as "progress." These contrasting beliefs not only cloud our analytic lenses, but also provide important ideological service for those who are organizationally and technologically powerful by masking the real-world distributional impacts of new technologies and projects. We must come to understand that at a minimum technologies and technological systems reflect (1) the natural and social environments in which they were developed, (2) their scale and corresponding organizational and resource prerequisites, and (3) their design goals and principles. Only by understanding the nonneutral characteristics and distributional impacts of specific technological systems can we evaluate whether or not they contribute to the health and regenerative capacity of a particular system (see Dahlberg, 1990a, for a discussion).

Regenerative Processes and Cycles

Biologists and ecologists use such terms as "fitness," "adaptiveness," and "resilience" to describe the net reproductive success of different species, populations, or habitats. As the history of Social Darwinism illustrates, there can be great risks in combining an important theoretical advancement with the dominant social assumptions of an era—where evolutionary theory was used to rationalize the racial biases of imperial Europe. The debate over sustainability suggests some parallels—whether the discussion is in terms of "sustainable development" or "sustainable agriculture." The original research and thinking on sustainability derived primarily from worries over the destruction of natural systems and their regenerative capacity, along with a concern for the loss of indigenous and traditional cultures. Today, however, there is the risk that only the language, but little of the substance of sustainability will be adopted. It would be much harder, but healthier for the agricultural establishments of the world to make a serious efforts to reexamine the basic assumptions and structures of conventional systems (Dahlberg, 1991b).

Thus, the debate over the meaning of sustainability is a crucial one at all levels of society. At the global level, as the debate over *Our Common Future* (WCED, 1987) suggests, there is a struggle between those who would redefine not only development, but the larger economic, resource, and trade relations of the rich and poor countries and those who seek to hold on to privilege by defining sustainability in terms of sustaining currents systems and life-styles. This is one of the important reasons why I prefer the term "regenerative" to "sustainable." That is, it points much more directly to the basic reproductive and generational questions that are crucial to the health of individuals, populations, and societies.[3]

It also suggests an ongoing and evolutionary process of change and continuity. Just as each of us as individuals carries the genes of our parents (continuity), they are combined in a new way (change). The process is similar between the generations. Each generation seeks to socialize the next into its social and cultural assumptions and values (continuity). Yet, each generation grows up in its own historical context where wars, depressions, etc. can strongly shape its particular values and views (see Marias, 1970).

By focusing on the health and regenerative capacity of living systems, one is also forced to consider how they are dependent on the fluctuations, availability, and purity of nutrients that flow through the great biochemical cycles of which they are a part. Thus, the term regenerative requires a consideration of the negative impacts of industrial societies in terms of pollution and the simplification and/or destruction of habitats. Efforts to recycle physical materials, to shift toward renewable sources of energy, and to reduce pollution and other impacts of fossil fuel intensive systems must become an integral part of searching for more regenerative systems—integral, but secondary in the sense of needing to be structured and evaluated in terms of how they best contribute to the regenerative capacity of living systems.

In terms of agriculture, there is no consensus on what *sustainability* means (see Harwood, 1990; Dahlberg, 1991b). Yet, all definitions suggest a systems approach and a longer time horizon than is typical. Conventional approaches still tend to be disciplinary based, reductionist, and employ narrow economic or production and productivity criteria to measure their "success." Broader approaches that seek to capture the full dimensions of regenerative systems will certainly include production and economics, but will stress three other "e's" as well: ecology, ethics, and equity. These correspond to the three broad approaches that Douglass (1984) identified: (1) Sustainability as long-term food sufficiency, i.e., food systems that are more *ecologically based* and that do not destroy their natural resource base. (2) Sustainability as stewardship, i.e., food systems that are based on a conscious *ethic* regarding humankind's relationship to other species and to future generations. (3) Sustainability as community, i.e. food systems that are *equitable* or socially just. As the turmoil in Central America demonstrates, food systems cannot be sustainable if there are gross maldistributions of land, wealth, and power.

REGENERATIVE FOOD SYSTEMS

The discussion so far has focused on the general and theoretical aspects of regenerative systems with examples from agriculture added here and there. We now need to turn to an examination of *food* systems. As indicated above, food systems can exist at a number of levels—from the household to the neighborhood to the regional, etc. This means we must examine the degree to which food systems at one level are independent of, symbiotic with, or in conflict with food systems at other levels. To do this we must first be able to describe the full characteristics of any given food system. What then, do we mean by a food system?

The very language we use to talk about food systems reflects our functionally specialized thinking as well as the functionally specialized structures of our society. Food systems include the following: (1) production processes and inputs, (2) food distribution, (3) food preparation and preservation, (4) food use and consumption, (5) the recycling and disposal of food wastes, and (6) the various support systems—which will vary by level—that are required for the viable operation of the food system. Examples include marketing systems, transportation and distribution systems, storage systems, and a range of government services, such as research and extension.

One crucial reason for seeking to understand food systems as systems is that it leads us to focus on their inherent cycles and variability—natural and social. Only when we understand the degree to which our current food structures are fragmented and not systems based on natural cycles and variability, will we be able to develop strategies and policies to more them in more regenerative directions (ecological, equitable, ethical, and economic). In this regard, we can see a serious weakness in current thinking and work on sustainable *agriculture*: it does not recognize that in the longer term, it can be successful only to the degree that other portions of the food system *and* the larger society also become more sustainable and regenerative.

Turning to each stage in the food system cycle, we can see that production processes and inputs typically receive the most attention under the label of "agriculture." One of the basic conceptual implications of a food systems approach is that agriculture—whether conventional or regenerative—must be understood as only one stage or portion of the full food system at that level, as well as linked to food systems at other levels. For example, household gardeners—whether on the farm or in the city—produce food and are part of household food systems, but are not seen to be part of agriculture—which focuses on the production of commodities on farm fields. The roughly $18 billion worth of vegetables and fruit produced each year in household and community gardens (NGA, 1989) is largely ignored by the USDA and the land-grant system. In part this is because of their orientation toward farmers and fields, and in part because the gardens are part of the informal economy—something generally neglected by agriculturalists, economists, and sociologists (except those few who have learned of its importance in Third World settings).[+]

Food distribution and access are important issues at all levels. While the physical aspects of distribution are typically not significant at the household and village levels in the Third world, many anthropological studies point out significant gender and age differences in access at the household level, plus economic (poverty) distortions at the village level. Within the major industrial countries, physical distribution again is typically not a major problem, although the age, gender, and economic disparities of access are very significant. Even more significant are the disparities between the rich and poor countries. These issues revolve around both cultural attitudes as well as the political economy of different countries and food systems—something covered in detail in several of the accompanying chapters.

Food preparation and preservation are crucial parts of any food system—traditional or modern. In modern systems it is labeled "the food processing industry." Yet at the state, national, and international levels, it remains only one component of the larger food system. It is important to recognize this conceptually, because the relative institutional power of the different components of a given level's food system may change over time—where, for example, U.S. farmers have become much weaker vis-à-vis both input suppliers and output purchasers in the post-World War II era. Of course, at the household and neighborhood levels, food preparation and preservation often remain part of the informal economy. Other aspects, such as storing food in cans, refrigerators, or freezers, are linked to the formal economy. Whatever the case, an important cultural gate is crossed when agricultural produce or commodities are transformed into "food." Agriculture as a field rarely deals with any of the myriad cultural and social dimensions of food. The costs of this are apparent particularly in efforts to "transfer" agriculture to the Third World. There production is much more closely linked—practically and culturally—to all the other aspects of the local food systems, so that the importation of Western "agriculture" rarely occurs without major disruptions or difficulties (see Dahlberg, 1979, 1990b).

Food consumption seen in a cross-cultural perspective involves everything from various cultural conceptions of food, diets, purity, and contamination to social status, rituals, cooking techniques and technologies, eating implements and habits, to the status of women and children in the home. The highly formalized and complicated long distance distribution and marketing systems of the industrial world—with all of their attendant advertising—have added a host of political, social, and health issues relating to food safety, food labeling, false advertising, etc. And, of course, there has been a massive shift from people fixing their own meals to eating out at restaurants or eating in various institutional settings (colleges, military mess halls, nursing homes, etc.) or "grazing" among fast food shops in shopping malls.

Food recycling and disposal are important, but neglected parts of food systems. In nonindustrial food systems, most foods wastes are recycled—mainly through the use of animals. Little is disposed of otherwise—although this will vary according to cultural definitions of "garbage" or "waste." In industrial countries, there has been little effort until recently to try to recycle food and dispose of food wastes in more than a haphazard manner. Until World War II, waste food and garbage were often fed to pigs. With the surpluses of the 1950s, grain growers seeking new markets challenged the practice by raising "health" questions. The out of sight, out of mind approach to garbage in cities—whether disposed of through a garbage collector or through a "garbage disposal"—has led to the waste of a valuable resource and the unnecessary use of landfills and the polluting of water ways. The cutbacks in various forms of public assistance since 1980 have led to a dramatic increase in food banks, soup kitchens, and food pantries. They recycle millions of tons of surplus, mislabeled, or otherwise unsalable, but edible foods. Also, as more and more states prohibit the landfilling of yard wastes, there is a greater interest in composting.

The pervasiveness of the national level food system in the United States is reflected in the fact that no city in the United States (to my knowledge) has a "food department," even though most will have water, transportation, health, sewage, and planning departments. It is ironic that while much of the literature on "sustainable agriculture" talks of the need to "localize" food systems, there is little discussion of how cities might improve their local food system. What discussion there is focuses primarily on ways to bring food produced by local farmers to urban consumers (Lockeretz, 1988). Not that this is not important; but it should be complemented with efforts to understand and encourage the food-producing potential of the urban populace itself. The benefits of this could be significant—in terms of empowering people, helping the poor improve their diets, building neighborhood groups and identity, and reducing a city's waste stream. The energy efficiencies gained can also be expected to be significant.

Of course, localizing food systems and encouraging city dwellers to produce more of their own food will tend to challenge those with vested interests in current systems. There is a rough parallel here to the situation with USDA's Low Input and Sustainable Agriculture Program (LISA), where many input suppliers (ranging from equipment makers to chemical companies to credit providers) are threatened by farmers becoming more resource-efficient and self-reliant.

Support systems can be broken down into direct and indirect supports. Direct supports are found mainly in the agricultural (production) portion of food systems. In the United States, such direct supports include the agricultural education, research, and extension provided by the land-grant system, the research done by USDA, as well as some 20 direct farm support programs. These range from supply control measures (such as acreage reductions, set-asides, and marketing quotas) to price support measures (such as those for butter and cheese) to income support measures (such as target prices for grains, marketing loans, and crop insurance) (Knutson et al., 1986). It should be noted that many of these programs, particularly the price and income-support programs, encourage crop specialization at both the farm and regional levels, thus discouraging the movement toward more sustainable approaches based on crop rotations (GAO, 1990).

Indirect support systems such as transportation can be localized as they are in most traditional food systems or they can be ocean spanning as they are in industrial food systems. The energy and other resource costs of transportation throughout the food system have never been fully calculated. While many of the energy studies done of the U.S. or the U.K. food systems in the 1970s clearly demonstrated their energy inefficiency, they did not include the transportation or energy costs of food disposal. Besides the longer term economic risks to national and international food systems posed by the escalating costs of fossil fuels as they are depleted over time (Gever et al., 1986), there is also increasing concern about the climate impacts of fossil fuel consumption. Parallel concerns can be raised about each of the other major "inputs" to our large-scale food systems: water, NKP, soils, genetic diversity, as well as capital and human resources. It should be noted that the inputs (and outputs—including "externalities") need to be examined at each level and each of the stages in the food system/cycle. Only by so doing can we get a full picture of the real world balance sheet of the pluses and minuses of our food systems.

Let us now turn from this general discussion of the nature of food systems to an overview of what such a broadening of analysis means at each level in terms of policy, research, and goals.

GLOBAL CHALLENGES

These face all societies, both industrial and developing. Historically, industrial development and its various associated "externalities" have been driven by a "frontier mentality" and have pushed into lesser used regions. Agriculture, forestry, and mining have expanded into forest lands and marginal regions. Urbanization has become a global phenomenon. Pollution has reached all parts of the oceans and the atmosphere. Since there are no more frontiers, we must now begin to deal responsibly with the consequences of exponential population, industrial , and pollution growth, rather than shoving the consequences off onto other groups, societies, or environments.

In dealing with any one of the following global threats and uncertainties, prudence requires us to seek to move toward more diverse and resilient systems. Internationally, this will require a new "global bargain" between the rich and poor countries, with a focus on sustainable development (WCED, 1987). Essentially, the rich countries will have to reduce their high levels of fossil fuel and resource consumption while they provide greater aid to the poor countries. For their part, the poor countries will have to reduce their population growth rates and redirect their development efforts to reach the needy. This will require a reinterpretation of the goals of economic growth, adding to them equity, ethics, and ecology, all set in a larger framework based on the long-term sustainability of both natural and social systems. Finally, the energy and production systems of industrial countries will need to shift from huge, but simple linear systems (mining, refining, manufacturing, distribution, consumption, and throw away) that push huge flows of energy and materials through each stage (with attendant waste, pollution, and social disruption) to more localized and complex systems based on the recycling of materials and the regeneration of natural and social systems.

Economic Uncertainties

One of the most important uncertainties is the "Third world debt crisis." A default by any major debtor country could provoke major economic disruptions, perhaps even a major international depression. Such a depression would have much more serious consequences than that of the 1930s, primarily due to the changed structure of the international system. This changed structure is characterized by much greater centralization of infrastructures, institutions, and trading patterns. States are much more dependent on these poorly understood systems than in the 1930s—something that has a double danger. First, these systems—which are structurally simple and have little redundancy (even if they are very complicated)—are more vulnerable to collapse (given a major disruption) than those of the 1930s. In addition, there will be a much greater international multiplier effect. The stock market crash of 1987 offers an example of both these weaknesses and their multiplier effect.

Second, given the dependency of states on these systems for basic food, resource, trade, and financial flows, any depression or collapse will hit much harder because domestic systems of self-reliance have been weakened at all levels. Issues that now face the poor—food accessibility, shelter, and employment—will face whole societies, as we can see in the republics of the former Soviet Union. For many Third World countries that have geared their food and industrial production to export markets, it will be difficult to shift them to meet domestic needs. Beside these national economic problems, try to imagine the difficulties of rapidly shifting agricultural production from an export commodity—say cotton—to food crops. Not only would great changes be needed in training, equipment, cultivation practices, etc., but there would be both less capital and lower profits available to farmers to finance such changes. All of this would be further complicated by the efforts of the multinational corporations and banks (backed by the industrial countries) to salvage as much of their power and capital as possible.

What this suggests is a need for states to develop national economic and trade strategies that are based on increasing the sustainability and self-reliance of their renewable resource systems (at all levels) and reducing their dependence on long-distance markets and fossil fuel-based production systems (Dahlberg, 1987). This means focusing on sustainable livelihoods for employment, sustainable food systems for food access, etc. across the various economic sectors—building in greater equity at each level. The fact that all of this goes against the grain of the currently powerful centralized institutions does not alter the larger global facts, pressures, and uncertainties that states need to address now if they are to survive.

Resource Uncertainties

As suggested above, there are unprecedented uncertainties regarding the basic resources underpinning industrial society, particularly fossil fuels. Since they are the energy base on which industrial societies have been built, uncertainties regarding their availability and price in the future are significant. At some point—the timing of which depends on the degree to which energy efficiency programs are seriously pursued—they will become too expensive for most current uses, suggesting the need to turn to renewable resource systems. Ironically, however, current fossil fuel uses are seriously undermining renewable resource systems, whether through simplification, pollution, or destruction. The other major concern regarding fossil fuels is their impact on climate. Levels of greenhouse gas emissions are increasing alarmingly and climatologists are recommending a 20% reduction in carbon dioxide emissions by the year 2005 if we are to avoid massive disruptions caused by global warming.

Genetic and Biological Diversity

The importance of genetic and biological diversity can perhaps best be understood through an analogy to language. Recognizing that the "letters" of nature's alphabet (species) are greatly more numerous than those in human languages, the loss of a letter or two—and the words that contain them—would severely restrict language, thought, and communication. Thus, the rapidly increasing losses of species and diversity in gene pools, populations, and habitats threaten to restrict severely the adaptability and evolutionary potential of these three levels of diversity. Thus, we must understand them not in an economic context of "resources," but in a broader and more fundamental way—as the very *source* of the regenerative capacity of the renewable resource systems on which we will become progressively more dependent (Dahlberg, 1987; Wilson, 1988).

While loss of species and habitat is recognized by most scientists and some policy makers to be important, few see it either as fundamental or as important as the threats of climate change. In terms of financial and program commitments, it is clearly given a much lower priority (NRC, 1991). Each of these global threats suggests a redefinition of "national security" and a resulting shift in budgetary and program allocations. There is a parallel within agriculture—where in spite of a long-standing concern regarding the collection and maintenance of crop and livestock germplasm, its long-term conservation, especially *in situ,* remains a relatively low priority in most ministries of agriculture.

Climate Uncertainties

Many conferences and research programs are devoted to evaluating these uncertainties. In spite of different assumptions, all models suggest increasing variability in the temperate zones. This is most significant for agriculture. Any increase in variability is likely to have overall negative impacts since most agricultural systems are based on historic levels of variation.[5] Any significant change in temperature and/or moisture patterns or in the frequency or location of damaging storms will produce overall negative effects. Also, changes in pH and/or ultraviolet radiation levels may differentially affect the reproductive rates of different species—from microorganisms on up, changing the basic population and community dynamics of agricultural production systems.[6]

At some point, the aggregate modeling done at the global and national levels needs to be converted into regional impact studies that map the distribution of potential impacts in real space and time. This is because averages are inadequate policy guides. As farmers know, 2 years with the same average precipitation can result in very different production levels. To go beyond averages requires many people with detailed contextual knowledge of the different aspects of a region to discuss together the feedthrough effects of various changes. Since the range of possible impacts will be great, the best approach would be to change our institutions so as to give us a "disaster response capability" to meet any of a number of the contingencies, rather than planning for one or two specific possible events—whose actual likelihood of occurrence would be low. Developing such a capability will require significant changes in our institutions to make them more flexible and responsive. The sorts of changes sought by those promoting sustainable agriculture would be adaptive in these terms since they would create more decentralized, more diverse, and more flexible and adaptable farming systems—from the field level to farm level to the regional and national levels.[7]

While a number of specific implications for food and agriculture have been drawn in discussing the above challenges, it is important to recognize that adequate global responses require major reconceptualizations and the employment of correspondingly broadened criteria. New research agendas and new policies are also needed. In terms of reconceptualization, it would appear that the relevant units of analysis at the international and global level have to be different than those historically employed at the national level. At these levels, it is necessary to more clearly define and gather data on our renewable resource systems—where one often finds an intermingling of agriculture, forestry, grazing, and fisheries. Regional renewable resource systems and regimes intertwine with land use, land tenure patterns, watershed and irrigation patterns, trade and corporate patterns, diet preferences, etc.

As one example, concern over the loss of genetic and biological diversity through the cutting of tropical forests to provide grazing land to raise beef for fast food franchises in the rich industrial countries led to the popular concept of "the hamburger connection." However, traditional disciplinary divisions have made the "mapping" of the various overlapping patterns and interests involved very difficult. While it is clear that various international resource regimes exist, we have few conceptual tools and few data to map out and explore their dimensions (Young, 1989; Dahlberg, 1992).

Since the global challenges and threats raise basic questions of survival, new evaluative criteria are also needed, both to evaluate these global threats and to offer guidance in meeting them. They must also be blended with the new global concepts and new goals and indicators developed. The concept of sustainable development as presented in the Brundtland Commission report (WCED, 1987) needs to be complemented with redefinitions of economics as a field as well as with new indicators of development. One modest step in this direction is the development of the "Human Development Indicator" by the United Nations Development Program. While superior to the dominant GDP indicator, it still does not itself get at issues of how its components (health, literacy, and poverty) are distributed within countries by race, gender, age, ethnic group, region, etc.

NATIONAL CHALLENGES

These need to be addressed from a perspective of how to move toward more regenerative (sustainable) systems given the global uncertainties and threats outlined above. The first step is to try to develop a broad overview of the basic structural factors of the country under consideration. Such factors include resource endowments, particularly the availability of renewable resources, urban–rural balances, and the political economy of the various national sectors: energy, transportation, industry, health, education, food, etc. In focusing on the national food system, factors such as trade dependency, basic institutional and organizational patterns and behavior, land tenure patterns, and farm and market structures need to be examined. These then need to be evaluated in terms of possibilities and difficulties they pose for moving toward regenerative systems. As emphasized above, adaptive responses to macrochanges will be manageable only if more diverse, flexible, and resilient institutions are created. They will have to facilitate the long-term health of the relevant interacting systems.

In turning to a brief overview of the types of food system policies and research needed at each level, it must be emphasized that the analysis and suggestions made relate primarily to the United States and other industrial countries—with most examples drawn from the United States. A rather different set of recommendations would be required for Third World countries. Throughout, there is an effort to analyze needs in terms of the four "e's" mentioned above, although the emphasis shifts somewhat between the sections on resources and those on policy and research in order to fill in those topics historically neglected in each.

National Resources and Needs

The primary need here is to move toward a society and a food system based more on a recycling of materials and the regeneration of biological and social systems. Such a society offers the best hope of surviving the global uncertainties outlined above, while doing so in a more equitable and ethical manner. Moving toward such a society will require a major restructuring of organizations, institutions, and infrastructures to make them more diverse, complex, and resilient (i.e., less centralized and standardized).

The policy challenges and opportunities here revolve around the need to overcome cultural and policy biases toward urban development. If, as suggested above, renewable resource systems are more fundamental than nonrenewable resource-based systems and are likely to take on more economic importance as fossil fuels become more expensive, then the health of the rural sector and how it relates to national and regional food systems needs to be stressed. This suggests not only a reexamination of urban policies in the United States, but the need for a major shift in priorities to create a rural development policy that includes not only food and agriculture, but the preservation of the genetic and cultural diversity of rural landscapes (Gall and Staton, 1989).

Implementation of any such policy changes will require major reforms in existing institutional frameworks and practices—which have allowed for a shocking impoverishment of rural communities, families, soils, and landscapes over the past century. Such reforms need to be structured (1) to encourage better soil conservation and water quality, (2) to encourage a greater integration and interaction between agriculture, pastures, grazing, forestry, and fisheries, (3) to give greater priority to rural (and urban) *support systems* such as health, education, extension, and research, and (4) to give greater priority to materials recycling, energy efficiency, and renewable energy systems.

Research is needed at the national level on each of the above topics. In addition, there is a need for three other types of research. First, global climate models need to be brought down to the national, regional, and local levels and contextualized through detailed impact analyses of different possible scenarios. Only in this way will data necessary for policy making be generated. Second, comparable general and localized scenarios need to be developed on fossil fuel prices, e.g., what would be the impacts of a doubling or a quadrupling of prices? Or of major supply short-falls? Such studies would also be helpful in the emerging debates over the utility of a "carbon tax" in reducing greenhouse gas emissions. Given the energy intensiveness of the industrial food systems—where it takes 10 calories of energy to deliver 1 food calorie on our plates—the multiplier effect of any energy price increases through each stage of the food system (production, processing, marketing, refrigeration, cooking, and waste disposal) can be expected to be very high and to lead to major changes and adjustments. In addition, research needs to be done on the long-term health and productivity of renewable resource systems—farms, forests, fisheries, pastures and grazing lands—and threats to their regenerative capacity, whether from acid rain, overuse, or conversion.

Third, these is a need to develop broader evaluative criteria. As indicated above, there are significant limits to what economic criteria tell us about the long-term health of systems. Research on both natural and social indicators needs to be encouraged if we are to have a more accurate and comprehensive reading of the real-world balance sheet when we compare conventional and regenerative food systems. In agriculture such work needs to relate to all of the "externalities" that are not addressed by economic criteria: the continued availability and quality of basic natural resources (air, water, and soil; genetic and biological diversity); the health costs of agricultural production systems, both to natural systems and to farmers and consumers; and the social costs of technological change. Overall, this would require more support for social scientists and public health analysts interested in rural and agricultural matters. Comparable research is needed throughout the food system.

Regional and/or State Resources and Needs

Contextual approaches are based on the recognition that food and agricultural systems will be both more sustainable and successful when they are adapted to local and regional variations in soils, species, weather and climate, culture, and social institutions. There is thus a corresponding need in industrial countries for greater regional research, planning, and coordination to conserve and regenerate the renewable resource and *support systems* (natural and social) within each region. This is a difficult challenge. On what basis should a region be constituted? In terms of watersheds, airsheds, climate zones, ethnic patterns, economic patterns, political boundaries, or what? Beyond that, thinking about regions in the United States has been greatly complicated by our federal structure and by the vast differences in the size of states—where in one case a state might include several "regions" and in others where a "region" might include all or parts of several states. While there is an emerging literature on bioregionalism, it does not take sufficient cognizance of the legal and political difficulties of trying to structure conservation efforts along regional lines.[8]

These are illustrated by the one example of regional planning and development in the United States—the Tennessee Valley Authority (TVA). The history of the TVA has been one of continual tensions and battles between the TVA, various federal agencies, states, local communities, and various corporate interests. The result of the particular political and economic circumstances of the Great Depression, it is unlikely that anything similar will be created in the future. And while there are multistate federal "regions," they are primarily administrative mechanisms. There are a few state-generated regional efforts, such as the Great Lakes Compact, where states seek to harmonize their policies and legislation on a particular issue and to build a political coalition in Congress. In spite of all these constraints, it is still important to think about what the relevant regions are for the conservation of renewable resource systems and how one might move in that direction.

In terms of policy and research, it is mainly the states that have any capacity distinct from the federal government. The remainder of this section briefly reviews the types of things states might do—either individually or possibly in cooperation with other states and/or the federal government. Obviously, the mix of federal and/or state jurisdiction varies from sector to sector (agriculture as compared to health) as well as from state to state. The discussion is therefore general.

Three general policy challenges and opportunities are discernible. First, there is a need to integrate soil conservation with water quality programs and land use planning. An important part of this should be to maintain and enhance both the genetic and biological diversity in a region and the landscape diversity throughout a region. The difficulties here—both theoretical and political—can be seen in the controversies over the definition and protection of wetlands. Second, there is a great need to think about how to regenerate farm families and farm labor. At a minimum, a healthy rural structure with good supporting rural services in terms of education, health, transport, and communications is required. Yet there is little systematic work on how a healthy agriculture depends on a healthy rural sector (as well as healthy natural resource systems). Another aspect here is the need to be aware of the great value of the indigenous contextual knowledge that farmers and other rural residents have about their regions. This knowledge, gained over decades through careful observation, can be lost if it is not passed on to the next generation.

Third, there is a need to shift new production support systems (resource, energy input, and waste disposal systems) from linear throughput structures to systems based on recycling. Such systems need to facilitate the diversification and intermingling of renewable resource systems on a regional/state scale, much as a small-scale diversified farm does with its pastures, fields, woodlots, ponds, hedgerows, garden, and fruit trees.

These policy challenges suggest corresponding research needs. These all must be based on better understands of the linkages between the various natural and social *support systems* (air and water quality; soil conservation; health, education, and basic research) and the *production systems*. In the past, the emphasis has clearly been on production systems rather than support systems. And as stressed throughout, agriculture has tended to receive more emphasis than either rural development or other portions of the food system. Two important, but not widely used concepts are crucial to being able to fully evaluate agricultural and food systems and their regenerative capacities. These are the informal sector and common property systems.[9]

As mentioned earlier, the value of produce from U.S. gardens—which has been ignored by USDA and the land-grant system—is large enough to rival that of the corn crop. Other informal systems provide people with a range of support services and assistance. Natural systems pose special conceptual problems. Land regimes are very different from water regimes, which are different from air regimes. Some may be privately owned, others publicly, others—like air—unownable (see Bennett and Dahlberg, 1990 for a discussion). The literature on common property resources seeks to understand complex interactions between local social and resource systems in terms of the ways in which multiple and variable access and use are managed, often through informal negotiating processes (NRC, 1986). Much of the legislation (federal and state) on "multiple use" on public lands would benefit by reinterpretation and broadening in light of this literature.

Landscape/Watershed Resources and Needs

This level needs particular attention since it is a crucial one for actually maintaining the regenerative capacities of natural supporting systems. In addition, it has been largely neglected in agricultural and rural research and policy. Just as at the regional level of analysis, there are few governmental or administrative structures that are congruent with watersheds or landscapes. Counties are typically too small, while substate regions are normally defined in terms of several counties—whose borders have little to do with those of watersheds. Also, as with regions and river basins, there are several different ways they can be defined.

While it is not suggested that special policy-making organizations be set up at this level, it is important that state and national policy makers develop specific sets of policies and operational capabilities for this level. Some policy and administrative capabilities might be built by creating new programs within and between the Soil Conservation Service, the Army Corps of Engineers, and various state natural resource agencies.

The research needed here should focus primarily on habitats and how to strengthen long-term conservation efforts. Landscape ecology (broadened to include agricultural land use patterns) can be used to increase the genetic and biological diversity of local habitats. The combination of remote sensing capabilities and geographic information systems offer the possibility of doing integrated data collection and analysis—greatly facilitating the development of watershed and landscape planning and scenario building.

Community Resources and Needs

Rather different approaches may be needed between rural and urban communities. Rural communities have always been more directly linked to surrounding farms— although their current decline reflects the decline in the number of farmers and the tendency of remaining farmers to seek their inputs and equipment from large volume, lower price dealers rather than local dealers. Rural communities thus need to be understood primarily as part of the larger rural support system on which a healthy rural sector and agriculture depends.

Urban communities have traditionally been conceived of as the "consumers" of the food produced on farms. This perception is carried over into much of the literature on alternative agriculture that calls for a localizing of food systems—something that otherwise has value in challenging our amorphous conceptions of an undifferentiated national food system. Besides building better and more localized urban–rural interactions and linkages, there is also the need mentioned above to examine the food production, processing, and recycling potential of communities and cities themselves.

The greatest policy challenge here is for communities and cities to begin thinking and developing policies that relate to their own food needs. As mentioned, no city has a department of food. The food programs that do exist are largely related to federal and state hunger programs, supplemented by various private efforts by churches and volunteer organizations to provide food assistance. A study by the U.S. Conference of Mayors (1985) of five cities that had set up food advisory committees or councils showed significant unmet potential for cities to improve the health and welfare of their poorer citizens through a range of local policies. These include encouraging household food production (by reexamining restrictions household gardens and the raising of small livestock, etc.), community gardens and canning facilities (by providing land and/or startup capital), designing bus routes to facilitate access to lower price stores, promoting school breakfast programs, encouraging nutrition education in the schools, encouraging farmers markets and the preservation of urban fringe farmland, etc.

The research needs in this area will once again vary somewhat between rural and urban communities. Rural research needs to begin with a consideration of rural community structures—seeking to broaden the pioneering work of Goldschmidt (1978) and his successors. In addition to examining ways in which farm structure and land tenure patterns affect political diversity, there needs to be greater analysis of how changes in banking and financial structures at the state and federal levels have affected not only the viability of rural communities, but the willingness of no longer locally controlled banks to make loans to farmers seeking to shift to more sustainable and regenerative approaches.

In terms of resources, research on watersheds and landscapes needs to include the demands and impacts of both rural and urban communities on them. Research on energy resources and use at all levels should do the same. Research on the food systems of urban communities, besides being almost nonexistent, has no conceptual base.[10] What one can see is that problems parallel to those of the absorption of rural banks exist in cities where local food processors, supermarkets, and restaurants are being displaced or absorbed by regional or national chains or franchises. What this means for jobs and local economic activity needs analysis (see Hochner, et al., 1988 for an example dealing with supermarkets).

Farm-Level Resources and Needs

Current approaches by national and regional bodies need to be redefined in terms of the systems perspectives required to move toward more sustainable and regenerative systems.

There are several longer term policy challenges and opportunities here: (1) to develop policies that encourage farmers in a transition to more regenerative and sustainable food systems, policies that need to be integrated with policies at other levels, especially those relating to rural development and the maintenance and enhancement of biodiversity through new landscape approaches; and (2) to rethink existing, or develop new policies that will provide greater protection to farmers from the vagaries of weather, climate change, and markets, while making them responsible for their choices.

Since it has been shown that the current commodity support system tends to discourage crop diversification and pasture-based crop rotations (GAO, 1990), it would seem to make much more sense to focus directly on supporting farm families through a negative income tax. In addition, such an approach would help small and medium-sized farmers, whereas the preponderance of current commodity program dollars goes to large corporate farms. Such approach would need to be combined with other measures (antitrust policy, banking reform, changes in tax policy on rapid depletion allowances, etc.) that would be designed to reduce the centralization and power of both input suppliers and the large processors and marketers—whose market share and power have steadily increased (Constance and Heffernan, 1989).

Much greater research is needed on all of the above—something that will require different priorities and allocations of research funds within USDA and the land-grant system. Particularly, there will be a need for much greater social science input into the systems research that is required. This research will necessarily also be more interdisciplinary (including research between the natural and social sciences).

Such research will require a new type of research model and structure. In addition to moving from the classical bivariate statistical analyses of the Experiment Station to ecosystems and human ecology models (Wilson and Morren, 1990), another dimension needs to be added. For lack of a better term, I call this "conditionality," an if/then type of research. Rather than making the typical assumption that there is one given goal—which historically has been to increase production and/or productivity—researchers need to be able to provide answers to questions based on the quite varied goals of different farmers and bioregions. *If* the goal is to conserve soil (or water, or energy), *then* this kind of research is needed. *If* the goal is to provide a secure inheritance for a farm family's children, *then* that kind of research is needed. *If* the goal is to make the farm more resilient to climate changes, *then* other factor need to be researched. Such an approach will not only complement systems-oriented approaches (in terms of helping to sort out basic system parameters), but can be an integral part of developing impact assessments on the natural, social, and technological aspects of agriculture.

Research on natural systems needs to focus more on population and community dynamics, rather than individual species or breeds. Research on integrated pest management, organic farming, and sustainable agriculture clearly is moving in this direction. Research on farming systems has generally been focused around economic criteria and needs to be linked and integrated with research on natural systems. As indicated above, there is a need to explore the potential of interacting agricultural, grazing, forestry, and fishery systems—especially as a feasible alternative for various small-scale operations. Permaculture approaches (Mollison, 1988) offer some interesting thinking along these lines. The rotational cropping and grazing patterns involved in organic systems might also be compatible with multistory approaches or with aquacultural systems.

As mentioned above, there is also a need to think about how farm-level systems interact with watershed–landscape level systems, particularly in terms of biodiversity. A recent workshop in the United States between agricultural production researchers and conservation biologists has outlined the main issue involved (Gall and Staton, 1989).

Much more social science research is needed on farm families, households, and their interactions with the rural community. As mentioned, other research is needed on demographics and basic structural trends in the political economy of food and fiber (for a useful example, see Le Heron, 1988).

Two levels of research are needed on technological systems. The first is to think of designing new technologies in terms of the natural and social systems within which they will operate. They should be not only "user-friendly," but "use-friendly" in terms of their environmental impacts. Second, research is needed to improve the capacity to do technology assessments. To do this properly, there will have to be a much greater awareness of the nonneutrality of technologies, especially agricultural technologies (Dahlberg, 1990a).

Just as conditional research is needed, so too is "conditional extension service." The former should facilitate the latter. It also complements another valuable farm level principle: end use analysis. That is, both research and consulting need to be developed in terms of the end use contemplated. Some extension is already done in this mode, for example, soil testing—which gives the farmer not only knowledge regarding his situation, but a range of options depending on what his own goals and objectives are.

Household- and Neighborhood-Level Resources and Needs

There has already been some general discussion of the need to include household and neighborhood food systems in any analysis of urban food systems. Because this is where the greatest numbers exist, this certainly should be the main focus, although they also need to be examined in rural contexts as well. Unfortunately, the annual farm census specifically excludes food grown for household use, so the only source of data we have is the annual Gallup survey which the National Gardening Association commissions. This, however, covers only garden produce, not livestock used for household consumption.

There would appear to be few legal limitations or constraints on rural household food production, processing, preservation, and recycling. This is not the case in urban areas where there are often limitations on the raising of small livestock, on composting, the use of gray water, the handling of garbage, etc. Many of these are rationalized in terms of "health" measures, but appear more often to relate to gentrification. The maze of rules and regulations that may be required to create community gardens or community canning centers can be discouraging to all but the most persistent—again in part because municipalities have no administrative machinery or tradition that addresses food issues in any larger sense.

As suggested earlier, the lack of any theoretical or conceptual framework for understanding food systems is especially visible in the community setting—although there is certainly a wealth of solid, practical experience embodied in a range of "how to" books. What theoretical work exists has tended to come from outside agricultural circles. *The Integral Urban House* (Olkowski et al., 1979), which details the ecological principles used in retrofitting a nineteenth-century San Francisco frame house for household food production and maximum resource efficiency, cycling, and recycling, has been out of print for some time now. The work on permaculture by Bill Mollison (1988) comes much more out of a landscape ecology perspective than an agricultural one. As mentioned above, there is virtually no research on the energetics of the disposal of food wastes, although Bill Rathje of the University of Arizona has done extensive "archeological" excavations of urban garbage.

CONCLUSIONS

It has been argued that new approaches—conceptual and organizational—are needed if industrial societies and their agriculture are to develop the flexibility and resilience needed to respond to the range of global uncertainties facing them. A general theory of regenerative systems is needed to address these rigidities and uncertainties. One important component of such a general theory—that dealing with food systems—has been outlined here along with some of the policy and research implications that flow therefrom.

In applying what I have termed "contextual analysis" several important, but neglected dimensions emerge. In terms of the different levels of analysis, it appears that much more work needs to be done on regional, landscape–watershed, community, and household levels. Much of this must include urban food systems, another neglected area. These as well as the national, state, and farm levels need to be mapped not only in terms of their production aspects (agriculture), but in terms of the total food system, including food processing, preservation, use, recycling, and disposal. In doing this the interactions over time of natural, social, and technological systems need to be traced at, and between levels. Fundamental shifts in methodology, theory, and evaluative criteria will be required to do all this. Specifically, "production and productivity" criteria need to be supplanted by criteria relating to the health and regenerative capacity of food systems.

The basic discontinuities between current industrial structures—which are complicated, but inflexible—and the adaptive institutions needed for sustainable and regenerative food systems hopefully become more visible using this approach. Also, it is hoped that this approach helps to clarify and sort out by and among levels the research and policy directions needed to move in these more regenerative directions. Ultimately, however, such changes in direction will require both political efforts and ratification— something that hopefully will emerge out of the fear and anger of family farmers over the injustices, health costs, and unsustainability current policies and production approaches force on them and the rather inchoate fear of consumers regarding the safety of their food and their search for healthier foods and environments. A regenerative food systems approach thus offers the possibility of building political bridges and alliances between farmers and consumers as well as conceptual ones.

NOTES

1. This is not to suggest that there is one "natural" way in which the various levels can or should be organized—only that there are mutual interdependencies that need to be recognized if communities and societies are to be able to adapt and survive in changing environments. As a wide range of anthropological literature has shown, various societies have organized the roles and functions of different social levels quite differently. One of the key questions facing us today is whether the ways we have chosen to organize modern industrial society—generally adopting the Hamiltonian model over the Jeffersonian one—will enable us to adapt to the dramatic social, environmental, resource, and population changes of this century, many of them brought on by industrial society itself. For a discussion, see Bennett and Dahlberg (1990).

2. There are, of course, many concepts that claim or assume "universality." They are, however, intellectual constructs that reflect the values and interests of the individuals, groups, and cultures using them. The key question is how much they also reflect or capture the realities that they are employed to seek. "Universal" concepts are rather like a particular gauge fishnet. It will catch all fish larger than that gauge, but will let smaller ones escape. It assumes that you are after a certain sized (or larger) fish. A more contextual approach is to use specific sized hooks and bait that are more precisely targeted at certain species. The fish net analogy, of course, suggests the need for a multiple level approach, as would a map analogy—where a "bounded universality" can be achieved at each scale of map. But as both analogies suggest, there are inevitable trade-offs between the scale of the map (or gauge of the fishnet) and the detail captured. World maps capture broad scope, but little detail while city maps capture great detail, but little scope. Equally, the "universality" is bounded temporally—as illustrated dramatically by the dilemma of map makers today in how to portray the changes going on in the Soviet Union.

3. There has been a long running debate over terminology. Robert Rodale championed the term "regenerative" over "sustainable" for a number of years (Rodale, 1983). By 1990, even he recognized that "sustainable" had become the popular label. Some biologists object to the term since they associate it with the capacity of certain creatures to regenerate appendages. The term is used much more broadly here—to refer to the adaptive reproduction and regeneration of individuals, populations, habitats, and social knowledge and systems under the selective pressures of changing environments.

4. In Third World settings, there is also an important gender aspect to household or "kitchen" gardens in that it is largely women who tend them. It is only recently that the many African women farmers growing subsistence crops have received any sort of official attention.

5. It is only recently that seasonal variability has gained any attention. Such variability includes not only variability in temperature and moisture patterns, but in the price of inputs, the markets for crops, and the availability and price of food (Sahn, 1989).

6. The geopolitical implications of all of this are also significant. Shifts in climate zones or

changes in monsoon patters could dramatically alter which crops are grown where, if at all. Naive speculation that crops and/or forests will simply migrate with changes in climate ignores all of the natural barriers (different soils, increased pest activity, natural diffusion rates, etc.) as well as the loss of massive capital investments. If climates favorable to corn growing "migrate" to Canada, that will mean that the Colorado River basin will largely dry up—leaving hundreds of billions of dollars of dams and irrigation canals behind as the modern equivalent of the pyramids—large monuments to a past way of life sitting in the desert. Even the supposed "winners" will have to make drastic organizational and infrastructural changes to take advantage of any climate "improvements." This will take a great deal of capital for agricultural investment exactly at a time when the dominant urban–industrial majorities will be pushing for capital investments to protect or move sea-level cities, relocate people from the new desert regions, etc.

7. Since the above implications of global warming clearly strengthen the many other arguments for moving to sustainable approaches and link it to other major societal concerns, the proponents of sustainable agriculture need to follow the debates and the literature on global warming. Besides "translating" and bringing the implications of climate change into debates over the future of agriculture, they also need to try to provide input into the climate debates since few of the global modelers are aware that there are proven alternatives to the highly energy consumptive and inefficient approaches of conventional agriculture.

8. Bioregionalism is an attempt to define regions in the United States primarily in terms of natural phenomena such as biomes and watersheds. For a discussion of bioregionalism, see Sale (1985).

9. It should be noted that these concept have been developed and employed primarily in analyzing the problems of Third World rural development and food and agricultural programs. It is curious that the great reservoir of knowledge, concepts, and experience gained in and about the Third World has been so little applied in the First World. One underlying reason is the cultural hubris that sees the First World as modern and "developed" and the Third World as traditional and backward.

10. Recently, I have begun work on this—exploring how household food systems relate to neighborhood and municipal food systems. This conceptual framework is being applied to the five cities that were involved in the U.S. Conference of Mayors food policy project.

REFERENCES

Bennett, J. W., and K. A. Dahlberg. 1990. Institutions, Social Organization, and Cultural Values. In *The Earth as Transformed by Human Action*. B. Turner II (ed.), pp. 69–86.Cambridge: Cambridge University Press.

Constance, D., and W. Heffernan. 1989. The Demise of the Family Farm: The Rise of Oligopoly in Agricultural Markets. Paper presented at the Agriculture, Food, and Human Values Society Meeting, Little Rock, Arkansas, November 2, 1989.

Conway, G. R. 1985. Agroecosystem Analysis. *Agricultural Administration* 20:31–55.

Dahlberg, K. A. 1979. *Beyond the Green Revolution: The Ecology and Politics of Global Agricultural Development*. New York: Plenum.

Dahlberg, K. A. ed. 1986. *New Directions for Agriculture and Agricultural Research: Neglected Dimensions and Emerging Alternatives*. Totowa, NJ: Rowman & Allanheld.

Dahlberg, K. A. 1987. Redefining Development Priorities: Genetic Diversity and Agroecodevelopment. *Conservation Biology* 1(4):311–322.

Dahlberg, K. A. 1988. Towards a Typology for Evaluating Conventional and Alternative Agricultural Systems and Resarch Strategies. In *Global Perspectives on Agroecology and Sustainable Agricultural Systems,* 2 volumes. P. Allen and D. Van Dusen (eds.), pp. 103–112. Agroecology Program. Santa Cruz, CA: University of California.

Dahlberg, K. A. 1990a. The Value Content of Agricultural Technologies. *Agricultural Ethics* 2(2):87–96.

Dahlberg, K. A. 1990b. The Industrial Model and Its Impact on Small Farmers: The Green Revolution as a Case. In *Agroecology and Small Farm Development,* M. Alteri and S. B. Hecht (eds.), pp. 81–88. Boca Raton, FL: CRC Press.

Dahlberg, K. A. 1991b. Sustainable Agriculture: Fad or Harbinger? *BioScience* 41(5): 337–340.

Dahlberg, K. A. 1992. Renewable Resource Systems and Regimes: Key Missing Links in Global Change Studies. *Global Environmental Change* 2(2):128–152.

Douglass, G. A. 1984. *Agricultural Sustainability in a Changing World Order.* Boulder, CO: Westview Press.

Gall, G. A. E., and M. Staton. 1989. *Integrating Conservation Biology and Agricultural Production: Executive Summary of International Workshops.* Public Service Research and Dissemination Program. Davis: University of California.

General Accounting Office (GAO). 1990. *Alternative Agriculture: Federal Incentives and Farmers' Options.* GAO/PEMD-90-12. Washington, DC: U.S. General Accounting Office.

Gever, J. R., R. Kaufman, D. Skole, and C. Vorosmarty. 1986. *Beyond Oil: The Threat of Food and Fuel in the Coming Decades.* Cambridge, MA: Ballinger.

Goldschmidt, W. 1978. *As You Sow: Three Studies in the Social Consequences of Agribusiness.* Montclair, NJ: Allanheld, Osmun.

Harwood, R. R. 1990. A History of Sustainable Agriculture. In *Sustainable Agricultural Systems.* (eds.), pp. 3–19. C. A. Edwards, R. Lal, P. Madden, R. H. Miller, and G. House. Ankeny, IA: Soil and Water Conservation Society.

Hochner, A., C. S. Granrose, J. Goode, E. Simons, and E. Applebaum. 1988. *Job Saving Strategies: Worker Buyouts and OWL.* Kalamzaoo, MI: W. E. Upjohn Institute for Employment Research.

Knutson, R. D., J. W. Richardson, D. A. Klinefelter, M. S. Paggi, and E. G. Smith. 1986. *Policy Tools for U.S. Agriculture.* Agricultural and Food Policy Center. B-1548. College Station: Texas A&M University.

Le Heron, R. 1988. Food and Fibre Production under Capitalism: A Conceptual Agenda. *Progress in Human Geography* 12(3):409–430.

Lockeretz, W., ed. 1988. *Sustaining Agriculture near Cities.* Ankeny, IA: Soil and Water Conservation Society.

Marias, J. 1970. *Generations: A Historical Method.* Alabama: University of Alabama.

Mollison, B. 1988. *Permaculture: A Designer's Manual.* Tyalgum, Australia: Togari Books.

National Gardening Association (NGA). 1989. *National Gardening Fact Sheet.* Burlington, VT: National Gardening Association.

National Research Council (NRC). 1986. *Proceedings of the Conference on Common Property Resource Management.* Board on Science and Technology for International Development. Washington, DC: National Academy Press.

National Research Council (NRC). 1991. *Managing Global Genetic Resources: The U.S. National Plant Germplasm System.* Board on Agriculture. Washington, DC: National Academy Press.

Olkowski, H., B. Olkowski, T. Javits, and the Farallones Institute Staff. 1979. *The Integral Urban House: Self-Reliant Living in the City.* San Francisco: Sierra Club Books.

Rodale, R. 1983. Breaking New Ground: The Search for a Sustainable Agriculture. *Futurist* 1:15–20.

Sahn, D. E., ed. 1989. *Seasonal Variability in Third World Agriculture: The Consequences for Food Security.* Baltimore: Johns Hopkins University Press.

Sale, K. 1985. *Dwellers in the Land: The Bioregional Vision.* San Francisco: Sierra Club Books.

United States Conference of Mayors. 1985. *Municipal Food Policies: How Five Cities Are Improving the Availability and Quality of Food for Those in Need.* Washington, DC: United States Conference of Mayors.

Wilson, E. O., ed. 1988. *Biodiversity: Proceedings of the Forum on Biodiversity.* Washington, DC: National Academy Press.

Wilson, K., and G. E. Morren, Jr. 1990. *Systems Approaches for Improvement in Agriculture and Resource Management.* New York: Macmillan.

World Commission on Environment and Development (WCED). 1987. *Our Common Future.* Oxford: Oxford University Press.

Young, O. R. 1989. *International Cooperation: Building Regimes for Natural Resources and the Environment.* Ithaca, NY: Cornell University Press.

CHAPTER 4

Vegetarianism and Sustainable Agriculture: The Contributions of Moral Philosophy

Tom Regan

A great deal of recent work by moral philosophers—much of it in environmental ethics, for example, but much of it also in reference to questions about obligations to future generations and international justice—is directly relevant to sustainable agriculture. Yet most of this work has been overlooked by most of the advocates of sustainable agriculture. Why this is so is unclear, but certainly the responsibility for this lack of communication needs to be shared. Like all other "specialists," moral philosophers have a tendency to converse only among themselves, just as, like others with a shared, crowded agenda, advocates of sustainable agriculture have limited discretionary time, thus little time to explore current tendencies in any discipline outside the canonical ones, even including moral philosophy. This chapter attempts to take some modest steps in the direction of better communication between moral philosophers and advocates of sustainable agriculture.

After a brief historic section, four tendencies in contemporary moral philosophy—utilitarianism, animal rights, holism, and ecofeminism—are described and some of their implications regarding sustainable agriculture are explained. Not all these tendencies can be true in every respect (for they

Food for the Future: Conditions and Contradictions of Sustainability, edited by Patricia Allen.
ISBN: 0-471-58082-1 © 1993 John Wiley & Sons, Inc.

contradict each other at crucial places), and perhaps none is true in any. Unquestionably, however, these four tendencies are among the most important options in moral philosophy today, so that if they can be shown to give the same answer to a given question, this unanimity is not only unusual, it is important. If I am right, just such unanimity obtains when each position surveys intensive animal agriculture (so-called "factory farming"). Though the reasons offered differ, each of the positions condemns this form of agriculture. Because intensive animal agriculture *is* a form of agriculture, advocates of sustainable agriculture cannot avoid asking whether it is sustainable. If it is sustainable, then these advocates will be obliged to answer the criticisms raised by the philosophers who reach a contrary judgment, while if intensive animal agriculture is not sustainable, then partisans of sustainable agriculture might look to these philosophers for their assistance in opposing this way of "growing animals." In either case, the arguments offered by utilitarians and ecofeminists, for example, have an obvious relevance to sustainable agriculture and should not be neglected in the future as they have been in the past.

Neglect, as it happens, may be the more comfortable option. For reasons I attempt to explain in what follows, not only have holists and ecofeminists, for example, roundly criticized intensive animal agriculture; they have also strongly defended a vegetarian life style, on ethical grounds. Now, vegetarianism seldom is mentioned let alone seriously discussed by proponents of sustainable agriculture. The plight of the family farm, worker safety, pollution, issues of race and gender, economic and others forms of exploitation—these and a host of related topics receive attention, as they deservedly should. But not vegetarianism, which, despite its recent rise in popularity (there are an estimated 15–20 million vegetarians in the United States), still retains something of its past reputation of being "weird" or worse. So perhaps a reason (though, as I have already noted, not the only reason) why both utilitarian and animal rights critiques of intensive animal agriculture have been largely ignored by advocates of sustainable agriculture has something to do with the distaste many people have for "the 'V' word," "vegetarianism"—the unsavoriness (so to speak) of the vegetarian life-style championed by these philosophers. Speculation about such matters to one side, what I hope to be able to explain is how, in addition to speaking with one voice in opposition to factory farming, the philosophies I describe can also speak with one voice in favor of ethical vegetarianism. In this way, then, the recent tendencies in moral philosophy I have mentioned, despite their many differences, may yet combine to force "the V word" onto the agenda of issues that need to be considered by advocates of sustainable agriculture.

In the nature of the case, of course, not all proponents of sustainable agriculture will be persuaded by the arguments advanced in the pages that follow. But none can be unpersuaded by these arguments if the arguments are not considered, and it is in the hope of having the issues I raise considered by informed people of good will that I raise them.

CHANGING TIMES

Philosophers once had little good to say about animals other than human beings. "Nature's *automata,*" writes Descartes (Descartes; Regan and Singer, 1976:61). Morally considered, animals are in the same category as "sticks and stones," opines the early twentieth-century Jesuit Joseph Rickaby (Rickaby; Regan and Singer, 1976:179) True, there have been notable some, throughout history, who celebrated the intelligence, beauty, and dignity of animals: Pythagoras, Cicero, Epicurus, Herodotus, Horace, Ovid, Plutarch, Seneca, and Virgil. Hardly a group of ancient world "animal crazies." By and large, however, a dismissive sentence or two sufficed or, when one's corpus took on grave proportions, a few paragraphs or pages. Thus we find Immanuel Kant, for example, by all accounts one of the most influential philosophers in the history of ideas, devoting almost two full pages to the question of our duties to nonhuman animals, while St. Thomas Aquinas, easily the most important philosopher-theologian in the Roman Catholic tradition, bequeaths perhaps 10 pages to this topic.

Times change. Today an even modest bibliography of the past decade's work by philosophers on the moral status of nonhuman animals would easily equal the length of Kant's and Aquinas's treatments combined,[1] a quantitative symbol of the changes that have taken place, and continue to take place, in philosophy's attempt to excise the cancerous prejudices lodged in the anthropocentric belly of Western thought.

With relatively few speaking to the contrary (St. Francis always comes to mind in this context), theists and humanists, rowdy bedfellows in most quarters, have gotten along amicably when discussing questions about the moral center of the terrestial universe: *Human* interests form the center of this universe. Let the theist look hopefully beyond the harsh edge of bodily death, let the humanist denounce, in Freud's terms, this "infantile view of the world," at least the two could agree that the moral universe revolves around us humans—our desires, our needs, our goals, our preferences, our love for one another. The intense dialectic now characterizing philosophy's assaults on the traditions of humanism and theism, assaults aimed not only at the traditional account of the moral status of nonhuman animals but at the foundations of our moral dealings with the natural environment, with Nature generally—these assaults should not be viewed as local skirmishes between obscure academicians each bent on occupying a deserted fortress.

At issue is the validity of alternative visions of the scheme of things and our place in it. The growing philosophical debate over our treatment of the planet and the other animals with whom we share it is both a symptom and a cause of a culture's attempt to come to critical terms with its past as it attempts to shape its future.

At present there are four major challenges being raised by moral philosophers against moral anthropocentrism. The first comes from utilitarians, the second from proponents of animal rights, the third from those who advocate a holistic ethic, and the fourth from certain ecofeminists. This essay offers brief summaries of each position with special reference to how they answer two questions: (1) Is vegetarianism required on ethical grounds? (2) Judged ethically, what should we say, and what should we do about the dominant forms of commercial animal agriculture, as practiced in affluent economies?

SOME PRELIMINARIES

A number of preliminary points need to be made before proceeding. The first concerns the meaning of vegetarianism. One thing seems clear: When it comes to what it is to be a vegetarian, different people mean different things. Thus, "ovo-vegetarians" do not eat meat or dairy products but do eat eggs, while "lacto-ovo-vegetarians" eat both eggs and dairy products but do not eat meat. And some people who describe themselves as vegetarians eat not only eggs and dairy products, they also eat chicken, turkey, and fish. For present purposes, I propose to understand vegetarianism in what I think is its most natural sense—namely, not eating meat (or flesh), whether "red" meat or "white" meat. Questions about eating eggs or cheese or other animal products will not be considered except by implication.

Second, to ask whether vegetarianism is required on ethical grounds is to ask whether there are reasons other than those of self-interest (for example, other than those that relate to one's own health or financial well-being) that call for leading a vegetarian way of life. Each of the theoretical tendencies mentioned above maintains that such reasons exist.

Third, as already indicated, this chapter focuses on ethical questions against the backdrop of the dominant forms of animal agriculture, as practiced in affluent economies. The ethics of other practices that routinely kill other animals (for example, hunting and trapping) will not be considered except in passing, not because they should be immune to critical moral reflection but only because space and time preclude my attempting it here. Space and time also preclude my offering sustained critical assessments of the important and divergent philosophical views I discuss.[2]

Fourth, despite their many differences, all moral philosophers agree that "ought" implies "can"; they agree, that is, that persons can be *morally obligated* to act in certain ways only if, without making extraordinary efforts or behaving in an uncommonly self-sacrificial manner, they can (they "are able") to act in the specified ways. This principle ("'Ought' implies 'can'") explains why, for example, blind persons are not obligated to warn others of the dangers posed by falling objects (for those who are blind cannot see such objects and so cannot be obligated to warn others about them). Moreover, this same principle also explains why people who act heroically or make uncommon sacrifices *do more than duty requires*. Given any reasonable account of what our duties are, there will be some acts that *exceed* them—acts that "go beyond the normal call of duty" (what philosophers call "supererogatory acts"). Our duties, whatever they are, are obligatory, not supererogatory. This much is common ground for all moral theoreticians. And so is the following, related point.

The same act can be obligatory for some people but supererogatory for others. For example, the range of one's income makes a difference to what one is morally obligated to do. The very rich have obligations to their children (providing financially for their educational and health needs, for example) that are not shared by the very poor. True, by making enormous sacrifices themselves, indigent parents might be able to set aside the money that is necessary to send their children to a good dentist or one of the better universities. And any parent who did so would deserve our moral praise. But they would deserve it precisely because they have acted in a supererogatory fashion. Whereas the wealthy parent has done only what duty requires, the indigant parent has done a good deal more.

The final preliminary point emerges from the previous ones. The question of the moral status of vegetarianism cannot be addressed in a real world vacuum. One of my primary interests in this chapter lies in asking whether people who *can* choose to live as vegetarians, without behaving heroically and without performing acts of great self-sacrifice, *ought* to do so. As I hope to explain, all the major philosophical tendencies to be examined arguably imply an affirmative answer to this question. But it is important to realize, for reasons already explained in the preceding, that it is not inconsistent to affirm that *these* people *are* morally obligated to forgo meat eating and to deny that *other* people (those who are chronically impoverished and malnourished, for example) *are not*. I return to this matter in my conclusion.

MORAL ANTHROPOCENTRISM

Aquinas and Kant speak for the anthropocentric tradition. This tradition does not issue a blank check when it comes to how humans may treat other animals. Positively, we are enjoined to be kind to them; negatively, we are prohibited from being cruel. But we are not enjoined to be the one and prohibited from being the other because we owe such treatment to these animals *themselves.* For we have no duties *to other animals,* according to the anthropocentric tradition; rather, it is because of human interests that we have those duties we do. "So far as animals are concerned," writes Kant, "we have no direct duties . . . Our duties to animals are merely indirect duties to mankind." "[H]e who is cruel to animals becomes hard also in his dealings with men," writes Kant. *That* is why cruelty to animals is wrong. As for kindness, Kant observes that "[t]ender feelings towards dumb animals develop humane feelings toward mankind." And *that* is why we have a duty to be kind to animals.[3]

So reasons Kant. Aquinas predictably adds theistic considerations but the main story-line is the same, as witness the following representative passage from his *Summa Contra Gentiles.*

> Hereby is refuted the error of those who said it is sinful for a man to kill dumb animals: for by divine providence they are intended for man's use in the natural order. Hence it is not wrong for man to make use of them, either by killing, or in any other way whatever. . . . And if any passages of Holy Writ seem to forbid us to be cruel to dumb animals, for instance to kill a bird with its young: this is either to remove men's thoughts from being cruel to other men, and lest through being cruel to animals one becomes cruel to human beings: or because injury to an animal leads to the temporal hurt of man, either of the doer of the deed, or of another: or on account of some (religious) signification: thus the Apostle expounds the prohibition against *muzzling the ox that treadeth the corn.*[4]

To borrow a phrase from the twentieth-century British philosopher Sir W.D. Ross, our treatment of animals fro both Kant and Aquinas is "a practising ground for moral virtue" (Ross, 1930:47). The real "moral game" is played between human players or, on the theistic view, human players plus God. The way we treat other animals is a sort of moral warmup—character calisthentics, as it were—for the real game in which these animals themselves play no part.

UTILITARIANISM

The first fairly recent spark of revolt against moral anthropocentrism comes, as do other recent protests against institutionalized prejudice, from the pens of the nineteenth-century utilitarians Jeremy Bentham and John Stuart Mill. In an oft-quoted passage Bentham enfranchises sentient animals in the utilitarian moral community by declaring "[t]he question is not, Can they *talk?*, or Can they *reason?*, but, Can they *suffer?*" (Bentham; Regan and Singer, 1976:130). And Mill goes even further, writing that utilitarians "are perfectly willing to stake the whole question on this one issue. Granted that any practice causes more pain to animals than it gives pleasure to man: is that practice moral or immoral? And if, exactly in proportion as human beings raise their heads out of the slough of selfishness, they do not with one voice answer 'immoral' let the morality of the principle of utility be forever condemned" (Bentham; Regan and Singer, 1976:132). Some of our duties are *direct duties to other animals,* not indirect duties to humanity. For utilitarians, these animals are themselves involved in the moral game.

Viewed against this historical backdrop the position of the influential contemporary moral philosopher, Peter Singer, can be seen to be an extension of the utilitarian critique of moral anthropocentrism.[5] In Singer's hands utilitarianism requires that we consider the interests of everyone affected by what we do, and also that we weight equal interests equally. We must not refuse to consider the interests of some people because they are catholic, or female, or black, for example. Everyone's interests must be considered. And we must not discount the importance of equal interests because of whose interests they are. *Everyone's* interests must be weighed *equitably.* Now, to ignore or discount the importance of a woman's interests *because she is a woman* is an obvious example of the moral prejudice we call sexism, just as to ignore or discount the importance of the interests of African or Native Americans, Hispanics, etc. are obvious forms of racism. It remained for Singer to argue, which he does with great vigor, passion, and skill, that a similar moral prejudice lies at the heart of moral anthropocentrism, a prejudice that Singer, borrowing a term first coined by the English author and animal activist Richard Ryder, denominates *speciesism* (Ryder, 1975).

Like Bentham and Mill before him, therefore, Singer denies that humans are obliged to treat other animals equitably in the name of the betterment of humanity, and also denies that acting dutifully toward these animals is a warm-up for the real moral game played between humans or, as theists would add, between humans and humans and God. We owe it to those animals who have interests to take their interests into account, just as we also owe it to them to count their interests equitably. In these respects we have direct duties to them, not indirect duties to humanity. To think otherwise is to give sorry testimony to the very prejudice—speciesism—Singer is intent on silencing.

UTILITARIANISM AND ETHICAL VEGETARIANISM

Singer believes that one of speciesism's most obvious symptoms is that we eat other animals, and his utilitarian case for vegetarianism gains strength from the radical changes that recently have taken place in commercial animal agriculture in the affluent world. Increasingly animals raised for food never see or smell the earth. Instead they are raised permanently indoors in unnatural, crowded conditions—raised "intensively," to use the jargon of the animal industry—in structures that look for all the world like factories. Indeed, it is now common to refer to such commercial ventures as factory farms. The inhabitants of these "farms" are closely confined in cases, or stalls, or pens, living out their abbreviated lives in a technologically created and sustained environment: automated feeding, automated watering, automated light cycles, automated waste removal, automated what-not. And the crowding: as many as nine hens in cages that measure 18 by 24 inches; veal calves confined to 22 inch wide stalls; pregnant hogs confined in tiers of cages, sometimes two, three, four tiers high. Many of the animals' most basic interests are ignored, and most are undervalued.

Add to this sorry tale of speciesism on today's factory farms the enormous waste that characterizes the animal industry, as we find it in affluent nations, waste to the tune of six or seven pounds of vegetable protein to produce a single pound of animal protein in the case of beef cattle, for example; and add to the accumulated waste of nutritious food the chronic need for just such food throughout the countries of the Third World, whose populations characteristically are malnourished at best and literally starving to death at worst—add all these factors together and we have the basis on which utilitarians can answer our first two questions. In response to the first question, Is vegetarianism required on ethical grounds?, utilitarians can reply that it is, noting that it is not for self-interested reasons alone that we should stop eating meat but for reasons that count the interests of *other* humans and *other* animals. And as for our second question, the one that asks what we should think and do about commercial animal agriculture as it exists today, in affluent nations, utilitarians can use these same considerations to support their moral condemnation.

THE RIGHTS VIEW

An alternative to the utilitarian attack on anthropocentrism, and one that also issues its own severe critique of intensive commercial animal agriculture, is the rights view. Those who accept this view hold that (1) factory farming methods are wrong because they violate the rights of animals and that (2) those of us who *can* adopt a vegetarian way of life morally *ought* to do so. How might one defend what to many people will seem to be an "extreme" view? This is not a simple question by any means; but something by way of a sketch of this position needs to be presented here.[6]

The rights view rests on a number of factual beliefs about the animals humans eat, use in science, hunt, trap, and exploit in a variety of other ways. These animals are not only in the world, they are aware of it—and also of what happens to them. And what happens to them matters to them. Each has a life that fares experientially better or worse for the one whose life it is. As such, all have a life of their own that is of importance to them apart from their utility to us. Like us, they bring a unified psychological presence to the world. Like us, they are somebodies, not somethings. They are not (our) "tools," not (our) "models," not (our) "resources," not (our) "commodities."

The life that is theirs includes a variety of biological, psychological, and social needs. The satisfaction of these needs is a source of pleasure, their frustration or abuse, a source of pain. And the untimely death of the one whose life it is, whether this be painless or otherwise, is the greatest of harms since it is the greatest of loses: the loss of one's life itself. In these fundamental ways the nonhuman animals in labs and on farms, for example, are the same as human beings. And so it is that, according to the rights view, the ethics of our dealings with them, and with one another, must rest on the same fundamental moral principles.

At its deepest level, an enlightened human ethic is based on the independent value of the individual: The moral worth of any one human being is not to be measured by how useful that person is in advancing the interests of other human beings. To treat human beings in ways that do not honor their independent value—to treat them as "tools" or "models" or "commodities," for example—is to violate that most basic of human rights: the right of each of us to be treated with respect.

As viewed by its advocates, the philosophy of animal rights demands only that logic be respected. For any argument that plausibly explains the independent value of human beings, they claim, implies that other animals have this same value, and have it equally. And any argument that plausibly explains the right of humans to be treated with respect, it is further alleged, also implies that these other animals have this same right, and have it equally, too.

Those who accept the philosophy of animal rights, then, believe that women do not exist to serve men, blacks to serve whites, the rich to serve the poor, or the weak to serve the strong. The philosophy of animals rights not only accepts these truths, its advocates maintain, it insists on and justifies them. But this philosophy goes further. By insisting on the independent

value and rights of other animals, it attempts to give scientifically informed
and morally impartial reasons for denying that these animals exist to serve
us. Just as there is no master sex, and no master race, so (animal rights
advocates maintain) there is no master species.

ANIMAL RIGHTS AND ETHICAL VEGETARIANISM

To view nonhuman animals after the fashion of the philosophy of animal
rights makes a truly profound difference to our understanding of what we
may do to them. Because other animals have a moral right to respectful
treatment, we ought not reduce their moral status to that of being useful
means to our ends. That being so, the rights view calls for the total dissolution
of the dominant form of commercial animal agriculture, as it exists in the
affluent world. Factory farming, in other words, must cease. In this respect,
this philosophy adds its voice to that of utilitarianism. Not suprisingly, moreover, its answer to our second question, Is vegetarianism ethically required?,
also is the same. According to the rights view, all those who can become
vegetarians, without acting in a supererogatory manner, have an obligation
to do so.

HOLISM

The "radical" implications of both the utilitarianism's and the rights view's
critique of factory farming and their joint call to adopt a vegetarian diet, for
ethical reasons, suggest how far some philosophers have moved from the
anthropocentric traditions of theism and humanism. Like utilitarian attacks
on this tradition, however, the rights view seeks to make its case by working
within the major ethical categories of this tradition. For example, utilitarians
do not deny the moral relevance of human pleasure and pain, so important
to our humanist forbearers; rather, they accept this and seek to extend our
moral horizons to include the moral relevance of the pleasures and pains of
other animals. For its part, the rights view does not deny the moral importance of the individual, a central article of belief in theistic and humanistic
thought; rather, it accepts this moral datum and seeks to widen the class of
individuals who are thought of in this way to include nonhuman animals.

Because both the positions discussed in the preceding use major ethical
categories handed down by our predecessors, some influential thinkers argue
that these positions, despite all appearances to the contrary, remain in bondage to anthropocentric prejudices. What is needed, these thinkers believe,
is not a broader interpretation of traditional categories (for example, the

category of "the rights of the individual"), but the overthrow of these very categories themselves. Only then will we have a new vision, one that liberates us from the last vestiges of anthropocentrism.

Among those whose thought moves in this direction none is more influential than Aldo Leopold's.[7] Leopold rejects the individualism so dear to the hearts of those who build their moral thinking on "the welfare (or rights) of the individual." What has ultimate value is not the individual but the collective, not the part but the whole, meaning the entire biosphere and its constituent ecosystems. Acts are right, Leopold claims, if they tend to promote the integrity, beauty, diversity, and harmony of the biotic community; they are wrong if they tend contrariwise. As for individuals, be they humans or other animals, they are merely "members of the biotic team," having neither more nor less value in themselves than any other member—having, that is, no value in themselves. What value individuals have, so far as this is meaningful at all, is instrumental only: They are good to the extent that they promote the welfare of the biotic community.

Traditional forms of utilitarianism, not just the rights view, go by the board given Leopold's vision. To extend our moral concern to the pleasures and pains of other animals is not to overcome the prejudices indigenous to anthropocentrism. One who does this is still shackled to these prejudices, supposing that those mental states that matter to humans must be the measure of what matters morally to the world at large. Utilitarians are people who escape from one prejudice (speciesism) only to embrace another (what we might call "sentientism"). Animal liberation is not nature liberation. To forge an ethic that liberates us from our anthropocentric tradition we must develop a holistic understanding of the community of life and our place in it. "The land" must be viewed as meriting our equal moral concern. Waters, soils, plants, rocks—inanimate not just animate existence—must be seen to be morally considerable. All are equal members of the same biotic team.

Holists face daunting challenges when it comes to determining what is right and wrong. These are to be determined by calculating the effects of our actions on the life community. Such calculations will not always be easy. Still, the situation is not hopeless. While it is true that we often lack detailed knowledge about how the biosphere is affected by human acts and practices, we sometimes know enough to say that some of the things we are doing are unhealthy for the larger community of life. For example, we do not know exactly how much we are contaminating the waters of the earth by using rivers and oceans as garbage dumps for toxic wastes, or exactly how much protection afforded by the ozone layer is being compromised by our profligate use of chlorofluorocarbons. But we do know enough to realize that neither situation bodes well for marine and other life forms as we know them. Those scientists and policymakers who insist on having "more data" before they raise their hands to vote in favor of serious social change are like the passers-by who could have prevented a flood if they would have but lifted a finger.

Let us assume, then, what I believe is true, that we sometimes are wise enough to understand that the effects of some human practices act like insatiable cancers eating away at the life community. From the perspective of holism, these practices are wrong, and they are wrong because of their detrimental effects on the interrelated systems of biological life.

It is important to realize that holists are aware of the catastrophic consequences toxic dumping and the ever widening hole in the ozone layer are having on seals and dolphins, for example. And it would be unfair to picture those who subscribe to holism as taking delight in the suffering and death of these individual animals. Holists are not sadists. What is fair and important to note, however, is that the suffering and death of these animals are not morally significant according to these thinkers. Morally, what matters is how the diversity, sustainability, and harmony of the larger community of life are affected, not what happens to individuals.

To make the holists' position clearer, consider the practices of trapping fur-bearing animals for commercial profit and hunting for sport. Holists find nothing wrong with these economic and recreational ventures so long as they do not disrupt the integrity, diversity, and sustainability of the ecosystem. Trappers can cause such disruptions, if they overtrap a particular species. The danger here is that the depletion of a particular species will have a ripple effect on the community as a whole and that the community will lose its diversity, sustainability, and integrity. The overtrapping (and hunting) of wolves and other predatory animals in the northeastern United States often is cited as a case in point. Once these natural predators were removed, other species of wildlife—deer in particular—are said to have overpopulated, so that today these animals actually imperil the very ecosystem that supports them.[8] All this could have been avoided if, instead of rendering local populations of natural predators extinct by overtrapping and overhunting, they had been trapped or hunted more judiciously, with an eye to sustainable yield. Although a significant number of individual animals would have been killed, the integrity, harmony, and sustainability of the ecosystem would have been preserved. When and if commercial trappers and sport hunters achieve these results, holists believe they do nothing wrong. From the perspective of holism, the inevitable suffering and untimely death of individual fur-bearing and other wild animals do not matter morally.

HOLISM AND ETHICAL VEGETARIANISM

Holism's position regarding the ethics of vegetarianism is analogous to its position regarding the ethics of our interactions with wildlife. There is nothing wrong with raising animals for food if doing so is good for the large life community. But it is wrong to do this if the community suffers. For example, some cattle ranchers claim that raising cattle on nonarable lands adds to the diversity of the ecosystem, and as long as the lands are not overgrazed, the system remains sustainable. However, to destroy delicately balanced communities of life in order to create new grazing lands for commercially raised beef, as is being done in the Amazon rainforest, is wrong. For holists, then, the ethics of meat eating must be judged on a case-by-case basis. What matters most is where the meat comes from, not the pain and death of the animals involved. If these animals are raised in an ecologically sensitive way, we do nothing wrong when we eat them.

While this may seem like good news to all meat eaters, it is not to some. Speaking generally, commercial animal agriculture, as practiced in affluent countries, is an ecological disaster. Or, rather, it is part of a more general ecological disaster, one that begins with grain production. Almost all the grain grown in affluent countries (over 90% in the case of oats, barley and rye, for example, and more than 60% in the case of soy beans)[9] is used as animal feed. The agriculture that produces these massive amounts of animal feed is literally killing the planet. Increasingly, knowledgeable people are reaching the conclusion that it is not sustainable. Because the same crop is grown on the same land year after year after year (this is what is meant by saying the system is "monocultural"), and because of the heavy use of toxic synthetic chemicals such as herbicides, nematicides, and fungicides (this is what is meant by saying the system is "chemically intensive"), the future fertility of the land is being compromised. For every bushel of corn from America's heartland, two bushels of topsoil are lost; that works out to an inch of topsoil the size of the state of Kansas every year. Even worse is the toil on the land directly attributable to the grazing of livestock, a practice that, both in the United States and throughout the world, is the principal cause of desertification.

Rainforest destruction is tied to commercial animal agriculture by way of the "hamburger connection." During the past 30 years fully 25% of the forests of Central America have been cleared in order to create pastures for grazing cattle. That hamburger at the local fast food eatery has an unadvertised price: five square meters of jungle have been destroyed to make room for a quarter pounder.

And it is not just the land that is a casualty of modern agriculture. The quality and availability of our water also are at risk. Nearly 50% of the water used in the United States goes for crops that are fed to farm animals. Remarkably, one serving of chicken takes over 400 gallons of water to produce, while an average size steak requires over 2,600 gallons. Moreover, once applied to crops, deadly chemicals do not disappear. Often they run-off into neighboring creeks or rivers, or trickle down through the earth into underground lakes only to surface again, at some other time and place. We do not speak loosely when we say that chemically intensive, monocultural agriculture is far and away the greatest single cause of the deteriorating quality of the earth's water supply. Because commercial animal agriculture is the largest consumer of a system of grain production that has these deleterious ecological consequences, holists can—and holists are[10]—speaking out in favor of a vegetarian way of life. On their view, those who can lead a vegetarian way of life, without acting supererogatorily, ought to do so.

There is a longer story to be told about the ecological carnage attributable to intensive animal farming—a much longer story than can be told here, one that would detail, for example, the environmental degradation that is the direct result of factory farms.[11] Suffice it to say on this occasion that the more we learn about why and how intensive commercial animal agriculture is implicated in environmental degradation, the stronger holism's case in favor of vegetarianism becomes, and this—paradoxically—despite the fact that what is done to the animals meateaters eat is not morally relevant.

ECOFEMINISM

Among those philosophers and other writers who are most actively distancing themselves from the anthropocentric tradition are ecofeminists. As is true of any other "ism," ecofeminism is not the name of just one idea; there are, in other words, important differences that separate some ecofeminists from others on specific issues. Even so, there is enough common ground shared by ecofeminists to enable a general characterization of this important tendency in contemporary moral philosophy. Perhaps no one offers a better characterization than Greta Gaard. "Ecofeminism is a theory, she writes,

> which has evolved from various fields of feminist inquiry and activism: peace movements, labor movements, women's health care, and the anti-nuclear, environmental, and animal liberation movements. Drawing on the insights of ecology, feminism, and socialism, ecofeminism's basic premise is that the ideology which authorizes oppressions such as those based on race, class, gender, sexuality, physical abilities, and species, is the same ideology which sanctions the oppression of nature. Its theoretical base is a sense of self most commonly expressed by women and various non-dominant groups—a self that is interconnected with all life. (Gaard, 1993)

The "ideology that authorizes oppressions" of which Gaard speaks, is patriarchy, an ideology that valorizes traditional "male" capacities (rationality and objectivity, for example) while minimizing traditional "female" characteristics (for example, intuition and subjectivity). In a fundamental sense, then, ecofeminists go beyond the critique of anthropocentrism that is common to the three other views described earlier—utilitarianism, animal rights, and holism. It is *androcentrism,* not anthropocentrism, that needs to be recognized and dismantled. Moreover, as Gaard and other ecofeminists view the moral landscape, androcentric approaches to ethics, grounded as they are in a conception of the self as separate from and in competition with others, emphasize individual rights and the following of abstract rules; a feminist approach to ethics, by contrast, emphasizes individual responsibility and care. Citing the pioneering work of developmental psychologist Carol Gilligan,[12] Gaard explains these important differences in the following passage:

> It is now common knowledge that rights-based ethics (most common of dominant-culture men, although women may share this view as well) evolve from a sense of self as separate, where a society of individuals must be protected from each other in competition for scarce resources. In contrast, Gilligan describes a different approach more common to women, one in which "the moral problem arises from conflicting responsibilities rather than competing rights and requires for its resolution a mode of thinking that is contextual and narrative rather than formal and abstract. This conception of morality as concerned with the activity of care centers moral development around the understanding of responsibility and relationships, just as the conception of morality as fairness ties moral development to the understanding of rights and rules." (Gaard, 1993)

In the hands of Gaard and other ecofeminists, this contextual, narrative approach to the resolution of moral problems is extended beyond, not limited to, humans only. Nature and nature's nonhuman citizens, including the other animals, are morally considerable.

The fundamental differences that separate ecofeminism and both utilitarianism and the rights view are deep and transparent. For all its similarities with ecofeminism when it comes to the denunciation of such prejudices as sexism, racism, and speciesism, utilitarianism is inseparable from reliance on and appeals to rules—the principle of utility paramont among them. And notwithstanding the agenda it shares with ecofeminism, of enfranchising nonhuman animals within the moral community, in the eyes of ecofeminists the rights view, with its insistence on "the rights of the individual," is yet another vestige of the ideology of patriarchy. And as for holism, here again appearances are deceiving. For despite its celebration of the "life community," holists continue to valorize values that are distinctly macho.

In her brilliant review of such prohunting literature, Marti Kheel, herself an ecofeminist philosopher, examines the recurring theme of the *naturalness* of the desire to hunt. Leopold is representative of this view. "The instinct that finds delight in the sight and pursuit of game," he writes, "is bred into every fibre of the human race" (Leopold, 1946:227). As Kheel notes—and her point cannot be emphasized too strongly—Leopold is "describ[ing] hunting not as a necessary means of subsistence, but rather as a desire that fulfills a deep psychological need" (Kheel; Diamond and Orenstein, 1991). It is *instinctive* to the species *Homo sapiens* to want to hunt and kill. No feminist (and, so, no ecofeminist) will find this vision acceptable. For the gospel of sport hunting is the gospel of patriarchy. In Leopold's work, the familiar themes of power and domination remain intact, only it is wild animals, not wild women, who exist to be vanquished. Because any ecofeminist ethic must be an ethic of care, and because such an ethic must find its expression in the caring relationships one has with *concrete individuals,* the chasm that separates ecofeminism, not only from utilitarianism and the rights view, but also from holism, is both stark and deep.

ECOFEMINISM AND ETHICAL VEGETARIANISM

While not all ecofeminists reach this conclusion, it is difficult to see how a consistent ecofeminism can avoid recommending vegetarianism on moral grounds. The massive destructiveness and waste that characterize commercial animal agriculture, and the rapacious monocultural, chemically intensive plant agriculture that sustains intensive animal production, hardly bespeak an ethic of care for and responsibility toward the natural world. Just the opposite. Notwithstanding the many important, foundational differences that separate ecofeminism from utilitarianism, the rights view, and holism, many of the same considerations that lead proponents of these latter three positions to conclude that people who can be vegetarians, ought to be, are available to ecofeminists and lead to the same conclusion. But as Carol Adams has shown in her pioneering work, *The Sexual Politics of Meat,* there are also distinctively *feminist* reasons for drawing this conclusion (Adams, 1990).

Briefly, the reasons are these. How women are perceived (and, indeed, how they often are treated) in patriarchal cultures mirrors how nonhuman animals are perceived and treated. Those who are "superior" (men) are encouraged by patriarchal ideology to view and treat "others" (women and nonhuman animals) as "inferior," which makes the subordination of the latter by the former seem not only rational but natural. It is no accident, Adams argues, that men talk about women as if they were animals. Like these animals, women are objectified as "a piece of meat." If, then, ecofeminism, as Gaard characterizes it, "calls for an end to all oppressions," and in the face of the distinctively feminist reasons for recognizing meat eating as a symptom of patriarchy's oppression of the other animals, akin in its origins and effects to its oppression of women, then Adams seems unassailably correct in maintaining that vegetarianism not only is a feminist issue, it should become part of the way of life of feminists themselves.

CONCLUSION

Most people who live in affluent countries, and who eat meat, purchase the meat they eat at nationally franchised food stores. This meat is therefore the end-product of factory farms. How we know this is simple: economics. Wholesalers and retailers buy as cheap as they can, which means—in the case of meat—at the trough of factory farms. From the animals' point of view, this is no blessing. More than six billion are slaughtered annually, just in the United States. That works out to 16 million each day, 700,000 each hour, 11,500 each minute.

As I have tried to explain, despite their many differences, utilitarians, advocates of animals rights, holists, and ecofeminists can speak with one voice on some occasions. This is one of them. All can agree that factory farming is wrong—wrong because it violates the rights of individual animals (the rights views' position), wrong because it either does not count the interests of nonhuman animals at all, or does not count them equitably (the utilitarian's view), wrong because it is destroying the ecology of the planet (the holist's view), and wrong because it is a symptom of patriarchal oppressions (the ecofeminist's view).

If the system of supply is wrong, what about individual consumers? Despite their many differences, and viewed against the backdrop of factory farming, each of the views discussed above can give the same answer: Vegetarianism is ethically required; those people who can lead a vegetarian way of life, ought to do so. Does this mean that, situated in the world as they are, each and every individual human ought to adopt this way of life? No. For we must always remember that "ought" implies "can," so that what is obligatory for some, is not necessarily obligatory for all. In particular, those of us who are privileged to have enough financial resources and time to participate in conferences on such topics as sustainable agriculture, or to read about such issues, must keep in mind that what is obligatory for us may be supererogatory for others. Nothing disgraces a good idea more than the moral excesses of those who champion it. Clearly, what we who are privileged are obligated to do should not be transferred, *mutatis mutantis,* to those who are not.

Having said this, however, I conclude by reminding champions of sustainable agriculture of the neglect they have showered on the question of ethical vegetarianism and why, if they are not to practice the very sort of moral imperalism they rightfully condemn, when imposed by others, they need to avoid excusing their own lack of attention to ethical vegetarianism by appealing to the dire circumstances of, say, people in the Third World. Truly, it is not right of us to impose our ethic on these people. But neither is it right to excuse ourselves for failing to take questions of *our* moral obligation seriously by insisting that, for these peoples, these questions raise issues of supererogation. It is not, after all, heroic or stupendously self-sacrificial for us to stop eating meat, for ethical reasons, whether these reasons have a utilitarian, animal rights, holistic, or ecofeminist lineage. When all the dust of excuses and self-exceptions settles, it is simply a matter of breaking a culturally induced habit and, if current tendencies in moral theory are correct, doing the right thing. But whether right or not, the *question* of its rightness should no longer be ignored, especially by advocates of sustainable agriculture. To secure even this much agreement, at this point in time, and whatever the answer such advocates reach, would be progress, any way it is measured.

ACKNOWLEDGMENTS

This chapter incorporates parts of two previously published essays, "Abolishing Animal Agriculture" and "Irreconcilable Differences," originally published in Tom Regan, *The Thee Generation: Reflections on the Coming Revolution.* Philadelphia: Temple University Press. ©1991 Temple University. Reprinted by Permission of Temple University Press.

NOTES

1. See, in particular, Magel, C. 1989. *Keyguide to Information Sources in Animal Rights.* London: Mansell.

2. I discuss all the theories considered here, at greater length: Regan, T. 1991. *The Thee Generation: Reflections on the Coming Revolution.* Philadelphia: Temple University Press.

3. For the relevant passages from Kant, see his selections in *Animal Rights and Human Obligations,* op. cit., pp. 122–124.

4. Ibid., pp. 58–59.

5. See in particular, Singer, P. 1990. *Animal Liberation,* 2nd ed. New York: Random House.

6. See in particular, Regan, T. 1983. *The Case for Animal Rights.* Berkeley: University of California Press.

7. See in particular, Leopold (1946).

8. For an interpretation of the situation that significantly departs from the one offered by traditional "wildlife managers," see Baker, R. 1985. *The American Hunting Myth.* New York: Vantage Press.

9. For an illuminating account of these and related matters, see Robbins, J. 1987. *Diet for a New America*. Walpole, NH: Stillpoint Press.

10. See, for example, Callicot, J. Baird 1986. The Search for an Environmental Ethic. In *Matters of Life and Death*. T. Regan (ed.). New York: Random House.

11. Ibid.

12. The passage quoted from Gilligan by Gaard appears in Gilligan, C. 1982. *In a Different Voice: Psychological Theory and Women's Development*, p. 19. Cambridge: Harvard University Press.

REFERENCES

Adams, C. 1990. *The Sexual Politics of Meat* Crossroad Continuum: New York.

Bentham, J. 1990. In *Animal Rights and Human Obligations*. T. Regan and P. Singer (eds.), p. 130. Englewood Cliffs, NJ: Prentice Hall.

Descartes, R. 1976. Discourse on Method. Reprinted in *Animal Rights and Human Obligations*. T. Regan and P. Singer (eds.), p. 61. Englewood Cliffs, NJ: Prentice Hall.

Gaard, G. (ed.). 1993. *Ecofeminism: Women, Animals, Nature*. Philadelphia: Temple University Press.

Kheel, M. 1991. Ecofeminism and Deep Ecology. In *Reweaving the World: The Emergence of Ecofeminism*. In I. Diamond and G. Orenstein (eds.). San Francisco: Sierra Club.

Leopold, A. 1946. *A Sand County Almanac,* p. 227. New York: Ballentine Books.

Rickaby, J. 1976. Moral Philosophy. Reprinted in *Animal Rights and Human Obligations*. T. Regan and P. Singer (eds.), p. 179. Englewood Cliffs, NJ: Prentice Hall.

Ross, W. D. 1930. *The Right and the Good,* p. 47. Oxford: Carendon Press.

Ryder, R. 1975. *Victims of Science*. London: Davis-Poynter.

PART II

CONTRADICTIONS: SUSTAINABILITY BARRIERS TO OVERCOME

CHAPTER 5

Is Sustainable Capitalism Possible?

James O'Connor

The short answer is "no." In the 1992 U.S. presidential campaign, neither major presidential candidate made "environment" a main issue. Federal, state, and local governments are neglecting the environment in their competition to attract scarce capital. The definition of "wetlands" is being narrowed, as is that of "endangered species." National and state parks are becoming more commodified as managers search for ways to make them more profitable. Union leaders still oppose or are indifferent to most environmental demands. The mainstream environmental movements have compromised their positions, and have become less effective. International agreements on ozone layer depletion are weak and on global warming merely symbolic. Agreements with respect to the exploitation of the world's forests, oceans, and "commons" are honored in the breach. Oil as an instrument of economic wealth and expansion and of national power is more important than ever. Even nuclear power may make a comeback. In the South, most governments are eager to sell their national birthrights to transnational corporations, and small farmers in the countryside and the urban poor are forced to deplete nature and pollute the air and water more rapidly, respectively, in order to survive.

One basic condition for an ecologically sustainable capitalism is a national budget that puts high taxes on raw material inputs (e.g., coal, oil, gas, nitrogen, raw materials) and certain outputs (e.g., gasoline, cars, chemical building blocks, etc.), meanwhile slapping a value-added tax on environmentally un-

Food for the Future: Conditions and Contradictions of Sustainability, edited by Patricia Allen.
ISBN: 0-471-58082-1 © 1993 John Wiley & Sons, Inc.

friendly products (with a Green Label program that would exempt friendly green products). Another condition is an expenditure program that heavily subsidizes solar energy and other alternative energy sources, technological research that leads to eliminating toxins at the source and creating healthy work conditions, and a reorientation of science and technology priorities generally, e.g., in the direction of sustainable agriculture. Nowhere is this kind of budget being developed.

Finally, the whole discourse on sustainable development obscures the truth. While environmentalists and others spin their wheels trying to identify the ways that corporate environmental practices need to be reformed to make them consistent with preserving nature, corporate research and development departments are devising ways to reform and restructure *nature* to make *it* consistent with profitable accumulation—e.g., plantation forestry, biotechnological agriculture.

DEMAND CRISIS: EXPANSION AND CONSUMPTION

A longer answer is "probably not." An even longer answer is, "not until capitalism changes its face in ways that would make it unrecognizable to bankers and money managers looking at themselves in the mirror today." A sufficient albeit still terse answer requires a brief account of how capitalism works, why it works when it works, and why it does not when it does not.

Until the rise of ecological economics, economists discussed the "sustainability" of capitalism in purely economic terms, i.e., in terms of money capital, investment and consumption, profits and wages, costs and prices. The physical world appeared in models of economic growth in two disguises: first, in the form of rent and location theory; and second, in the concept of the "accelerator," or the amount of physical product that new productive capacity could be expected to produce.

From a strictly economic point of view, sustainable capitalism is an expanding capitalism. Productive capacity expands more or less steadily and purchasing power or market demand grows more or less in proportion to productive capacity—all without serious inflation. While there are many variations of economic growth theory, they all presuppose certain relationships between the real and money economy; physical production and incomes, i.e., increases in investment and consumption goods, on the one hand, and profits and wages, on the other. They also presuppose technical progress, or increases in labor productivity. And all of them theorize that capitalism cannot stand still, i.e., that the system must expand or contract, grow or shrink, "accumulate or die," in Marx's words.

The key to the growth of the economy defined in material or physical terms is technological change and investment in physical and "human" capital, or new production processes, new consumer products, and new technical, organizational, managerial, and other skills. The social and economic laws governing technological change are so complex or slippery that economists have made careers defending or attacking one theory of technology or another.

The secret of economic growth defined in monetary terms is clearly profits. Profits are the means of expansion, i.e., of new investment and technology, as well as the diffusion of existing technologies. Profits are also incentives to expand. Also, expansion is the means of making more profits. Individual capitals need to grow to defend profits when aggregate demand for commodities contracts, hence to retain their realized profits. Profits and expansion or growth are thus the means and ends of one another. They are each other's content and context and the average investor does not really see any difference between them.

In the simplest (and most simple-minded) model of capitalism, the rate of growth or rate of accumulation of capital depends on the rate of profit. The higher the profit rate (everything else being the same), the more sustainable is capitalism. A negative profit rate spells real trouble, a general crisis and depression. In this simple model, anything that interferes with profits, new investment, and markets threatens the sustainability of the system, or threatens to bring about an economic crisis with unknown and unknowable economic, social, and political consequences.

In the traditional Marxist discourse, capital is its own worst enemy. Capital threatens its own profits because of what Marx called the "contradiction between social production and private appropriation." This means that the greater the exploitation of labor, the more potential profits are produced, but, just for this reason, the more difficult it is to realize these profits in the market, i.e., to sell goods at prices reflecting costs and the average profit rate prevailing at a certain time. This "first contradiction of capitalism" states that when individual capitals attempt to defend or restore or expand profits by increasing labor productivity, speeding up work, cutting wages, and other time-honored ways of getting more production from fewer workers or with smaller wage advances, the unintended effect is to reduce demand for consumer commodities. The reason is that workers, technicians, and others in the labor process produce more, hence by definition are able to consume less. The greater are produced profits, or the exploitation of labor, the smaller are realized profits, or market demand, all other things equal. Of course, other things are never equal. The federal budget deficit, consumerism and consumer credit, business borrowing, and an aggressive foreign trade policy (among other things) buoyed up U.S. capitalism in the 1980s (although it is highly doubtful that the credit system can do the same job through the 1990s).

COST CRISIS: CONDITIONS OF PRODUCTION

Today, this kind of economic thinking, while still valid, is clearly one-sided and limited. The reason is that it presupposes limitless supplies of what Karl Marx called "conditions of production." Put another way, this type of model presupposes that capitalism does not have to face potential bottlenecks on the "supply side," or that growth is demand-constrained only. However, if the costs of "land and labor" increase significantly, capitalism faces a possible "second contradiction," namely, economic crises striking from the cost-side. Examples are the English "cotton crisis" during the U.S. Civil War, wage advances in excess of productivity in the 1960s, and the "oil shocks" of the 1970s.

Cost-side crises originate in two general ways. The first is when individual capitals defend or restore profits by neglecting work conditions (hence raising the health bill); degrading soils (hence lowering the productivity of land), or turning their backs on decaying urban infrastructures (hence increasing congestion costs)—to take three examples.

The second is when social movements demand better health care to defend the human body, protest the ruination of soils in the name of "environmental protection," and protect urban neighborhoods in ways that increase capital costs or reduce capital flexibility—to stay with the same three examples. Here we are talking about the potentially damaging economic effects of labor and women's movements, environmental movements, and urban movements—a subject that obsesses mainstream economists and capitalist management but that the leaders of labor and social movements do not wish to discuss in public.

In the real world, both types of cost-side crises combine or intermingle in complex and contradictory ways that no one has ever theorized or studied. For example, no one knows how much urban congestion costs are the results of capital's celebration of the automobile and neglect of urban mass transport that commuters look forward to using, and how much the unintended effects of community struggles to keep freeways from scarring their neighborhoods.

We need a more refined theoretical approach to the problem of "land and labor." Marx inadvertently supplied us with such an approach with his concept of "conditions of production."[1] Conditions of production are things that are not produced as commodities in accordance with the laws of the market (law of value) but that are treated as if they are commodities, i.e., they are "fictitious commodities" with "fictitious prices." According to Marx, there are three conditions of production: first, human laborpower, or what Marx called the "personal conditions of production;" second, environment, or what Marx called "natural or external conditions of production;" third, urban infrastructure (we will add "space"), or what Marx called "general, communal conditions of production." The fictitious price of laborpower is the wage rate, and that of environmental and urban infrastructure and space is rent. Given that wage and rent theory are not and cannot be based on costs of production, it is no accident that both bourgeois and Marxist economic accounts of wages and rents are the least developed and satisfying in the entire science of economics.

Sustainable capitalism needs all three conditions of production at the right time and right place and in the right quantities and right qualities, and at the right fictitious prices. As already noted, serious bottlenecks in the supply of laborpower, natural resources, and urban infrastructure and space threaten the sustainability of capitalism in the sense of driving up costs and impairing the flexibility of capital. "Limits to growth" thus do not appear as absolute shortages of laborpower, raw materials, clean water and air, and urban space, etc., but as *high cost* laborpower, resources, and infrastructure and space. This leads to capital's constant attempts to rationalize labor markets, supplies and markets for fuel and raw materials, and urban and rural land-use patterns and land markets to reduce costs of production.

Supply-side bottlenecks or shortfalls pose especially difficult problems when the economy is weak, or faces a demand-side crisis or fresh competition from other countries. Stagnant or falling profits force individual capitals to attempt to reduce the turnover time of capital, i.e., to speed up production and reduce the time that it takes to sell their products. This obsession with making money faster and faster to compensate for low or falling profits inevitably runs up against union-organized labor markets, OPEC-dominated oil markets, conventional agriculture's "inefficient" uses of the soil and water, etc. On the one hand, money capital seeks more of itself faster and faster; on the other hand, what Polanyi called "society" and what we can call out-of-date patterns of land and labor utilization, and land and labor markets, combined with the resistance of capitalist rationalization by labor and social movements, constitute themselves as obstacles, or "barriers to overcome." At the very least, capital faces massive social indifference and inertia.

Capital's solution to this dilemma, at least in the short-term, is as simple as it is economically self-destructive. Money capital abandons the "general circuit of capital," i.e., the long and tedious process of leasing factory space, buying machinery and raw materials, renting land, finding the right kind of laborpower, organizing and implementing production, and marketing commodities, and finds its way into speculative ventures of all kinds. Money capital, based on the expansion of credit, or money that cannot find outlets in real goods and services, jumps over society, so to speak, and seeks to expand the easy way—in the stock and bond markets. Hence the anomaly appears—the value of claims on the surplus or profits grows at the same moment that the real value of fixed and circulating capital declines. This makes a bad situation worse.

During earlier periods of capitalist development, there was plenty of pre-capitalist laborpower, untapped natural wealth, and space. This was true in fact, and also in the perceptions of the bourgeoisie at the time. Hence the (fictitious) prices of laborpower, natural resources, and space were held in check. Nor were there environmental movements or urban movements that raised political and social barriers to capital. Sooner or later, however, capital tends to capitalize everything and everybody, i.e., everything potentially enters into capitalist cost accounting. For millennia, human beings have been "humanizing" nature or creating a "second nature," e.g., deforestation and drought/flood cycles under the Roman plantation system, the devastating ecological consequences of the Punic Wars, and soil depletion and water scarcity in Mayan civilization. But in capitalist modes of production and social formations, this second nature is turned into capitalist nature, i.e., nature is commodified and valorized, e.g., multiple-use of national parks means multiple sources of profit. From the point of view of those who want capitalism to be ecologically sustainable, this is when problems start to appear. Labor markets become tight, and developed countries have to rely on imported labor from the South—with all attendant economic and social costs and problems. Examples are the cost of settling newcomers who use a different language and a resurgence of racism. Raw materials and unpolluted commons become scarce, driving up what Marx called the "costs of the elements of capital," e.g., domestic oil and gas, trees and lumber, and supplies of clear water. And, last but not least, urban infrastructure and space become scarce—creating rising congestion costs, ground rents, and pollution costs. Lost Angeles is a good example; Mexico City is a better one.

In sum, as Lewis Mumford wrote long ago, "In American during the [19th century], we mined soils, gutted forests, misplaced industry, wasted vast sums on needless transportation, congested population, and lowered the physical vitality of the community without immediately feeling the consequences of our actions. Today, we feel very little other than these consequences."[2]

Putting two and two together, a model of sustainable capitalism states the following: the capitalization of the conditions of production in general and environment and nature in particular may have the effect of raising the cost of capital and reducing its flexibility. There are two general reasons: first, a systemic reason, namely, that individual capitals have little or no incentive to use production conditions in sustainable ways, especially when faced with economic bad times of capital's own making. Second, precisely for this reason, labor, environmental, and other social movements challenge capital's control over laborpower, environment, and the urban (and, increasingly, the rural, as well, especially in the South). Examples are the National Toxics Campaign, occupational health and safety and right-to-know struggles, direct action to save wild rivers and original growth forests, antifreeway and development movements in the cities, and sustainable agriculture movement.

This line of thinking suggests that there are two, not just one, contradictions of capitalism, two, not just one, types of economic crisis, and two, not just one, types of crisis resolution. The "second contradiction" of capitalism results in economic crisis that strikes not from the demand side but from the cost side. Marx himself came close to developing this type of crisis theory in his account of the rising organic composition of capital due to increasing raw material prices. But, strangely enough, he never integrated the many passages in *Capital* on this subject into his theory of the falling rate of profit. Only in Marx's discussion of the cotton crisis of the 1860s can we find an explicitly cost-push theory of crisis. By contrast, since the 1960s, there have been many economists, Marxist and non-Marxist, who defend the thesis that cost-side (as well as demand-side) crises characterize the late twentieth century, e.g., the cost-push crisis of profitability due to wage increases, struggles against productivity, welfare state expansion, and high oil prices in the 1960s and 1970s.

In crude terms, the second contradiction states that when individual capitals attempt to defend or restore profits by cutting costs, the unintended effect is to reduce the "productivity" of the conditions of production hence to raise average costs. Cost may increase for the individual capitals in question, other capitals, or capital as a whole.

Some examples pertaining to the three conditions of production will illustrate this line of thought. Chemical pesticides in agriculture at first lower costs but ultimately increase costs as pests become more chemical resistant and also as the chemicals kill the soil. Permanent yield monoforests in Sweden were expected to keep costs down; but the loss of biodiversity over the years has reduced the productivity of forest ecosystems and the size of the trees. In the United States, nuclear power promised to reduce energy costs. But bad design, problems of finance, and, most of all, popular opposition to nuclear power have the effect of increasing costs. As for the "communal" conditions of production, new highways designed to lower the costs of transport and the commute to work tend to raise costs when they attract more traffic and create more congestion. And in relation to the "personal" production condition, it is clear that the U.S. education system, which is supposed to increase potential labor productivity, produces as much stupidity as learning, hence impairs labor productivity.

Before I discuss the various "solutions" to the second contradiction of capitalism, and also the political ramifications of these "solutions," I want to complicate the issue a little. Remember that the conditions of production are not produced in accordance with laws of the market. Nor does the market generally regulate capital's access to these conditions when and if they are produced. Therefore, there must be some agency whose task it is to either produce the conditions of production and/or regulate capital's access to them. This agency is the state. If we ignore the monetary and military functions of the state, every state activity, including every state agency and budgetary item, is concerned with providing capital with access to laborpower, nature, or urban space and infrastructure. For example, in the United States, there are the Department of Labor and the education bureaucracy; the Department of Agriculture, National and State Park Services, and the Bureaus of Land Management and Reclamation; and urban planning bodies and traffic authorities. Examples of specific functions related to the three conditions of production are first, with respect to laborpower, child labor laws and laws governing hours and conditions of work, work safety, and so on; second, in relation to environment, laws governing access to federal lands, regulating coastal development, pollution, etc.; third, with respect to urban infrastructure and space, zoning laws, traffic planning, and land-use regulation generally. I cannot think of one state activity or budgetary item that does not concern itself in different ways with one or more conditions of production.

MANAGING COST CRISES

What is the solution to these cost-side crises? I will discuss this question briefly, first, from the standpoint of individual capitals, and second, from the standpoint of capital as a whole.

The worst case is when individual capitals, faced with both higher costs and lower demand, cut costs even more, thus intensifying both the first and second contradiction. But this result is not the only possibility. In relation to the environment, there are many examples of individual capitals responding to green consumerism, e.g., the public demand to reduce waste and recycle, by finding new uses for waste products; and also examples of companies that upgrade their capital equipment when forced to reduce their pollutants and other companies that specialize in environmental clean-up.

The best solution for capital as a whole is to restructure the conditions of production in ways that increase their "productivity." Since the state either produces or regulates access to these conditions, restructuring processes are typically organized and/or regulated by the state, i.e., by politics. Examples are banning cars in cities to lower congestion and pollution costs, subsidizing integrated pest management in agriculture to lower food and raw materials costs, and shifting some emphasis from curative to preventative health (e.g., the fight against AIDS in the United States) to lower health care costs.

The ideal situation for capital is as follows: Huge sums of state monies are expended to restructure production conditions in ways that restore or increase their "productivity" and lower costs of capital. Long-term productivity is thus enhanced at the expense of short-term profits. New industries produce environmentally friendly products, urban transport, and education systems, and so on, which (like the examples cited above) effectively lower the costs of the elements of capital and the consumption basket, and also ground rent, and, at the same time, raise the level of aggregate demand, thus attacking the first contradiction in noninflationary ways. By contrast, if new systems of forest management, pollution control spending, urban planning, and so on have no effect on costs, the result is an increase in effective demand and also inflation, or a reduction in profits.

So much for theory. Practice is another question. In liberal democratic states, the normal political logic of pluralism and compromise prevents the development of overall environmental, urban, and social planning. The logic of the state administration or bureaucracy is undemocratic, hence insensitive to environmental and other issues raised from below. And the logic of self-expanding capital is antiecological, antiurban, and antisocial. All three logics combined are extraordinarily contradictory in terms of developing political solutions to the crisis of the conditions of production; hence the chance of a "capitalist solution" to the second contradiction is remote.

In fact, there is no state agency, or corporatist-type planning mechanism, in any developed capitalist country, that engages in overall ecological, urban, and social planning. The idea of an ecological capitalism, i.e., a sustainable capitalism, has no institutional infrastructure. Where is the state that has a rational environmental plan? Intraurban and interurban planning? Health and education planning organically linked to environmental and urban planning? Nowhere that I know of. Instead, there are piecemeal approaches, fragments of regional planning at best and irrational spoils allotment systems at worst.

Thus, every day new headlines announce another health care crisis, environmental crisis, and urban crisis. The ultraimage we have is of a largely illiterate labor force living in a totally polluted city, immobilized by gridlock (in any case, homeless because of high rents), and even unable to obtain clean water. This may not fit Rome or New York yet, but it comes close in Mexico City and Bombay, which are definitely parts of the capitalist world.

CONDITIONS IN THE SOUTH

The situation is especially bad in the South, as everyone knows. Thus, we are witnessing a new discourse on "sustainable development," which has become an ideological battleground of growing importance. Practically every one uses the expression, which has many meanings. Sustainability is usually defined by environmentalists and ecological economists to mean the use of only renewable resources and also low, nonaccumulating levels of pollution. The South is, in fact, closer to "sustainability" defined in this way than the North; but the North has many more capital and technological resources than the South to attain sustainability. Capital uses the term to mean sustainable profits, which presuppose long-run planning of the exploitation and use of renewable and nonrenewable resources. "Sustainability" is also defined in terms of natural systems: wetlands, wilderness protection, air quality, and so on. But these definitions have little or nothing to do with sustaining profitability. In fact, there may be an inverse correlation between ecological sustainability and profit. "Sustainability" of rural and urban life, the life of indigenous peoples, the conditions of life for women, the rights of labor, and so on are inversely correlated with sustainability of profits—if the history of the late twentieth century is any guide.

Independent of the question of the desirability of the South following the industrial, consumerist path of the North is that of the possibility of doing this. Indian, Brazilian, and Mexican (to take three examples) industrial capitalism occur at the expense of vast poverty and misery and also of ecological stability. East Asia is doing well economically, but has yet to prove that it can be industrial powerhouse and also pay good wages and provide good working conditions, progressive social policies, meaningful environmental protection, and stable forms of liberal democracy. Most of the rest of the South is an economic, social, and ecological disaster zone. In short, there are many barriers to capitalist development in the South, such as weak markets, in turn the result of a hugely unequal wealth and income distribution, the absence of agrarian reform favoring small and middle farmers, and instabilities in the demand for, and supply of, raw materials; foreign debt and balance of payments crises; and problems of maintaining ruling blocks of propertied interests and stable governments—among other reasons—independent of ecological conditions in particular and the state of the conditions of production in general. Needless to say, this situation creates permanent social and political instability, new migration patterns to the North, more and more economic and ecological refuges, and so on—which, in turn, spell continued trouble in the North.

If we are looking for institutions to blame for ecological decline and cost-push economic crises, we have to blame the state, as well as capital. Deep reforms within the state apparatus, as well as within the decision-making centers of capital, are required for a sustainable capitalism, or the sustainable production and reproduction of human labor, environment, and urban life, generally.

I do not think that the kinds of reforms that are needed, whether by the economic ruling class or the state power elite, are in the cards, especially in the South. I think that the attempts to rationalize the conditions of production are weak—even from the standpoint of profitability. It is true that capital and states the world over are trying to restructure the conditions of production with the aim of making capitalism sustainable, i.e., remake nature. Examples are biotechnology, plantation forestry, low-input agriculture, saving the Amazon for long-term profit, imposing a kind of state environmentalism on the Third World, and restructuring cities and transport systems. I do not think that most of these attempts will succeed. Capitalism is too irrational and contradictory and so is the capitalist state, equally or more so in the South or Third World.

POLITICAL POSSIBILITIES

I would like to end with more stress on the political aspects of the issue. As I have suggested, I do not think that most of the center-right governments that have been ruling most of the world since the late 1970s and early 1980s are incapable of steering capitalist development in ways that improve the conditions of life, labor, the cities, or the environment. These governments are too intent on expanding the international division of labor, deregulating and privatizing industry, forcing economic "adjustments" on the South, hence marginalizing up to half the population of some Third World countries, and pretending that neoliberalism generally will solve the growing economic crisis. By and large, things will get worse before they get better, especially in the South. Global warming will grow, as will ozone holes, ocean pollution, and world deforestation.

Meanwhile we have seen the growth of various "red–green" movements in different countries. At the local level, some labor unions are beginning to address environmental issues. Environmental movements are addressing economic and social issues that 5 or 10 years ago they ignored or downplayed. Labor and feminist movements, urban movements, and environmental movements have organized themselves around the general issues of the conditions of production or conditions of life. While I have little or no hope for a sustainable capitalism, there might be hope for some kind of ecological socialism—a kind of society that pays close attention to ecology along with the needs of human beings in their daily life, as well as to feminist issues, antiracism, and issues of social justice and equality generally. Globally, it is around these issues that there is movement and organization, agitation and action, which can be explained in terms of the new contradictions of capitalism and the nature of the capitalist state discussed earlier.

Politically, this means that labor, feminist, urban, environmental, and other social movements must be combined in a single powerful, democratic force—one that is both politically viable and also capable of radically reforming the economy, polity, and society. This suggests the need for three general and related strategies.

The first is the self-conscious development of a public sphere, a political space, in which labor, women, urban, and environmental organizations can work politically, a kind of dual power, if you like. Here there could be developed not the temporary tactical alliances between movements and movement leaders as we have today, but strategic alliances, including electoral alliances. A strong civil society, defining itself in terms of its struggles with capital and the state, as well as of democratic impulses and forms of organization within alliances and coalitions of movement organizations—and within each organization itself—is the first prerequisite of sustainable society and nature. The second is the self-conscious development of economic and ecological alternatives within this public sphere—alternatives such as green cities, pollution-free production, biologically diversified forms of silvaculture and agriculture, and so on, which are very well known today. The third is to organize struggles to democratize the workplace and the state administration so that substantive contents of an ecological progressive type can be put into the shell of liberal democracy. This presupposes that the movements not only use political means to economic, social, and ecological goals, but also agree on political goals themselves, especially the democratization of the state apparatus.

These ideas may seem to be as unrealistic as an ecological capitalism. Perhaps they are. But I keep on reminding myself that while the existing structures of capital and the state do not seem to be capable of anything more than the occasional reform, social movements worldwide get bigger every day—hence that at some point there will be a general social and political crisis as the demands of these movements clash with existing rigid economic and political structures. This will create a general crisis, at which point, no doubt, all kinds of "morbid forms" will appear. Some will say that this is precisely what is happening today—that the social, political, and economic fabric is unraveling, and that the resurgence of racism, nativism, discrimination against foreign workers, male backlash, and other reactionary trends and tendencies are becoming a much greater danger than capital and the state themselves. Others link the revival of right wing populism and reaction to the rightward shift in the political and economic mainstream. There are many other types of analysis of the current world political situation—including the line that the well-off are rebelling against the demands of the poor, the welfare state, redistributive economic policies, and the like. Whatever the case(s), from the standpoint of progressives, red or left greens and feminists, the last thing in the world we need is factionalism, sectarianism, "correct lineism"—instead, we need an ecumenical spirit, and to "celebrate our commonalities as well as our differences."

NOTES

1. "Inadvertently" because Marx used the concept of "conditions of production" in different and inconsistent ways; he never dreamed that the concept would or could be used in the way that I use it in this chapter; and no one could have used the concept in this way until the appearance of Karl Polanyi's *The Great Transformation*. New York: Farrar and Rinehart, 1944.
2. Quoted in Ramachandra Guha, Lewis Mumford: The Forgotten American Environmentalist. In *Capitalism, Nature, Socialism* II, 3, October 1991, p. 72 (from Lewis Mumford, The Theory and Practice of Regionalism. *Sociological Review* 20:10–12, 1992).

CHAPTER 6

Sustainable Agriculture in the United States: Engagements, Silences, and Possibilities for Transformation

Patricia Allen and Carolyn Sachs

S ustainable agriculture discourse—definitions, priorities, strategies—has been produced by numerous individuals, groups, and institutions concerned about the future of food and agriculture. We use the term "discourse" to refer to the body of statements made about sustainable agriculture that represents these groups. In this chapter we consider dominant perspectives of sustainable agriculture, i.e., the ones created and promulgated by those whose legitimacy to speak within and for agriculture is most widely recognized, such as farmer, scientific, and governmental organizations. We know, of course, that not every sustainable agriculture group or individual advocate agrees with all of the primary themes of the predominant discourse. Nonetheless, this discourse is the most powerful agent shaping sustainable agriculture today.

Sustainable agriculture has attracted a larger, broader following than previous agrarian movements, such as those in support of family farms or agricultural labor. It is unclear, however, if this is because sustainability is thought to be more comprehensive and crucial (e.g., because it is fueled by concerns of exhausting our ability to produce food) or because it is so vaguely defined that it can mean anything one decides it means. In one way, it would appear that the efforts of the early sustainable agriculture proponents have met with success. Their original struggle—to have their issues listened to rather than summarily discounted by the agricultural establishment—seems to be over. Nearly all agricultural groups, from those composed of local farmers to major commodity organizations, have adopted the use of the term "sustainable agriculture."

This universal appropriation of the term, however, has created its own problems. For example, some have defined sustainable agriculture as "competitiveness" and hole that its central tenets are specializations, exports, and vertical integration (Youngberg, 1992), notions either not included by or antithetical to the original ideas of sustainable agriculture advocates. Thus, now that this term has been effectively integrated into conventional agricultural discourse, the new struggle of the sustainability movement is to ensure that the *meaning* of the term is not subverted to equate with "business as usual." This chapter is about finding the conceptual space for the sustainability movement to challenge, rather than reproduce, the conditions that led to nonsustainable agriculture in the first place.

While sustainable agriculture cannot be classified as a single, articulated movement, those working to achieve sustainability do espouse common themes and hold common objects. Sustainable agriculture therefore can be considered an identifiable, nascent movement. Sustainable agriculture shares many characteristics with what have been called "new social movements," such as the environmental, cultural-diversity, and bioregionalist movements. These movements are sometimes characterized as "not-in-my-backyard" in orientation, i.e., that what compels them is less likely to be large social issues than concerns about the participants' own life conditions and personal identities. Perhaps because of this immediacy, these movements have become quite powerful. According to Frank and Fuentes (1990), new social movements are increasing in strength and importance; they inspire and mobilize people more than did previous social movements, such as those for workers' rights.

There are various perspectives on the relevance and efficacy of new social movements. One is that they are based on the "fractured subjects" of identity politics, not analytically coherent, and do not have an understanding of the fundamental causes of the problems they address. For example, Boggs (1986: 220) says, in discussing the Greens, that their characteristic calls for decentralization and citizen empowerment ignore "the structural and material sources of domination that lie at the heart" of environmental crises. Others, such as Scott (1990), conclude that new social movements are distinct from traditional worker movements in that they are primarily cultural rather than political, do not directly challenge the state, and focus less on class relations and more on life-style changes. The implication is that the issues of the new social movements are often not central and do not support analysis or solutions that address the roots of problems. As a consequence, the methods for dealing with the problems identified can be ineffectual.

Another way of viewing new social movements is that they have evolved as appropriate forms under the current flexible and fragmented mode of the economic system, characterized by an accelerating globalization and mobilization of resources. New global divisions of labor and new labor processes (e.g., in computers and biotechnology) have reconfigured people's relationships with each other and their experience of time, space, and consciousness (Harvey, 1989). In this context, new circuits of capital and social forms have generated and required new circuits of organization and resistance. According to O'Connor (1988; this volume), new social movements involve struggles over protecting and restructuring "conditions of production," i.e., the personal (human beings and their labor power), the environmental ("external nature" and natural resources), and the communal (urban and rural space, including social and physical infrastructures). While new social movements do break with previous forms of movement solidarity, they also represent and create new forms that cannot be dismissed. We hold open the possibility that new social movements can indeed challenge the basis of current social problems and create improved social forms.

Our goal in this chapter is to create an opening for ensuring that sustainable agriculture can achieve this and become an even more effective social movement than it is in its present form. We believe that certain theoretical and practical changes are necessary if the sustainable agriculture movement is not to be relegated to a marginal reformist movement, but instead is to have a significant role in the social, economic, and policy reconstruction needed for agricultural sustainability. In this chapter we discuss current efforts toward agricultural sustainability in the United States, focusing on the issues they address and those they leave out, what we have termed their "engagements" and their "silences." Next, we discuss the contradictions inherent in these sustainable perspectives, i.e., the omissions that threaten the effectiveness of such efforts. Finally, we suggest ways of transforming how sustainable agriculture is conceptualized and practiced.

CONVENTIONAL APPROACHES TO SUSTAINABLE AGRICULTURE

The large number of organizations and interest groups involved in the sustainable movement has resulted in a plethora of definitions for sustainable agriculture. Despite this diversity, however, there are identifiable central themes that express the main criteria of sustainable agriculture.

The predominant approach to sustainable in the United States is one that emphasizes farm-level resource conservation while maintaining production and profits. The USDA (1988), in a brochure describing its low-input sustainable agriculture (LISA) program,[1] stated that "LISA helps keep farmers profitable by improving management skills and reducing the need for chemicals and other purchased inputs. It helps sustain natural resources by reducing soil erosion and groundwater pollution and by protecting wildlife." The focus on farm production environment, and profits is echoed in the scientific literature on sustainable agriculture: "There is a growing awareness about the need to adopt more sustainable and integrated systems of agricultural production that depend less on chemical and other energy-based inputs. Such systems can often maintain yields, lower the cost of inputs, increase farm profits, and reduce ecological problems" (Edwards, 1990:xiii). Francis (1988) defines sustainable agriculture as a "management strategy," the goals of which are to reduce input costs, minimize environmental damage, and provide production and profit over time. Ruttan (1988) emphasizes that enhanced productivity must be a key factor in any sustainability definition. Finally, the National Academy of Sciences (National Research Council 1989:4) states that

> Alternative agriculture is any system of food or fiber production that systematically pursues the following goals:
>
> - More thorough incorporation of natural processes such as nutrient cycles, nitrogen fixation, and pest-predator relationships into the agricultural production process;
> - Reduction in the use of off-farm inputs with the greatest potential to harm the environment or the health of farmers and consumers;
> - Greater productive use of the biological and genetic potential of plant and animal species;
> - Improvement of the match between cropping patterns and the productive potential and physical limitations of agricultural lands to ensure long-term sustainable of current production levels; and
> - Profitable and efficient production with emphasis on improved farm management and conservation of soil, water, energy, and biological resources.

The National Academy of Sciences report is especially important because of the role it has played in bringing discussion of sustainability to the forefront in circles where it had previously been ignored.

Several organizations (e.g., the International Alliance for Sustainable Agriculture, the California Action Network, the Center for Rural Affairs, and the Wisconsin Rural Development Center) and writers (e.g., Altieri, 1988; Gips,

1988; Freudenberger, 1986) include social concerns in their priorities for and discussions of sustainable agriculture. In general, however, predominant efforts in sustainable agriculture ignore its human face or constitute its human issues too exclusively. Concern with the beneficiaries of sustainability is often limited to certain categories of people, such as farmers or those who can afford organic food. Social problems and priorities, as seen above, are usually not included in sustainability definitions. In a review of the literature on sustainability, Lockeretz (1988) identifies the problems addressed as environmental contamination due to pesticides, plant nutrients, and sediments; loss of soil and degradation of soil quality, vulnerability to shortages of nonrenewable resources, such as fossil energy; and low farm income resulting from depressed commodity prices in the face of high production costs. One sustainable agriculture leader stated that "The fundamental social responsibility of organic agriculture is improving the health of the soil—there is universal consensus that farmers and agriculture systems have to take care of the soil. But there is no consensus on the nature of justice and what equity is and how the state should intervene in the structure of agriculture" (Benbrook, 1992:8).

The points made in dominant sustainability discourse—preserving ecological conditions of production in farming—are relevant, but partial. Social issues, when raised, are often safely vague and framed in terms of "socially acceptable" as it refers to environmentally and economically sustainable institutions and practices. This begs the question, "socially acceptable for *whom?*" Not clearly specifying the social subject for sustainability leads advocates to prescribe future visions that tend not to consider significant alternatives to current social arrangements.

As demonstrated by Douglass (1984), in sustainable agriculture discourse there is a sense that as long as the present agricultural system can be maintained, sustainability can be said to be operant. Combining several perspectives, Douglass produces a composite definition for agricultural sustainability: "Agriculture will be found to be sustainable when ways are discovered to meet future demands for foodstuffs without imposing on society real increases in social costs of production and without causing the distribution of opportunities or incomes to worsen." The goal is to have an agricultural system the results of which are "no worse than the existing system's" (Douglass, 1984: 25).

These perspectives do not question the inequities many people experience in current structures of family farms, rural communities, or agricultural labor. The economic, social, and production systems of a sustainable agriculture may look entirely different depending on whether the goal is sustaining the current world economic order, an individual nation's agricultural economy, a farm family's income, a rural Ethiopian woman's life, or a middle-class American's access to pesticide-free food.

ISSUES ON WHICH THE U.S. SUSTAINABLE AGRICULTURE MOVEMENT IS SILENT

Sustainability is an inherently regressive term if used to mean only the sustaining of the existing "system," since the benefits of maintaining the current system are uneven, accruing not to everyone, but to those currently in privileged positions. While there is variation in the goals of those in the sustainable agriculture movement, the goals, at least implicitly, tend to reinforce the socioeconomic status quo—maintaining the benefits of the food and agriculture system for those who currently possess them, rather than securing benefits for everyone. The emphasis is on upholding the structure rather than improving the conditions for the majority of people living within the structure. For example, dominant sustainable agriculture discourse advocates safe, organic food, but does not address hunger. It is interested in fair returns to farmers, but has little to say about equitable conditions for hired agricultural labor. Similarly, it promotes the preservation of a family farm-based agricultural structure, but does not complement this with a focus on reconfiguring problematic gender and racial relations that have been part of this structure. These are examples of the "silences" of sustainable agriculture, which we elaborate briefly in the following sections.

Hunger

One of the major contradictions of the twentieth century has been the persistence of hunger despite unprecedented growth in agricultural production. Even though it would seem that the primary purpose of agriculture is to feed people, this has not been fully achieved, nor does it look likely to be. In the 1980s the number of hungry people increased five times faster than in the previous decade, despite record-level food stocks (U.N. World Food Council, 1990) and despite the fact that global food production has more than kept up with population growth (World Bank, 1986). Although cereal production has drawn ahead of population growth in developing countries (U.N. World Food Council, 1990), 40,000 people die every day of hunger and hunger-related causes primarily due to poverty (Speth, 1992). In Mexico, for example, increased production has made little or no difference in improving the situation of the malnourished: "Rapid growth in the productivity of the economy and rapid growth in the productivity of agriculture have not been accompanied by an improvement in the lives of most Mexicans but rather by continued immiseration and the creation of scarcity" (De Walt, 1985:54).

Although hunger is concentrated in the rural areas of impoverished countries, the problem is not only overseas—20 million Americans do not have regular access to sufficient food (UN Food Council, 1990). Women, children, and the elderly are at greatest risk of hunger (Nestle and Guttmacher, 1992); 76% of the hungry in the United States are people of color (Bread for the World Institute on Hunger and Development, 1992). In California, the wealthiest state in the nation, 1.4 million children are hungry or at a risk of hunger (True, 1992). A particular irony is that many of the hungry in the United States are those who harvest and process our abundant food supply. Hunger has been documented recently in the heart of America's most abundant agricultural region, California's Central Valley (California Rural Legal Assistance, 1991).

Yet sustainable agriculture has largely excluded the equitable distribution of food as a category of concern. Many U.S. sustainable agriculture advocates, in fact, see no connection between sustainable agriculture and hunger (see Clancy, this volume). This is in contrast to the perspective of many international nongovernmental organizations. For instance, the International Movement for Sustainable Agriculture's declaration on ecological agriculture *begins* with the issue of hunger (International Movement for Ecological Agriculture, 1991). Where consumption issues are discussed in the context of U.S. sustainability, they are usually in terms of toxin-free food, a possibility primarily for the affluent. Visions that include factors such as who should have a fundamental right to eat or access to land on which to grow food are not presented in dominant sustainability discourse. In one of the earliest discussions of sustainability perspectives Douglass (1984) provides a clear picture of how hunger issues are approached in sustainability discourse. Douglass discusses what he sees as the three primary sustainability groups writing on agricultural sustainability: those who see sustainability as food sufficiency, those who see sustainability as community, and those who see sustainability as stewardship.

According to Douglass (1984:5), the sustainability-as-food-sufficiency school takes a "relatively short-term view of sustainability, asking how much needs to be produced for a hungry world." Yet the problem of hunger is not actually addressed, and it is clear that this school sees production, not distribution and access, as the primary cause of hunger. Douglass (1984:22) states that advocates of food sufficiency consider agriculture to be sustainable when "farmers produce enough food to meet reasonable projections of global market demand." Meeting market demand and meeting food needs, however, are not the same thing—in the market system those who lack the resources to purchase food go hungry. For the sustainability-as-stewardship school,

"The ecological view of agricultural sustainability arise from the belief that nature in the long run imposes definite limits on humankind's collective capacity to provide food for the people of the world. These limits are primarily physical in nature" (Douglass, 1984:11). Those in this school advocate population controls as an integral component of achieving sustainability. They view those who are hungry as a major source of environmental degradation, but fail to acknowledge the links between poverty, hunger, and environmental destruction. At its extreme, this viewpoint essentially proscribes the right of the hungry to exist, as when stopping population growth is advocated over ending hunger and poverty. In the sustainability-as-community school, there is the beginning of a focus on hunger in that "acceptable standards of nutrition" is a goal. Still, their goals are always referred to in the context of the "community," which seems to refer to their local, rural agricultural communities and thus excludes nutritional goals for those outside this framework.

Labor

Farm workers have received few of the benefits of profitable and abundant agriculture. Compared to farm owners (and workers and in other industries), their incomes are much lower, their living conditions are worse, their working conditions are more physically arduous, their control of the production process is minimal, and they are more often exposed to pesticides which results in higher incidences of health problems related to pesticide use.

In the United States, agricultural workers have traditionally low standards of living—they are often hungry, poorly housed, and have limited or no access to health care. Half of U.S. farmworkers and their families have incomes below the poverty level, with the median family income between $7,500 and $10,000 per year (U.S. GAO, 1992). Housing is available for only about one-third of hired farmworkers in the United States and what does exist is often inadequate (U.S. GAO, 1992). In Santa Cruz County, California, one of the richest agricultural areas in the world, adequate housing shelters only 11% of farmworkers; many workers and their families crowd into rentals too costly for one family to afford; many camp out or live in caves for which they must pay rent (Beebe, 1991).

Every year in the United States hired farmworkers suffer up to 300,000 acute illnesses and injuries from exposure to pesticides (U.S. GAO, 1992). They are most likely to suffer from acute pesticide illnesses such as dizziness, vomiting, and systemic poisoning; the chronic effects of pesticide exposure, such as cancer and reproductive problems, have barely been researched and remain largely unknown (Perfecto, 1992). Pesticides are exempt from the hazardous substances regulations that require employers to notify workers about the substances being used, potential risks, and first-aid procedures. Children, who often must work in the fields to earn income to contribute to family survival, are also exposed to pesticides. A study in New York found that 40% of farmworker children studied had worked in fields still wet with pesticides (Global Pesticide Campaigner, 1992).

Despite these obvious hardships, sustainable agriculture does not address the specific needs of farmworkers and does not recognize that farmers and farmworkers have different interests. For example, the *Asilomar Declaration for Sustainable Agriculture* states that "Healthy rural communities are attractive and equitable for farmers, farmworkers, and their families" (Committee for Sustainable Agriculture, 1990). We consider this document to be representative of major sustainability perspectives, since it was produced through an intensive meeting of U.S. sustainable agriculture experts.[2] This statement assumes the necessity of present social relations in the production process. It recognizes no inherent problem with an economy based on farm owners who hire landless laborers and does not question the existing structure of land tenure. where the National Research Council (1989) discusses farm labor on alternative farms, labor is viewed only as a cost of production. There is no discussion of who the workers are, their working conditions, or their wages; people are objectified and treated primarily as economic inputs, along with equipment and fuel.

Much of the work in the agrifood sector occurs not in the fields, but in factories. Working conditions in food processing vary with each particular industry, but workers in vegetable, fruit, and chicken processing industries are often poorly paid, seasonally laid off, have no benefits, and work in miserable conditions. In their search for cheap labor, these food-processing industries prefer to hire women and have moved to regions and localities where wages are lower. For example, chicken processing industries are located predominantly in the Southern United States and rely on the labor of black women who often work in unsafe conditions. Fruit and vegetable processing industries are moving across the border from California to Mexico where labor costs are lower and environmental regulations are few. The poor conditions of workers in the agrifood sector, whether on farms or in food processing, have not yet been addressed by the sustainable agriculture movement.

Race and Gender

Conventional food and agriculture systems have developed with highly skewed power distributions that marginalize women and people of color. While the success of these systems has depended on the labor of women and people of color, women and ethnic minorities have not had equal access to land, capital, or decision making in the food and agriculture system. Sustainable agriculture reinforces this distorted distribution of power and opportunities partly through its emphasis on keeping rural communities the same. For example, the *Asilomar Declaration for Sustainable Agriculture* (Committee for Sustainable Agriculture, 1990) states: "The continuation of traditional values and farming wisdom depends on stable, multi-generational population." This implies that current rural values, which include the patriarchal family and Christian religious beliefs, are ideals we should advocate and preserve. In addition, the call for a return to traditional rural values fails to challenge the racist attitudes that have historically permeated the rural United States and have provided a major arena for the subjugation of people of color.

People of color.   People of color have been integral to the functioning of American agriculture, but in subordinate roles. African-Americans, Latinos, and Asian-Americans have historically provided and continue to provide much of the labor in U.S. agriculture, but are much less likely than European-Americans to be farm owners. Even in California, an ethnically diverse state, only 9.2% of farm owners are ethnic minorities; in contrast, 75% of farmworkers are ethnic minorities (Vaupel, 1988). It is significant that the impetus for sustainable agriculture was generated in part by the level of public distress about farmers losing their land during the 1980s when the crisis affected mostly European-American farmers. In contrast, little concern has been raised in sustainability discourse about the nearly complete separation of African-American farmers from their land. Black farm ownership has declined 94% in this century (Belden, 1986). In 1920, one U.S. farmer in seven was black, whereas by 1987 the ratio of white to black farm operators was 91 to 1 (Demissie, 1992).

As farmworkers, people of color are much more likely than Caucasians to experience both chronic and acute pesticide poisoning. Perfecto (1992) emphasizes the inherent racism and classism in our concerns with the impact of pesticides. The major impetus for regulating pesticides has been persistence of the chemicals in the environment or residues on food. Thus, chlorinated hydrocarbons such as DDT that were persistent in the environment have largely been replaced with organophosphates, which are nonpersistent but acutely toxic. DDT, for example, has only 2 to 5% the toxicity of parathion. While the banning of organophosphates has improved conditions for wildlife and the environment, there has been a trade-off with the health of people who work in the fields. Farmworkers, particularly people of color, are exposed directly to these new chemicals with high acute toxicity.

In addition, people of color have not traditionally been involved in Western agricultural science, where the technologies that affect both workers and the structure of agriculture are developed. Separate and poorly funded agricultural colleges were established for blacks in the United States in 1890. Most dominant agricultural technologies are generated in well-funded agricultural institutions staffed overwhelmingly by whites—in the late 1970s 95.5% of U.S. agricultural researchers were Caucasian (Busch and Lacy, 1983).

Women.   Throughout the world, women are poorer, own less property, do more work, hold less power, are less educated, and suffer more hunger than men. According to Lipman-Blumen (1986:54), "Control over resources lies at the very heart of all power relationships—between nations, races, socioeconomic and ethnic groups, generations, and individuals—most particularly between women and men. . . . The paradigmatic power relationship between women and men, with its intransigent inequality mapped on all other relationships, across all nations, is the most crucial and fundamental issue underlying social justice." In the U.S. food and agriculture system, including the majority of family farms, men control land, capital, and women's labor (Sachs, 1983), while women contribute significantly to the sector's profitability as wage workers and as unpaid labor on family farms.

Traditional gender roles are reinforced in the sustainable agriculture discourse. Populist visions of sustainable agriculture see the family farm as the ideal organizational structure for sustainable agriculture, but generally do not discuss gender roles within the farm family. An exception is Berry (1977), who explicitly discusses differences between men and women on farms and suggests that both women and men suffer when nurturing is the sole purview of women. However, he advocates a return to traditional values associated with the home without questioning the patriarchal privilege that underlies many of these values. The fat that family farms are based on historically inequitable relations between men and women is often not problematized by the sustainable agriculture movement.

Although women on farms are resisting their roles as "farm wives" and insisting on access to land and recognition of their roles as decision makers or farmers, most sustainability discourse does not recognize or incorporate women's demands for change. The "farmer" is generally assumed to be male, as illustrated by the frequent use of masculine pronouns. Women's roles in agricultural production remain largely unacknowledged except as they support the male farmer. Also, women have been virtually absent in the agricultural scientific community, where as of the late 1970s 99.6% of agricultural scientists were male (Busch and Lacy, 1983). This results in few women being included in the scientific discourse on sustainability and an absence of feminist viewpoints in the shaping of sustainable agriculture, which takes place in scientific forums.

The gender bias of conventional agriculture carries over into the sustainable agriculture movement. In the food safety sector of the movement, for
example, women are often targeted specifically for the part they can play in
developing this nutritional/nurturing aspect of agricultural sustainability.
This is an essentialist role, however, since women are appealed to in their
capacity as food purchasers, family caretakers, and child-care providers. Although women are key forces in grassroots efforts organizing around issues
of organic farming, food safety, and pesticide poisonings, they are underrepresented in leadership positions. At conferences on sustainable agriculture
women are much more likely to be listening in the audience than speaking
on the program, which will be overwhelmingly composed of men. This is
despite the fact that women are prominent in organizing these conferences,
coordinating community endeavors, and creating linkages among different
sustainable agriculture groups.

The partial vision of dominant discourse on sustainable agriculture does
not include improving conditions for everyone, but promises benefits primarily for those who already have them. What we find missing in these perspectives on sustainability is an attempt to improve food and agriculture systems
for *all* people, regardless of class, race, gender, or national origin. Advocating
the preservation of family farms, for example, can mean reproducing the
uneven race, gender, and class divisions that have historically existed in
agriculture (although some would argue that family farms are still preferable
to the alternative of corporate-controlled farming). The call for preserving
rural communities does not ask how farmworkers, people of color, women,
and Third World people could benefit from an alternative form of agriculture.
Similarly, advocating food safety, certainly a worthwhile objective, tends to
exclude issues of corporate control of the food system, poor people's access
to high-quality food, and, most importantly, their access to *enough* food for
a productive and healthy life.

INTERNAL CONTRADICTIONS OF SUSTAINABLE AGRICULTURE

In addition to silences on issues such as hunger, labor, race, and gender,
there are a number of contradictions implicit in conventional approaches to
agricultural sustainability. That is, there are issues that, if neglected, threaten
to prevent us from achieving even purely environmental priorities in agriculture. The subjects we consider here are the units of analysis used in sustainability discourse and science, the causes attributed to nonsustainability,
the analysis of economic structures, the reification of science, and the role
of technology.

Overly Narrow Unit of Analysis

Like conventional agriculture, sustainable agriculture focuses its attention largely on the farm-level production aspects of agriculture without proportionate emphasis on the other aspects of the system—distribution, consumption, and exchange (Allen, 1991). For example, Edwards (1989), arguing the importance of integrating parts of the whole in sustainable agricultural systems, limits the main components of these systems to agronomic elements: fertilizers, pesticides, and cultivations (his other components include machinery, crop breeding, and rotations, still at the farm level). Edwards states that increasing knowledge of the main inputs and how these practices interact will form the basis of developing agricultural systems that increase profitability for the farmer and reduce environmental problems. This perspective is also predominant in the USDA's Sustainable Agriculture Research and Education program and the National Academy of Sciences' (National Research Council, 1989) approach. Sustainable agriculture is thought of almost solely in terms of farms and farmers, which is not an accurate representation of today's food and agriculture system.

Even though the on-farm transformation of resources into food and fiber is a core process of the food and agriculture system, it is but one of many components. Interactions among the large environmental, social, and economic systems in which agriculture is situated directly influence agricultural production and distribution. For example, farmers' choices of production strategies are frequently dictated by food manufacturing industries. Large agroindustrial firms buy the vast majority of farm produce and often specify inputs such as feeds, pesticides, and fertilizers and also set quality standards for produce and livestock that often require heavy chemical applications. Sustainable agriculture needs to be conceptualized in a way that includes not only the production process itself, but all of the related backward and forward linkages, i.e., the whole of the food and agriculture system. Lowrance et al. (1986) describe a model that approximates a whole-systems approach. They see four different loci or subsystems of sustainability: (1) individual farm fields where agronomic factors are paramount, (2) the farm unit wherein microeconomic concerns are primary, (3) the regional physical environment where ecological factors are central, and (4) national and international economies where macroeconomic issues are most important. Their model, though lacking a social component, demonstrates that focusing on only one level of the agricultural system neglects others that are equally essential. A whole-systems perspective is essential because it fosters an understanding of complex interactions and their diverse ramifications throughout agriculture and the systems with which it articulates (Allen et al., 1991). Yet, the context of problems identified in dominant U.S. sustainability texts is usually First

World oriented and treats U.S. agriculture as if it were independent rather than part of a global food system.

The existing global division of labor and global market by definition create a global economy and society. The system includes not only production of agricultural products, but also their distribution and the infrastructure that affects production and distribution at regional, national, and global levels. Farming is but one aspect of a larger, transnational agrifood system that is controlled predominantly by large corporations such as Philip Morris, Con Agra, Tyson Foods, and Cargill. Transnational restructuring of the agricultural sector has blurred national and sectoral boundaries, intensified agricultural specialization for both enterprises and regions, and created large agro-industrial complexes (Friedmann and McMichael, 1989). Therefore a global unit of analysis is an essential beginning framework for explanation and solution. As Friedmann (this volume) illustrates, we can no longer view agriculture as a separate or distinct sector of the economy but must more accurately understand it as an agrifood sector in a world economy, a series of commodity chains linked together into food complexes. In this way we move beyond the present, limiting farm-centric focus for sustainability efforts and consider the entire food and agriculture system as the unit of analysis. From within this, point-specific components can be studied and addressed as appropriate. It is epsitemologically accurate to include larger social and economic issues in sustainability if we want to use relevant units of analysis in conceptualizing, studying, and creating agricultural sustainability.

Nonsustainability Goes Unexplained

The sustainable agriculture movement has, of necessity, documented and described environmental problems in agriculture. Numerous reports have called our attention to soil erosion, groundwater depletion, the use of fossil-based inputs, and destruction of the genetic base. This is an essential first step. The transition from problem description to problem solution, however, has skipped the intermediate step of problem explanation (Allen, 1991). Little effort has been focused on answering the "why" questions of ecological destruction in agriculture.

An analysis of the root causes of nonsustainable agriculture is not apparent in conventional scientific texts on sustainability. Some groups outside formal academic institutions do, however, attribute causes for the development of nonsustainable agricultural systems. For example, Jackson (1984) states that the lack of an ecological approach has led to an unsustainable agriculture characterized by soil loss, fossil fuel dependence, and chemical dependence. Berry (1984) decries the industrialization and mechanization of corporate agriculture and asserts that the current U.S. agricultural system is unsustainable because of the continual attempt to get the highest possible production with the smallest number of workers. Particularly important for Berry is the erosion of cultural values associated with family farming, such as hard work, respect for place, respect for nature, and commitment to home and community. For food safety advocates, primary causes cited in regard to food contamination are the failure of government to adequately regulate pesticides (Natural Resources Defense Council, 1989) and lack of consumer awareness.

But in offering these explanations of nonsustainability—corporate agriculture, inadequate government regulation, and loss of respect for nature—we have not analyzed why these very conditions, in turn, exist. While sustainable agriculture discourse delineates proximate causes of sustainability problems, it does not get at the "causes behind the causes." The reasons for the current condition of agriculture, such as the forces that have led to resource-intensive farming practices, are rarely if ever explained. For example, sustainable agriculture advocate want to reduce pesticide use because pesticides cause groundwater contamination and remain as residues on food. But they do not ask how and why pesticide use has become so common and entrenched in agriculture. Sustainable agriculture discourse addresses the directly visible, surface cause of environmental problems. It does less well at addressing their less visible, structural causes, i.e., those that reflect the deeper, systemic operating principles of the food and agriculture system.

Failure to Examine Economic Structures Conducive to Sustainability

While twentieth-century, scientific agriculture has achieved unprecedented yields, the drive to maximize profits and production has been accompanied by a tendency toward ecodegradation. In both "existing socialist" and "existing capitalist" countries, agricultural practices have been driven by economic imperatives, since production for the market has been essential for economic survival. This system has been based on the commodification of both the product and the means of production in agriculture. Natural resources, labor, capital, technology, and food have all become commodities that are sold and bought for a price. This has required certain forms of appropriation from nature that have resulted in negative ecological externalities. The problem is particularly evident in agriculture, since, even in its industrialized form, agriculture remains dependent on natural resources and processes such as soil, water, and weather. Conventional agricultural systems have been self-negating, producing scarcity of the natural resources on which agricultural processes depend and thus producing barriers to long-term ecological sustainability.

Current economic systems have limited choices in the agrifood sector, subordinating environmental rationality and ethical priorities to economic rationality. Rarely are these congruent. Organic farmers may have a broad commitment to the environment, but they cannot remove themselves from the economic system in which they must purchase inputs and sell their products. As with conventional farmers, organic farmers will be driven by economic efficiency rather than ecological rationality once market competition among organic farmers develops (Bird, 1988). Similarly, wages for agricultural labor will also be dictated by the market as will other social relations in organic agriculture.

Dominant sustainable agriculture perspectives do not recognize the need to examine the political, macroeconomic, and structural context that defines profit and economic efficiency in agriculture. Although the U.N. FAO acknowledges the causal sustainability link between socioeconomic and agroecological factors (Consultative Group on International Agricultural Research, 1989), this is usually not considered in predominant U.S. sustainability efforts. In the U.S. Department of Agriculture's (1988) "10 Guiding Principles of Low-Input, or Sustainable, Agriculture," for example, the first principle is that "If a method of farming is not profitable, it cannot be sustainable." The National Academy of Sciences states that "Successful alternative farmers do what all good managers do—they apply management skills and information to reduce costs, improve efficiency, and maintain production levels" (National Research Council, 1989). A brochure publicizing an organic farming field day has the headline, "Boost Profits $50 Per Acre With [organic farming techniques]."

As the organic market expands, large-scale agribusiness is entering the industry, either through converting small amounts of their vast acreages to organic production, thereby competing directly with small organic producers, or through contracts with or outright purchase of successful organic enterprises, including processing and distribution. Farmers have little choice but to sell their products on commodity markets that are increasingly controlled by agrifood industries. Organic food producers and distributors must also seek larger and larger market shares. An article in a journal for the organic foods industry asks, for example, "In an age when marketing is reaching new heights of sophistication, how are retailers building sales of organic products?" (Snyder, 1990).

These strategies are necessary for sustainable agriculture businesses to survive in the current economic situation. They illustrate why the present forms of agricultural production, which are primarily for the market rather than for need, should be examined for their role in agricultural problems. Even the environmental aspect of sustainability cannot be understood outside of the larger economic context, since the social and economic structure of agriculture (e.g., land tenure, resource allocation and regulation, and terms of trade) affects environmental quality. Thus, there is a need to examine the basic premises, social and political structures, and processes at the root of agriculture's nonsustainable aspects.

Reification of Natural Science

In conventional agriculture a major emphasis has been on developing profitable production techniques and systems through science. As Danbom (1986) notes, the policy makers who created the publicly funded agricultural research system were primarily concerned with increasing production. Historically in the United States, agricultural science has been called on to resolve agronomic, economic, and ecological problems in agriculture, through institutions and agencies such as the land-grant colleges, Cooperative Extension Service, and Soil Conversation Service. Science has developed the highly capitalized, chemical-intensive agricultural system in the United States, and is being uncritically called on to develop sustainable agriculture as well. In neither case have socioeconomic inequities been addressed by, nor are they resolvable by, science. Nonetheless, or perhaps because of this, science serves as an ideology that supports the current organization of agriculture.

Western science is based on the exploitation of nature. The Baconian conception of science was developed explicitly to enable humans to dominate and master nature through knowledge. Following this Baconian imperative, publicly supported agricultural science has attempted to subdue and conquer nature to serve human interests through mechanistic, reductionist, and fragmented approaches to understanding and shaping the natural world. The approach to solving problems in agriculture at this level is fundamentally flawed. Nature is both objectified and essentialized. Feminist critics of science have led the way in pointing out how the objectification of nature has enabled scientists to construct false dualisms that set humans apart from and above nature (Harding, 1986; Keller, 1986).

Sustainable agriculture has tended to accept this reified perspective, constructing itself in the image and likeness of conventional agricultural science in terms of its lack of social focus and insistence on "objectivity." Sustainability problems are framed almost exclusively in terms of natural science and there is an implicit assumption that social relations in food and agriculture have a fixed quality and are beyond human control (Allen, 1991). Thus, problems in agriculture are viewed as technical rather than social and resolvable only through the efforts of natural scientists. Where sustainable agriculturists have recognized how environmental problems have resulted in part from the approaches of conventional agricultural science, they have typically not viewed these problems as related to problems in natural-scientific epistemology. Rather they maintain scientists' claims of objectivity—that scientists are separate from their objects of investigation—and propose that such problems will be solved by studying "farming systems" rather than individual components.

Sustainable agriculture research and education programs are placed in conventional agricultural departments at traditional agricultural universities and are staffed by scientists with expertise in traditional agricultural fields. Only casual attention is paid to integrating underrepresented disciplines or groups into these programs. The resulting research and education projects have focused largely on environmentally friendly versions of what agricultural science already does: improve production practices on farms. While research on biological pest control, rotational grazing, intercropping, cover crops, and nutrient management have been necessary and have provided farmers with less environmentally damaging production strategies, social research has lagged far behind. Research on social issues in sustainable agriculture has been limited primarily to evaluating how farmers' values and attitudes encourage or block the adoption of sustainable technologies. These projects have been useful in examining the possibilities of developing more environmentally sound farming operations. However, there has been little serious investigation of the social, political, and economic relations that are needed to encourage sustainable agriculture. This natural- and social-science research has generally assumed that environmental changes take precedence over social changes, and that achieving agricultural sustainability is possible without changing social relations. The biological and physical agricultural sciences, while important in providing agronomic and technical alternatives for agricultural problems, are not equipped to deal with the fundamental causes and corresponding solutions to agricultural sustainability problems.

Technological Determinism

Dominant sustainability discourses manifest a tendency toward technological determinism—that if the right technologies were developed, sustainability would result. This follows logically, if unfortunately, from the tendency to adopt a too-narrow unit of analysis, to overlook the fundamental causes of nonsustainability (including failure to examine the role of economic structures), and to reify natural science. The National Research Council's (1989) report on alternative agriculture embodies a lack of concern with factors other than technical production. In this report "alternative" refers to only biological and technological alternatives to conventional agricultural practices, not to alternative social and economic arrangements.

The major strategies employed in the effort toward sustainable agriculture are providing more information to farmers and consumers through better communication, gathering more knowledge about agroecological processes, and developing better technology. The *Asilomar Declaration for Sustainable*

Agriculture, for example, states that, "Given scientifically validated techniques, farmers will adopt sustainable agricultural practices" (Committee for Sustainable Agriculture, 1990:1). Offering a similar viewpoint, Strange (1988:206) concludes that "the most serious environmental problems in agriculture are those caused by technologies that make large-scale farming possible, and that sever the rewards of farming from the rewards of stewardship and husbandry." In these perspectives, an agricultural production system that is environmentally sound will automatically be achieved as less environmentally damaging technologies are developed and substituted for existing chemical technologies.

Although most sustainable agriculture research programs at universities have changed their research foci from "chemical" to "natural" technologies (e.g., crop rotations and biological pest management), they generally retain the emphasis on technological development as the solution to agricultural problems. The major strategies employed in the effort toward sustainable agriculture are providing more information to farmers and consumers through better communication, gathering more knowledge about agroecological processes, and developing better technology. We agree that research effort is needed to develop new technologies and management systems that do not degrade the natural environment. We argue not with their inclusion in a package of strategies for sustainability, but with the emphasis placed on them to the exclusion of other strategies.

Such an emphasis ignores the overarching structural forces that have contributed to the adoption of resource-intensive farming practices. Technologies and social relations are inseparably linked, both in terms of their inspiration and their consequences. This is clearly seen in the way that agricultural technologies are developed in publicly funded universities. Although agricultural research universities ostensibly operate on the Baconian model of the atomistic scientist pursuing "pure knowledge," implicitly or explicitly, their priorities in research and technology development have been driven by economics and politics. According to an ex-dean of agriculture at the University of California at Davis, because funding availability is a primary determinant of agricultural research priorities, private supporters—generally agricultural industries—can determine many agricultural research priorities by making relatively small financial contributions (McCalla, 1978).

Commodifiable technologies are demanded by business and thus produced. The development of chemical, rather than cultural, pest management techniques, for example, is not accidental. Within the existing economic system, commodifiable solutions—those that can be profitably manufactured and sold—are the solutions to agricultural problems that will be developed, marketed, and used (Bird, 1988). Maximizing profits depends on repeated sales of inputs, not products that can be reproduced by the farmer or which are self-reproducing under proper environmental conditions. Thus, the tendency to overrely on technology, combined with business leveraging of the kind of technology developed, constitutes a contradiction to agricultural sustainability.

It is clear that neither science nor new technologies can by themselves solve larger food and agriculture problems, as witnessed by the problems associated with the scientifically based Green Revolution. And, of course, the overarching causes of nonsustainability are not primarily the absence of proper technology or sufficient information about technology. As de Janvry and Le Veen (1986:83) affirm, "Although we acknowledge that both types of agriculture [socialist and capitalist] have become increasingly dependent upon productivity-increasing, industrial technologies, our position is that the nature of agricultural development is determined much more by prevailing economic, political, and social relations than by the requirements of these technologies." New technologies and policy reforms must be accompanied by structural transformations if sustainability is to be achieved.

MOVING TOWARD TRANSFORMATION

While the U.S. sustainability movement has raised important issues that have begun to change some aspects of agriculture, a broader perspective is needed if we are to avoid reproducing many of the fundamental ecological and social problems of conventional agriculture. To proceed with achieving agricultural sustainability, discourse and action must become more comprehensive, both conceptually and strategically. What changes in approach will help to bring this about? Based on the position that transformation toward a more sustainable agriculture necessarily involves both environmental and social changes, we must begin with strategies for incorporating social issues into the current production and environmental foci of research, education, policy, and citizen action. While a blueprint for developing a sustainable food and agriculture system is beyond the scope of this chapter, we suggest several areas in which the transformation of sustainability can begin.

Composing Sustainability Based on Human Need

First, the sustainability movement needs to develop a proactive vision of what "should be" to add to what is now primarily a reaction to food and agriculture problems. Such a vision should include a comprehensive answer to the question, "Who and what do we want to sustain?" (Allen and Van Dusen, 1990). Attention to *how* agricultural products are produced should be supplemented as, Altieri (1988) proposes, by addition to *what* agricultural products are produced and *for whom*. Without the resolution of social equity issues, the structures of domination that led to environmental degradation and social inequity in the first place will be reproduced. Thus, we need to address issues such as poverty and hunger as well as those of preserving the environment and family farms.

In fundamental terms, sustainable agriculture is a struggle over life and death. The concept of sustainable agriculture therefore should be based on meeting basic human needs. We consider these needs to be consumptive (food, water, fuel), protective (clothing, shelter), and regenerative (dignity, self-determination, freedom from exploitation) (Allen and Sachs, 1992). These needs should be met both for generations to come and for generations that are here now. Thus, sustainable agriculture should maintain the ecological conditions of production *and* provide the means for everyone to live and work with dignity, including securing adequate, safe food. This, in turn, is predicated on developing nonexploitative relations in terms of race, class, gender, and nation.

To guide action, a socially equitable vision for sustainable agriculture must provide guidelines for translating abstract principles into specific plans for action in local circumstances. Although this vision cannot fully articulate which actions are appropriate for a specific agent at a given time and place, it can provide a starting point for deliberation and action. for example, rather than assuming that family farms are ideal workplaces or that male control of decision making and women's and children's labor is unproblematic, sustainability advocates should promote strategies for enhancing equitable relations in families. Relations between wage workers, a large proportion of whom are people of color, and farmers should be questioned with an eye toward developing more participatory and less exploitative working conditions. Also, as land ownership becomes more concentrated, alternative land ownership arrangements that provide access to land for women, African-Americans, Latinos, and working-class people must be developed. To be effective, sustainability strategies must include all those who are affected by the food and agriculture system and must be accessible to and usable by those who have the power to end ecological destruction and human degradation at all levels. A vision that catalyzes consideration of these issues within localized centers of action is crucial.

A Chorus of Voices on Sustainability

This vision of sustainability must include equitable distribution of power and democratic decision-making structures. The secretariat of the United Nations Conference on Environment and Development (1991) recognized this need when it pointed out that unequal access to natural resources and other opportunities can lead to poverty and environmental degradation. Resolving this can be achieved only if there is equal access to decision making for those involved in all aspects of food and agriculture. Characteristics such as these must be built into standard conceptualizations of sustainability if it is to be a meaningful departure from conventional agriculture.

It is essential that we increase the participation of underrepresented groups in order to develop a broader range of possibilities for transforming the global food system. Presently, agricultural policy makers, farm-group leaders, development planners, and researchers and educators in the United States are predominantly male and European-American. Sustainability groups, whether focused on food safety, family farmers, agricultural science, or other concerns are, of necessity, involved in day-to-day struggles for sustainability within their own spheres. In the process of these daily struggles, the interests of the most disadvantaged groups in society fall through the cracks, and the poor, women, and people of color remain invisible and continue to be exploited. For example, policy efforts in sustainable agriculture have focused largely in improving farmers' ability to farm profitably in an environmentally sustainable manner. These efforts are critical, but a broader agenda that considers nonfarmers and people in other regions of the world is also necessary. An agricultural system cannot be sustainable unless its leadership represents all of its constituents.

A related point is that we need less reductionist and more participatory science. Research on sustainable agriculture has recognized that narrow, reductionist scientific approaches are inadequate to solve agricultural problems at the farm level (MacRae et al., 1989), let alone at the complex macrolevels that constitute today's agriculture. There are other important limitations to research as well, and they must be addressed. Individual perspectives, based on background and experiences, play a pivotal role in what people see as problems and solutions. We must therefore broaden the kinds of perspectives held by researchers. Continued efforts to conduct interdisciplinary research in agriculture should proceed with renewed efforts to include voices of farmworkers, the poor, women, farmers, people of color, environmentalists, and consumers in defining research problems and their resolution.

Action in Concert for Creating a Sustainable Future

To be successful, sustainable agriculture advocates must seek solutions through common political action. Achieving sustainability requires a political realignment to change relations of resource control and power. This, in turn, requires subordinating particular interests at some levels, while effective coalitions and broad agendas are formed.

The subordination of particular interests to universal interests requires an expansion of our frames of reference along both vertical and horizontal axes. A *vertical* analysis involves recognizing the links between and among conditions in food and agriculture, such as that ecological and social problems have similar roots. A *horizontal* consciousness involves a redefinition of who we see as "us" and "them," such as when consumers are interested not only in pesticide residues on food, but also farmworkers exposure to pesticides. Mies (1986) points out that once all people are considered inviolable as human beings and no longer considered objects or natural resources for others to use, it will no longer be possible for one group to expect progress for itself at the expense of another. Thus, ending ecological and human degradation requires that we make transparent the connections among political economic, ecological, patriarchal, racist, and nationalist forms of domination. We need to find ways to encourage integration of different interest groups in developing a sustainable food and agriculture, ways that include environmentalists, farmers, welfare organizations, consumers, farmworkers, and feminist groups working together.

In the sustainability movement there are diverse platforms for different groups, all of which have as their basis their own material interests: organic farmers (market premiums, concern for their land, sometimes their health), agricultural universities (maintaining a constituency base for research), consumers (their health and the low price of food), and family farm activists (preserving their traditional rural way of life). Each of these groups is engaged in tensions within their spheres—i.e., organic farmers vs. conventional farmers, sustainable agriculture researchers vs. conventional researchers, family farmers vs. corporate farms. Even when the goals of sustainable agriculture organizations are quite broad, in their day-to-day activities., members have often been reluctant to work on projects that go beyond their immediate interests. But sustainable agriculture advocates must recognize the common foundations of their particular interests, become aware of the broader contradictions among sustainability groups, seek common interests, and increase the participation of underrepresented groups in order to develop a politically powerful coalition for transforming the global food system.

Of utmost importance is development of international, national, and local food policies, the objectives of which are to provide everyone with adequate, safe, and nutritious food while guaranteeing fair returns to those who earn their living in food and agriculture. The Toronto Food Policy Council started such a project in 1990. In its goal, "to promote food production and distribution systems which are equitable, nutritionally excellent, and environmentally sound," it recognizes the imperatives of developing both a viable agricultural system and ensuring food security for all people (Toronto Food Policy Council, nd).

In designing food and agriculture policies we must also think beyond our own shores to recognize the global impacts of our actions and develop strategies to address their ramifications on people in other regions of the world. For instance, we should revise export/import policies that currently permit exports of toxic pesticides and imports of crops raised with pesticides that are banned domestically, something on which many sustainable agriculture groups are working. Policy makers should consider the impact of such policies on Third World workers and residents as well as on workers, consumers, and farmers in the United States.

CONCLUSION

As a relatively new concept, sustainable agriculture does not yet reflect a coherent vision of possible and preferable modes of agricultural production and distribution. In the current moment we have the opportunity to create conceptual and practical forms that eliminate exploitative conditions in agriculture, both for the environment and for human beings. Our ideas for creating pathways to an equitable and environmentally sustainable agricultural system are only a beginning. More comprehensive, creative, and locally appropriate agendas must necessarily arise through the many-faceted efforts of sustainable agricultural organizations.

In this, sustainable agriculture faces the same challenges as do other new social movements. Rather than focusing only on the immediate and personal, to be effective these movements must also address broad issues, seek fundamental causes of problems, and develop concerted strategies for change. This kind of reorientation is necessary in the sustainable agriculture movement in order to effect a transformation within food and agriculture that eliminates environmental as well as human degradation. To articulate the agendas of sustainable agriculture advocates with the issues of women, people of color, and working-class movements, we propose that as sustainable agriculture researchers, educators, policy makers, and activists, we assess our current concepts and strategies. We must continually ask both how we can bring together social and ecological concerns and to what extent we have incorporated a diversity of participants—men, women, people of color, working-class people—and their concerns into our efforts.

We have an active role to play in producing the history of sustainability. Although inertia is a strong force in determining history, so too are the concerted efforts of groups and individuals. History is the product of the "creative, and conscious, activity of real individuals" (Sayer, 1987:108) and people are the active producers of their concepts and possibilities. Each of us—whatever role we play in food and agriculture—needs to examine aspects of our daily life that have the potential to be forces for or against sustainability. An understanding of the complexities of the global food and agriculture system must not dissuade us from taking action, but rather inform our actions in influencing international and national policies as well as initiating grassroots, local efforts for sustainability.

Sustainable agriculture remains a contested concept—it can lead to a food and agriculture system that privileges the few, or one that transforms existing social relations and ecological conditions in the global food and agriculture system. We must all recognize that we make decisions every day to use our abilities, positions, and resources either in the direction of keeping things the same or changing them for the better.

NOTES

1. This program has been renamed "Sustainable Agriculture Research and Education" (SARE).
2. Sustainable agriculture groups represented included the Committee for Sustainable Agriculture, the Department of Agronomy at Washington State University, Healthy Harvest, the International Alliance for Sustainable Agriculture, Keyline Agricultural Consultants, the Land Institute, Madden Associates, Molino Creek Farming Collective, National Organically Grown Week, Pavich Family Farms, the Rodale Institute, the Rocky Mountain Institute, the Working Land Fund, the U.C. Santa Cruz Agroecology Program, the U.S. Sustainable Agriculture Research and Education Program, and the Whole Earth Review.

REFERENCES

Allen, P. 1991. Sustainable Agriculture at the Crossroads. *Capitalism, Nature, Socialism* 2(3):20–28.

Allen, P. L., and C. E. Sachs. 1991. The Social Side of Sustainability: Class, Gender, and Ethnicity. *Science as Culture* 2(13):569–590.

Allen, P., and C. E. Sachs. 1992. The Poverty of Sustainability: An Analysis of Current Discourse, *Agriculture and Human Values* 9(4):30–37.

Allen, P., and D. Van Dusen. 1990. *Sustainability in the Balance: Raising Fundamental Issues*. Santa Cruz, CA: University of California.

Allen, P., D. Van Dusen, J. Lundy, and S. Gliessman. 1991. Integrating Social, Environmental, and Economic Issues in Sustainable Agriculture. *American Journal of Alternative Agriculture* 6:34–39.

Altieri, M. A. 1988. Beyond Agroecology: Making Sustainable Agriculture Part of a Political Agenda. *American Journal of Alternative Agriculture* 3:142–143.

Beebe, G. 1991. Migrant Farmworkers Struggle to get By. *Santa Cruz Sentinel*, September 6.

Belden, J. N. 1986. *Dirt Rich, Dirt Poor*. New York: Routledge & Kegan Paul in association with Methuen.

Benbrook, C. 1992. Quoted in Dose Organic Mean Socially Responsible?: A Conversation. By Grace Gershuny and Thomas Forster. *Organic Farmer* 3(1):7–11.

Berry, W. 1977. *The Unsettling of America: Culture and Agriculture.* New York: Avon.

Berry, W. 1984. Whose Head Is the Farmer Using? Whose Head Is Using the Farmer? In *Meeting the Expectations of the Land.* W. Jackson, W. Berry, and B. Colman (eds.). San Francisco: North Point Press, 19–30.

Bird, E. 1988. Why "Modern" Agriculture Is Environmentally Unsustainable: Implications for the Politics of the Sustainable Agriculture Movement in the U.S. In *Global Perspectives on Agroecology and Sustainable Agricultural Systems.* P. Allen and D. Van Dusen (eds.) Santa Cruz, CA: University of California, 31–37.

Boggs, C. 1986. *Social Movements and Political Power: Emerging Forms of Radicalism in the West.* Philadelphia: Temple University Press.

Bread for the World Institute on Hunger and Development. 1992. *Hunger 1992: Second Annual Report on the State of World Hunger.* Washington, D.C.

Busch, L., and W. B. Lacy. 1983. *Science, Agriculture, and the Politics of Research.* Boulder, CO: Westview Press.

California Rural Legal Assistance Foundation. 1991. *Hunger in the Heartland: Community Childhood Hunger Identification Project In California's Central Valley.* San Francisco, California.

Committee for Sustainable Agriculture. 1990. *Asilomar Declaration for Sustainable Agriculture,* January 12. Asilomar, California.

Consultative Group on International Agricultural Research. Food and Agriculture Organization of the United Nations. 1989. *Sustainable Agricultural Production: Implications for International Research.* FAO Research and Technology Paper 4. Rome, Italy.

Danbom, D. B. 1986. Publicly Sponsored Agricultural Research in the United States from a Historical Perspective. In *New Directions for Agriculture and Agricultural Research.* K. A. Dahlberg (ed.). Totowa, NJ: Rowman and Allanheld, 107–131.

De Janvry, A., and E. P. LeVeen. 1986. Historical Forces that Have Shaped World Agriculture: A Structural Perspective. In *New Directions for Agriculture and Agricultural Research.* K. A. Dahlberg (ed.). Totowa, NJ: Rowman and Allanheld, 83–104.

Demissie, E. 1992. A History of Black Farm Operators in Maryland. *Agriculture and Human Values* 9(1):22–30.

De Walt, B. R. 1985. Mexico's Second Green Revolution: Food for Feed. *Mexican Studies* 1(1):29–60.

Douglass, G. K. 1984. The Meanings of Agricultural Sustainability. In *Agricultural Sustainability in a Changing World Order.* G. K. Douglass (ed.). Boulder, CO: Westview Press, 3–29.

Edwards, C. A. 1989. The Importance of Integration in Sustainable Agricultural Systems. *Agriculture, Ecosystems, and Environment* 27:25–35.

Edwards, C. A. 1990. Preface. In *Sustainable Agricultural Systems.* C. A. Edwards, R. Lal, P. Madden, R. H. Miller, and G. House (eds.). Ankeny, IA: Soil and Water Conservation Society, xiii–xiv.

Farmworkers not Protected in U.S. *Global Pesticide Campaigner.* February 1992:16.

Francis, C. A. 1988. Research and extension for sustainable agriculture. *American Journal of Alternative Agriculture* 3:123–126.

Francis, C. A., C. Butler Fora, and L. D. King. 1990. *Sustainable Agriculture in Temperate Zones.* New York: John Wiley.

Frank, A. G., and M. Fuentes. 1990. Civil Democracy: Social Movements in Recent World History. In *Transforming the Revolution: Social Movements and the World System.* S. Amin, G. Arrighi, A. G. Frank, and I. Wallerstein (eds.). New York: Monthly Review Press, 139–180.

Freudenberger, C. D. 1986. Value and Ethical Dimensions of Alternative Agricultural Approaches: In Quest of a Regenerative and Just Agriculture. In *New Directions for Agriculture and Agricultural Research.* K. A. Dahlberg (ed.). Totowa, NJ: Rowman & Allanheld, 348–364.

Friedmann, H., and P. McMichael. 1989. Agriculture and the State System: The Rise and Decline of National Agricultures: 1870 to Present. *Sociologia Ruralis* 29(2): 93–117.

Gips, T. 1988. What Is Sustainable Agriculture? In *Global Perspectives on Agroecology and Sustainable Agricultural Systems.* P. Allen and D. Van Dusen (eds.). Santa Cruz, CA: University of California, 63–74.

Harding, S. 1986. *The Science Question in Feminism.* Ithaca: Cornell University Press.

Harvey, D. 1989. *The Condition of Postmodernity.* Oxford: Basil Blackwell.

International Movement for Ecological Agriculture. 1991. Declaration: From Global Crisis Towards Ecological Agriculture. Drafted January 10–13, 1990. *Lokayan Bulletin* 9(1):78–87.

Jackson, W. 1984. A Search for the Unifying Concept for Sustainable Agriculture. In *Meeting the Expectations of the Land.* W. Jackson, W. Berry, and B. Colman (eds.). San Francisco: North Point Press, 208–229.

Keller, E. F. 1986. *Reflections on Gender and Science.* New Haven: Yale University Press.

Lipman-Blumen, J. 1986. Exquisite Decisions in a Global Village. In *New Directions for Agriculture and Agricultural Research.* K. A. Dahlberg (ed.). Totowa, NJ: Rowman and Allanheld, 42–62.

Lockeretz, W. 1988. Open Questions in Sustainable Agriculture. *American Journal of Alternative Agriculture* 3:174–181.

Lockertez, W. 1989. Comparative local economic benefits of conventional and alternative cropping systems. *American Journal of Alternative Agriculture* 4(7):75–83.

Lowrance, R., P. F. Hendrix, and E. P. Odum. 1986. A Hierarchical Approach to Sustainable Agriculture. *American Journal of Alternative Agriculture* 1:169–173.

MacRae, R. J., S. B. Hill, J. Henning, and G. R. Mehuys. 1989. Agricultural Science and Sustainable Agriculture: A Review of the Existing Scientific Barriers to Sustainable Food Production and Potential Solutions. *Biological Agriculture and Horticulture* 6:173–219.

McCalla, A. 1978. The Politics of the U.S. Agricultural Research Establishment: A Short Analysis. *Policy Studies Journal* 6(4):479–483.

Mies, M. 1986. *Patriarchy and Accumulation on a World Scale.* London: Zed Books.

National Research Council, Board on Agriculture. 1989. *Alternative Agriculture.* Washington, DC: National Academy Press.

Natural Resources Defense Council. 1989. *Intolerable Risk: Pesticides in Our Children's Food.* Washington, DC: National Resources Defense Council.

Nestle, M., and S. Guttmacher. 1992. Hunger in the United States: Rationale, Methods, and Policy Implications of State Hunger Surveys. *Journal of Nutrition Education* 24(1):18S–22S.

O'Connor, J. 1988. Capitalism, Nature, Socialism: A Theoretical Introduction. *Capitalism, Nature, Socialism* 1:11–38.

Perfecto, I. 1992. Pesticide Health Effects on People of Color. Paper presented at "Diversity in Food, Agriculture, Nutrition and Environment" conference, East Lansing, Michigan, June 4–7.

Ruttan, V. W. 1988. *Sustainability Is Not Enough. American Journal of Alternative Agriculture* 3:128–130.

Sachs, C. 1983. *Invisible Farmers: Women in Agricultural Production.* Totowa, NJ: Rowman & Allanheld.

Sayer, D. 1987. *The Violence of Abstraction.* Oxford: Basil Blackwell.

Scott, A. 1990. *Ideology and the New Social Movements.* London: Unwin and Hyman.

Snyder, T. 1990. Creativity, Credibility Build Organic Sales. *Organic Times* Spring: 1.

Speth, J. G. 1992. A New U.S. Program for International Development and the Global Environment. *WRI Issues and Ideas.* World Resources Institute, Washington, D.C.

Strange, M. 1988. *Family Farming: A New Economic Vision.* Lincoln: University of Nebraska.

Toronto Food Policy Council. No date.

True, L. 1992. *Hunger in the Balance: The Impact of the Proposed AFDC Cuts on Childhood Hunger in California.* San Francisco: California Rural Legal Assistance.

United Nations Conference on Environment and Development. 1991. *Earth Summit News* 1.

United Nations World Food Council. 1990. *The Global State of Hunger and Malnutrition.*

U.S. Department of Agriculture. 1988. *Low-Input/Sustainable Agriculture: Research and Education Program.*

U.S. General Accounting Office. 1992. *Hired Farmworkers: Health and Well-Being at Risk.*

Vaupel, S. 1988. *Minorities and Women in California Agriculture*. University of California Agricultural Issues Paper No. 88-2. University of California, Davis.

World Bank. 1986. *Poverty and Hunger—Issues and Options for Food Security in Developing Countries*. The International Bank for Reconstruction and Development, Washington, D.C.

Youngberg, G. 1992. Telephone conversation with author, 7 April.

CHAPTER 7

Sustainable Development: Concepts, Contradictions, and Conflicts

Michael Redclift

B oth "sustainable development" and "environmental management" have become buzzwords in development policy circles, but the discussion surrounding these terms pays scant attention to the way in which people in developing countries participate in the management of their resource base and, through their participation, help to transform the practice of environmental management. This chapter, in addressing these issues, seeks to correct two kinds of bias that exist in much of the sustainable development debate. First, there is a bias toward "managerialism" rather than resource management, stemming from a "top-down" approach to local-level development. Second, there is a tendency to treat "sustainable development" as merely a variation of the prevailing approaches to development and to see sustainability as a goal that can be attained through making adjustments to the standard development models.

This chapter, in contrast, will argue that the concept of sustainable development needs to be recognized as an alternative to the prevailing view, rather than a modification of it. The approach taken here reflects a way of examining resource conflicts through political economy. The emphasis is placed on the structural determinants of local-level decision making, at the local, national, and international levels, rather than on a more "human resources" or interactional approach. At the same time, the analysis emphasizes that what distin-

Food for the Future: Conditions and Contradictions of Sustainability, edited by Patricia Allen.
ISBN: 0-471-58082-1 © 1993 John Wiley & Sons, Inc.

guishes environmental concerns in the North from those of poor areas of the South is not simply material conditions, but different epistemologies and different systems of knowledge.

The first sections of this chapter analyze the concept of sustainable development, and seek to enlarge the conceptual discussion on this topic in order to take more account of some of the inconsistencies and limitations of the definitions currently available. The current thinking in environmental economics, which has gained favor within some international development agencies, and which emphasizes the use of calculations of the environment's value, is critically discussed. The economists' rather technical treatment is compared with a more thoroughgoing account of the economic, political, and epistemological dimensions of sustainable development. In this context, some of the new approaches that outside development agencies are currently taking toward local-level environmental management are briefly discussed.

Next, the chapter examines some instances of conflicts over resource use that have prompted popular participation and struggles to gain greater local control over the environment. The analysis focuses on situations in which natural resources are highly valued and have been heavily contested politically.

The final section of the chapter outlines an approach to contested environments that departs radically from the analysis of most development agencies by focusing attention on power and political mediation in the resolution of environmental conflicts at the local level. In this section the chapter tries to incorporate some experiences of poor people's participation in resource management in order to set out a framework for analysis that takes into account both the need for popular participation and the utility of local-level environmental management as complementary facets of the problem. It is hoped that through addressing the political problems associated with local resource management, as well as through developing a more rigorous analysis of the terms under which poor people and their environments are incorporated within development policy, we will begin to identify the potential for determining better policy interventions that is contained in the struggles and resistance of the rural majorities in the South.

SUSTAINABLE DEVELOPMENT: CONCEPTS AND CONTRADICTIONS

The problem with using the term "sustainable development" is that it has proven difficult to formulate a definition of it that is comprehensive but not tautological, and that retains analytical precision. In this it is similar to many terms in the development lexicon, whose very appeal, it can be said, lies in their vagueness. "Sustainable development" means different things to ecologists, environmental planners, economists, and environmental activists, although the term is often used as if consensus exists concerning its desirability. Like "motherhood" and "God," sustainable development is invoked by different groups of people in support of various projects and goals, both abstract and concrete.

One of the sources of the conceptual confusion surrounding the term "sustainable development" is that no agreement exists regarding what exactly is to be sustained. The goal of "sustainability" sometimes refers to the resource base itself, and sometimes to the livelihoods that are derived from it. Some writers refer to sustaining levels of production, while others emphasize sustaining levels of consumption (Redclift, 1987). This divergence in emphasis is important since what makes continued "development" unsustainable at the global level is the pattern of consumption in the rich countries, while most policies designed to tackle development problems, including those that fit within the "sustainable development" idiom, are essentially production oriented.

The different uses made of the concept of sustainable development reflect varying disciplinary biases, distinctive paradigms, and ideological disputes. In my view there are also at least two sets of contradictions that soon become evident when sustainable development is discussed.

First, embedded in much of the "sustainability" thinking is an important difference of emphasis. For some writers, the principal problem to be addressed is that "human progress" carries implications for nature itself, and should cause us to reexamine the "ends" of development, as well as the means (Devall and Sessions, 1985). Others view sustainability as a serious issue because nature is a major constraint on further human progress. They are concerned, basically, with the constraints that will be imposed on the conventional growth model if the warnings we receive from the environment, the "biospheric imperatives," are ignored. The solution, according to this view, is either to develop technologies that avoid the most dire environmental consequences of economic growth, or to take measures to assess and "price" environmental losses in a more realistic way, thus reducing the danger that they will be overlooked by policymakers.

Second, when "sustainable development" is considered within a North/South framework, attention must be paid to the contradictions imposed by the structural inequalities of the global system (WCED, 1987; Redclift, 1987). Green concerns in the North, such as alternatives to work and ways of making work more rewarding, can often be inverted in the South, where the environment is contested not because it is valued for its amenities or aesthetic value, but primarily because its exploitation creates economic value.

In the North, natural resources are also a source of value, and conflict between those who want to exploit them for commercial gain and those who wish to conserve the "countryside" is often highly charged. However, the very fact that conservation issues are given increasing weight in planning decisions in the developed countries bears witness to the shift in priorities that occurs in the course of "development." In urbanized, industrial societies, relatively few peoples' livelihoods are threatened by conservation measures. The "quality of life" considerations that play such a large part in dictating the political priorities of developed countries surface precisely because of the success of industrial capitalism in delivering relatively high standards of living for the majority (but by no means all) of the population in these countries.

In the South, on the other hand, struggles over the environment are usually about basic needs, cultural identity, and strategies of survival, rather than about providing a safety valve within an increasingly congested urban space. Under these circumstances, when the individual and household are forced to behave "selfishly" in their struggle to survive, there is no point in appealing to idealism or altruism to protect the environment.

Sustainable Development Perspectives

Of the two major trends in sustainable development thinking, one, exemplified by the economic approach taken by Pearce et al. in *Blueprint for a Green Economy* (1989), fails to take into consideration the contradictions discussed above. "Sustainable development," in this view, is treated as a modification of traditional development strategy, rather than an alternative to it, and this approach is therefore limited in scope and application. The second major trend, exemplified by the Brundtland Report, *Our Common Future* (1987), treats sustainable development as alternative concept of development, and, therefore, in the end, shows more promise.

The managerialist approach of ecological economics. A common point of departure of a discussion of sustainable development is to define it as what Barbier (1989) refers to as *sustainable economic development*. This is an optimal level of interaction between three systems—the biological, the economic, and the social—which is achieved "through a dynamic and adaptive process of trade-offs" (Barbier, 1989:185). Many economists, notably David Pearce, also emphasize the *trade-offs* between systems, or between present and future needs, as the key issue (Pearce et al., 1989). In similar terms it is argued that "sustainable economic development involves maximizing the net benefits of economic development, subject to maintaining the services and quality of natural resources over time" (Pearce et al., 1989), and that "[sustainable development] is development that maintains a particular level of income by conserving the sources of that income: the stock of produced and natural capital (Bartelmus, 1987:12). For economists interested in the environment, then, procedures such as environmental accounting, which aim to give a numerical value to the environment and to environmental losses, are essential instruments for the achievement of greater sustainability.

In Chapter Three of *Blueprint for a Green Economy* Pearce and his colleagues argue, from a declared interest in environmental quality, that environmental improvements are equivalent to economic improvements "if [they] increase social satisfaction or welfare" (p. 52). The resolve of these economists is to demonstrate that there are economic costs in ignoring the environment. This approach is growing in influence within international development agencies such as the World Bank, the United Nations agencies, and the Overseas Development Administration (ODA) (see World Bank, 1987, 1988a,b). Although all of these organizations have been strongly criticized in the past for funding development projects with very damaging ecological effects, such as cattle ranching in Central America, their new approach has, in a relatively short space of time, become almost synonymous with effective environmental management in many people's estimation.

One of the main problems with the view of environmental management is that it works better for developed than for developing countries. Most neoclassical economists use the "willingness to pay" principle as a means of assessing environmental costs and benefits, and Pearce et al. argue that the emphasis in environmental policy should be shifted toward this principle to avoid future damage to the environment (Pearce et al., 1989:55). It is not hard to appreciate some of the difficulties in applying the new environmental economics when we consider developing countries. As Pearce et al. (1989) demonstrate, there is widespread popular concern about the environment in the North, where environmental quality is often placed before economic growth in surveys of public opinion. In the South, on the other hand, immediate problems of acquiring subsistence needs preclude extensive and expensive efforts to improve the environment. In this sense, it is not useful to attempt to quantify the developing countries' "willingness to pay" for improved environmental quality, when their access to merely the basic livelihood essentials typically requires the sacrifice of environmental quality for short-term economic gain. Their ability to pay, or effective demand, for environmental quality is so limited under these circumstances that attempts to construct a level of "willingness to pay" must be speculative at best.

These uncomfortable facts have important implications for the ultimate utility of efforts to quantify assessments of environmental value in the Third World. No matter how complex and sophisticated the price imputation techniques, for instance, the revaluation of tropical forest to include its "full" environmental value would do little directly to prevent forest destruction, although it might serve to highlight the scale of the problem. Colombia's foreign debt, which requires the country to obtain foreign currency, enables the transnational companies buying valuable hardwoods in protected area like the Choco to pose as national saviors, rather than national vandals.

Equity considerations, in this context, are not necessarily a minor element in total utility, as Pearce et al. suggest (1989:48), but are often the driving force behind indiscriminate resource degradation, and must be recognized as such. The process of environmental degradation, including the wanton destruction of primary tropical forest, needs to be viewed within the context of highly unequal landholding, which forces poor men and women to colonize the tropical forests and other untitled land. In situations like those of tropical Colombia and Brazil we need to specify greater equity, or the reduction of poverty, as the *primary objective* of sustainable development, before the question of environmental quality can be fully addressed.

It is also essential that we widen the discussion of sustainable development to include the immediate influences of national and regional policies on environmental management decisions taken at the local level. It is at this level that we are least able to provide a clear framework of policy interventions, although a start has been made (IUCN, 1988). There is considerable evidence, much of it drawn from the experience of people living within fragile environments, about alternative, or sustainable uses to which resources can be put. In addition, largely because of the work of Pearce and other economists who take the environment seriously, we now have a much better basis from which to conduct environmental accounting within such environments.

These important advances, however, do not imply that the reformulation of environmental policy in developing countries should be confined to an assessment of environmental and economic "trade-offs," for to do so would mean ignoring other essential points of reference. These include the regional and national political economy of resource use, as well as dimensions of social justice that provide the backcloth against which much environmental degradation occurs. On its own, resource accounting also tacitly endorses a highly ethnocentric and "North-biased" view of the development process. Without attention to the analysis of resource use decisions, and the way these are influenced by structures of power and social relations at the community even within the South, we are unlikely to be able to influence the behavior of people who cut down primary forests in order to make a living.

The human needs approach.   An approach that is ultimately more successful than these primarily economic views of sustainable development is that taken by the Brundtland Commission's report, *Our Common Future* (Brundtland Commission, 1987). Although the economic concept of discounting plays a key role in the report, Brundtland immediately enlarges the compass of the debate about sustainability to include consideration of noneconomic factors. *Our Common Future* places the emphasis of the discussion of sustainable development on human needs, rather than on the trade-offs between economic and biological systems. While the future effects of present economic development is a central concern of the report, costs and benefits (both present and future) are assessed not only on economic grounds, but also in political, social, and cultural terms. In fact, Brundtland mapped out a very political agenda for shifting the emphasis of development, for the North as well as the South, without departing form the language of consensus. According to the Commission, "sustainable development is a process in which the exploitation of resources, the direction of investments, the orientation of technological development and institutional change are all in harmony, and enhance both current and future potential to meet human needs and aspirations" (Brundtland Commission, 1987:46).

One of the important things to notice about the approach taken by the Brundtland Commission is that it regards sustainable development as a policy objective, a methodological approach, *and a normative goal*, quite properly the end-point of development aspirations. Many economists acknowledge that normative considerations are important, but few would be prepared to state as unequivocally as does Brundtland that, without normative goals of this kind, improved methodologies and more well-designed policies will prove unworkable. Brundtland places the responsibility for environmental problems, and for mobilizing the political will to overcome them, firmly in the hands of human institutions and interests. Although the report remains committed to convergence and consensus, rather than divergence and conflict as a means of achieving sustainable development, the clear implication of Brundtland (and one that has broad appeal in the South, if not the North) is that unless the political and economic relations that bind the developing countries to the developed are redefined, sustainable development will prove a chimera.

It is worth noting that some authors, including people like Robert Chambers, who contributed to the Brundtland process, take an even more "human-focused" approach than that reflected in the report. Chambers argues for using "sustainable livelihood security" as an integrating concept (Chambers, 1988). For Chambers, the sustainability of the resource base makes little sense if it is separated from the human agents who manage the environment. Gordon Conway similarly emphasizes human actors in development. In a series of very influential papers, he argued that "sustainability (is) the ability to maintain productivity, whether of a field, farm, or nation, in the face of stress or shock" (Conway and Barbier, 1988:653). Originally, Conway had been thinking primarily in ecological terms, about the ability of natural systems to cope with system disturbance. His broader commitment to human needs, however, led him to seek to define a concept that retained the idea of system disturbance but incorporated a concern for the context of decision making within which poor rural households operate.

This does not exhaust the possibilities for defining "sustainable development," but it does point to a number of significant areas of both convergence and divergence. There is little agreement about what needs to be sustained, present or future population levels. It is also unclear whether a population should be sustained in terms of its minimum needs or at a particular level of consumption, or if this level of needs and consumption requires changing. Further, there are different "levels" at which sustainability is important, e.g., the farm level, the field level, and the village level in Conway's agroecological analysis, or the level of the village, region, and nation, according to other accounts. These distinctions are important because what is sustainable at one level may not be sustainable at another (and vice versa). An example is that of the Santa Cruz area in Eastern Bolivia, where farming systems are "sustainable" in agroecological terms, but the existence of important contraband trafficking, and a buoyant market for coca leaves, the raw material for cocaine, serve to undermine these systems (Redclift, 1987). Finally, some writers refer to sustaining levels of production and others to levels of consumption. Again, this is important since it can be argued that what makes development unsustainable at the global level are the patterns of consumption in the rich countries, while most efforts to tackle development problems are essentially production oriented. Sustainable development, then, is either about meeting human needs or maintaining economic growth or conserving natural capital, or about all three.

What is certain is that human beings are self-conscious actors in the development process. It has been let to the sociologists and anthropologists to take further the discussion of the human agency in sustainable development. In this context, both the participation of people in environmental management at the local level and the relationship between the implementation of empowering strategies an successful sustainable development are essential issues to explore.

The Multiple Dimensions of Sustainable Development

To establish an adequate conceptual framework within which to explore the issue of participation in sustainable development, we need to identify the multiple dimensions of the concept. There are three dimensions that require our attention: the economic, political, and epistemological.

The economic dimension. As we saw in the discussion of environmental accounting, much of the economic argument has been conduced at the level of present and future anticipated demand, assessing the costs, in terms of foregoing economic growth, of closer attention to environmental factors. It was John Stuart Mill, in his *Principles of Political Economy* (1873), who emphasized the idea that we need to protect nature from unfettered growth if we are to preserve human welfare before diminishing returns begin to set in. Malthus had earlier stressed the limits of the carrying capacity of the environment, although his emphasis was on the adverse effects that population pressure would have on consumption, rather than on the impact of environmental degradation itself.

Mill's concern with the environment, which today we would identify as part of the alternative, sustainable tradition of thought, has not been integrated into the mainstream of economic theory during this century. Following Ricardo's much more optimistic assessment of the potential of technology to overcome the limitations of existing resources, the more recent tradition has been to rely on humankind's promethean spirit and ingenuity to enable society to make scientific and technological advances capable of "putting back" the day in which population growth would begin to overtake available resources.

This optimism was shaken, although not destroyed, by the publication of *Limits to Growth* in the early 1970s (Meadows et al., 1972). This influential book argued that natural resources were indeed in short supply, undermining the assumption that humankind could continue to overcome the obstacles placed in its path by nature. The 1970s was a time in which—particularly following the oil price shocks—economic growth endangered the planet, primarily because the clamor for growth had meant the neglect of the environment on which growth was dependent. Twenty years later, the situation in the developed world is different: today we are beginning to be aware that it is the damage to our environment, caused by a heavy dependence on fossil fuels to drive industrial growth, that potentially imperils our ability to continue to support industrial society. The global externalities today, notably the greenhouse effect and the depletion of the ozone layer, are not the product of scarcity but of reckless and unsustainable production systems.

The sociopolitical dimension. The political dimension of the concept of sustainability comprises two separate but related elements: the weight to be attached to human agency and social structure, respectively in determining the political process through which the environment is managed; and the relationship between knowledge and power in popular resistance to dominant world views of the environment and resources. In both cases it is useful to draw on a body of emerging social theory that has evolved and gained currency while environmentalism has risen to prominence.

The problem of human agency in relation to the environment is well recognized in the literature, especially by geographers (O'Riordan, 1989). It is also a central concern of sociologists, although rarely linked to environmental concerns per se. The British sociologist Anthony Giddens has devoted considerable attention to what he describes as a theory of "structuration," which would enable us to recognize the role of human beings within a broad structural context, in seeking to advance their individual or group interests (Giddens, 1984). Giddens notes that "human agents . . . have as an inherent aspect of what they do, the capacity to understand what they do while they do it." (Giddens, 1984:xxii). It is their knowledgeability as agents that is important. Although Giddens does not apply his ideas specifically to environmental questions, they have clear utility for any consideration of the political and social dimensions of sustainability.

An examination of the ways in which power is contested helps us to explain human agency in the management of the environment, as well as the material basis of environmental conflicts. In this sense it is useful to distinguish between the way human agents dominate nature—what has been termed "allocative resources"—and the domination of some human agents by others, or "authoritative resources" (Giddens, 1984:373). Environmental management and conflicts over the environment are about both processes: the way groups of people dominate each other, as well as they way they seek to dominate nature. Not surprisingly, the development, or continuation, of more sustainable livelihood strategies carries important implications for the way power is understood between groups of people, as well as for the environment itself. The "Green" agenda is not simply about the environment outside human control; it is about the implications for social relations of bringing the environment within human control.

The second question of importance in considering the political dimension of sustainability is the relationship between knowledge and power, a dimension often overlooked by observers from developed countries when they turn their attention to poorer societies. As we shall see in a moment, the consideration of epistemology in sustainable development carries important implications for our analysis, since it strikes at the cultural roots of quite different traditions of knowledge. It is also important to emphasize, however, that knowledge and power are linked, as Foucault observed in much of his work (Smart, 1985; Sheridan, 1980).

We can, following Foucault, distinguish three fields of resistance to the "universalizing" effects of modern society, and these fields of resistance are particularly useful in delineating popular responses, by the rural poor in particular, to outside interventions designed to manage the environment in different ways. The first type of resistance is based on opposition to, or marginalization from, production relations in rural societies. This is resistance against *exploitation,* and includes attempts by peasants, pastoralists, and others to resist new forms of economic domination, which they are unable to control or negotiate with. The second form of resistance is based on ethnic and gender categories, and seeks to remove the individual from domination by more powerful groups whose ethnic and gender identity has conferred on them a superior political position. In many cases the only strategy open to groups of people whose environmental practices are threatened by outsiders, and whose own knowledge, power, and identity are closely linked with these practices, is to seek to distance themselves from "outsiders" by, for example, reinforcing ethnic boundaries between themselves and others. Finally, poor rural people frequently resist *subjection* to a world view that they cannot endorse, in much the same way as people in developed countries often confront "totalizing" theories, such as psychoanalysis or Marxism. In the South, development professionals frequently have recourse to a body of techniques for intervening in the natural environment that are largely derived from developed country experience. "Environmental managerialism" is one way of describing these techniques. The refusal to be subordinated to a world view dominated by essentially alien values and assumptions marks resistance against subjection. This does not imply that such resistance should necessarily be equated with political struggle, whatever the basis of the resistance itself. Frequently people who are relatively powerless, because their knowledge systems are devalued, or because they do not wield economic power, resist in ways that look like passivity: they keep their own counsel, they appear "respectful" toward powerful outsiders, but they simply fail to cooperate.

The epistemological dimension.   Sustainable development is usually discussed without reference to epistemological issues. It is assumed that the system of acquiring knowledge in the North, through the application of scientific principles, is a universal epistemology. Anything less than "scientific knowledge" hardly deserves our attention. Such a view, rooted as it is in ignorance of the way we ourselves think, as well as of other cultures' epistemology, is less than fruitful. Goonatilake (1984) reminds us that large-order cognitive maps are not confined to Western science, and that in Asia, for example, systems of religious belief have often had fewer problems in confronting "scientific" reasoning than has the Judaeo-Christian tradition. The ubiquitousness of Western science, however, has led to traditional knowledge becoming "fragmented" in the South, increasingly divorced from that of the dominant scientific paradigm.

The philosopher Feyerabend, in his influential book, *Farewell to Reason,* has distinguished between two different traditions of thought, which can usefully be compared with "scientific" and "traditional" knowledge. The first tradition, which corresponds closely to scientific epistemology, is the *abstract tradition*. This enables us "to formulate statements (which are) subjected to certain rules (of logic, testing and argument) and events affect the statements only in accordance with the rules. . . . It is possible to make scientific statements without having met a single one of the objects described" (Feyerabend, 1987:294). He gives as examples of this kind of tradition elementary particle physics, behavioral psychology, and molecular biology. In contrast, the kinds of knowledge possessed by small-scale societies Feyerabend would label as *historical traditions*. In these epistemological traditions "the objects already have a language of their own," and the object of enquiry is to understand this language. In the course of time much of the knowledge possessed by people outside mainstream science, especially in developing countries, becomes encoded in rituals, in religious observations and in the cultural practices of everyday life. In societies that make an easy separation between "culture" and "science" such practices can easily be ignored, although they are frequently the key to the way environmental knowledge is used in small-scale rural societies.

It is evident from some of the cases discussed briefly in the later sections of this chapter that any view of epistemology that rests solely on Northern experience will often fail to galvanize opinion among people such as the Brazilian rubber tappers or the Indian women involved in the Chipko movement. What is required is the admission that we are dealing, when we observe local resource management strategies, with multiple epistemologies possessed by different groups of people. Furthermore, the existence of global environmental issues, and the reporting of these issues by the media, forces us to consider the links between local epistemologies (all of which have evolved from their own encounter with other systems of thought, and are not fixed, "traditional" systems) and global systems of knowledge.

THE RURAL POOR AND SUSTAINABLE DEVELOPMENT: OUTSIDE INTERVENTION, INSIDE KNOWLEDGE

The first part of this chapter has sought to extend the definition of "sustainable development" by enlarging the compass of debate, and considering the dimensions of sustainability that usually lie outside the parameters of most Northern environmental policy interventions. As such it represents a contribution to the still small body of work that has begun to examine the links between local environmental knowledge, political processes, and the management of resources (McNeely and Pitt, 1985; IUCN, 1989; Norgaard, 1985). By enlarging the discussion it is hoped that we can begin to get at the texture of "actually existing" sustainable practices, and thus to make more qualified decisions about the direction that future policy should take. The remainder of the chapter employs the framework of sustainable development outlined above in order to consider the role of external agencies and local knowledge in a more genuinely participatory view of resource management.

The Crucial Component: Popular Participation

Because environmental management in the North utilizes a scientific epistemology, development "experts" frequently devalue the contribution of local knowledge to environmental planning and policy and, simultaneously, assume that local people should "participate" in Western sustainable development. However, it is not clear why or how poor people can retain their knowledge systems, and put them to practical use within development activities, while "participating" in other people's projects.

Rural people are unlikely to perceive the problems that face them in everyday life as "environment problems." Nevertheless the "answers" arrived at by the state, and other outside institutions, make assumptions about what is beneficial for people, and ways in which the environment can be more effectively managed (Blauert, 1990). In fact, the approaches of outside agencies frequently address the problems of the agencies themselves, rather than those of the rural poor or their environments. To most poor people in rural areas, for whom daily contact with the environment is taken for granted, it is difficult, if not impossible, to separate the management of production from the management of the environment, and both form part of the livelihood strategy of the household or group. It is increasingly recognized by many development agencies, notably NGOs working in developing countries, that the sectoral, "single problem" approach to policy and planing undertaken by most official bodies prevents a workable assessment of sustainable development options.

The current call for more participatory approaches to local-level environment management stems from the failure to recognize the importance of popular participation in influential reports such as those of Brandt (1980) and Brundtland (1987) as well as the original *World Conservation Strategy* document (1980). It also reflects the acknowledgment that national governments are less likely to ignore international opinion when it is buttressed by popular, grassroots support.

The call for more participation also reflects a third important variable: during the 1970s and 1980s an influential body of knowledge, along with new methodological interventions, stressed the importance of capturing the knowledge of poor people themselves—through farming systems research, agroecology, and "rapid rural appraisal" techniques. However, the cultural and political aspects of these gains in understanding received almost no attention. Social structure and political action remained essentially outside the map of development policy at the microlevel, and were given scarcely any attention in discussions of the natural environment.

The problem of rural poverty and the environment has frequently been posed in terms of available and appropriate technologies, while more reflexive, more iterative ways of working with rural people in developing countries were confined to the relatively "marginal" concerns such as community development. Anthropologists, for example, frequently found unlikely allies in ecologists, whose negative experience of large-scale development projects echoed their own (Ewell and Poleman, 1980).

If often appeared as if the larger the financial commitment of an organization to "development" goals, the smaller was the commitment to discovering how to assist the empowerment of the poor, drawing on their knowledge, their priorities, and their politics. One of the consequences, with which we grapple today, was that most environmental knowledge, like environmental management, is handed down from the First World to the Third, from large development agencies to the supposed beneficiaries of change.

The Report of the World Commission for Environment and Development, *Our Common Future* (1987), served to set the agenda for recent thinking about the environment and development. Despite its trenchant analysis, accessible style and clear exposition of the issues, the Brundtland Commission has relatively little to say about popular participation in environmental management at the local level. Other than a few short, but useful, sections on participation the Commission's Report has little to say about local empowerment until the conclusion, in which, after a long account of the international measures required to achieve more sustainable development, a short section on popular participation is included:

> progress will also be facilitated by recognition of, for example, the right of individuals to know and have access to current information on the state of the environment and natural resources, the right to be consulted and to participate in decision making on activities likely to have a significant effect on the environment, and the right to legal remedies and redress for those whose health or environment has been or may be seriously affected. (WCED, 1987: 330)

Despite the fact that these points are not elaborated in the Report, and popular involvement in environmental management gets only the most cursory treatment, these few phrases represent a commitment of immense value, which deserve to be taken seriously by the international community and national governments. Suddenly the issue of sustainable development is linked to human rights, and these rights are specified in terms of "their" right to know and be consulted. Participation, it is implied, is not simply a means of ensuring the efficacy of "our" development (via more attention to factors such as the creation of employment) but a means of ensuring their sustainability through the possession of the rights without which it cannot be achieved.

Evidence for greater attention to participation, and with its poor peoples' rights in the environment, can be gleaned from the first draft of the *World Conservation Strategy for the 1990s,* prepared by IUCN, UNEP, and the WWF and currently being circulated. This document, which exists only in draft form, goes some way to redressing the lack of attention to people in the original *World Conservation Strategy* (1980). The discussion of "policy, planning, legislation and institutions" (pp. 137–144) pays particular attention to the obligations that a more sustainable development strategy places on governments, to consult them, to facilitate their participation in decisions, and to make information available to them. It also recognizes that "special attention should be given to participation by women and indigenous peoples," which should be provided for by governments and intergovernmental agencies (p. 138).

The final section of the document gives considerable attention to local strategies for sustainable development, arguing that local communities should be given the opportunity to prepare their own sustainable development strategies "expressing their views on the issues, defining their needs and aspirations, and formulating a plan for the development of their area to meet their social and economic needs sustainably" (p. 156). This should be undertaken, like the regional and national strategies to which it would contribute, on the basis of consensus. Achieving "a community consensus on a future for an area" would require consultation and agreement with other, non-community interests, as well as "a forum and process through which the community (itself) can achieved consensus on the sustainable development of the area" (p. 157).

In practice, however, in most developing countries local-level environmental management will be left to understaffed, underfunded, and underesteemed enforcement agencies. The new *World Conservation Strategy* recognizes that legislative changes will be necessary before sustainable development strategies can be implemented with any success, but it attempts no analysis of the forces at the local, national, and international level that would need to be pressed into service to ensure that legislation is enforced and local management decisions are implemented. This document, in fact, shares the assumptions of much discussion of "participation," which is predicated on the presence of a social consensus that, in practice, rarely exists, especially in the most threatened parts of developing countries. Unless we analyze specific power structures in relation to the environment, we are in danger of being far too sanguine about the potential of negotiation and agreement. We are in danger, in fact, of drowning in our own rhetoric rather than identifying the underlying political processes whose understanding would facilitate the formulation of better environmental policy.

Conflicts over Resource Management: Forms of Resistance

Table 1 sets out some of the important variables for an analysis of conflicts over resource management at the local level. It must be emphasized that in the cases described the resources in question are heavily contested, and the conflicts surrounding them have drawn in both national and international interest groups. Many conflicts over local resource management in developing countries lack the heavily politicized nature of the Chipko or Brazilian rubber tappers' disputes, which have attracted media attention and become the focus for alternative development agendas. Nevertheless, these conflicts, and others such as the case of Bolivian frontier colonists, and freelance logging in the Choco of Colombia do illustrate the inadequacy of environmental interventions that proceed on the assumption of existing consensus, and in ignorance of the social and political struggles that lie behind environmental disputes.

The conflicts between Chipko activists in India, logging companies, and the Indian Government are well-known and have been exhaustively discussed in the literature (Guha, 1989; Shiva and Bandyopadhyay, 1986; Kunwar, 1982). Similarly, the struggle of the Brazilian rubber tappers in the Amazon to establish their rights to use the forest in a sustainable way has received extensive coverage, notably since the murder of the rubber tappers' leader, Chico Mendes. The struggles of the rubber tappers have reached the world stage, especially through the press and television, but the precise

TABLE 1.   Conflicts over Resource Management: Forms of Resistance

	Choices for Resource Utilization	Political Demands	Points of Tension and Resistance	State/External Intervention
Chipko (Shiva, 1989; Guha, 1989)	Forest conservation, commercial logging	Respect for traditional forest uses	Peaceful noncooperation (satyagraha)	Indian government intervention
Brazilian rubber tappers (Lab, 1989; Hecht and Cockburn, 1989)	Sustainable forest extraction, ranching	Conservation reserve	Forest clearing, federal government support	Brazil-wide solidarity growth, international ecological awareness
Tropical colonists (Bolivia) (Redclift, 1987)	Sustainable farming system; commercial rice cultivation/land engrossment	Land titles, institutional support	Disputed land ownership, migration, economic policy	Land reform, cocaine surveillance
"Freelance" logging (Choco, Colombia)	Contracted "logging" for transnational corporations, community stewardship		Individual livelihood strategy v. INDERENA	INDERENA/military base

circumstances of the conflict require some explanation (Schwartzman, 1989; Hecht and Cockburn, 1989; Hecht 1989).

According to Schwartzman (1989) there are approximately 1.5 million people in the Brazilian Amazon who depend on the forest for their living. Of these, about 300,000 are engaged in the sustainable harvesting of wild rubber. In fact, most rubber tappers, like other sectors of the forest population, are involved in several activities other than their main cash-earning occupation: they cultivate small gardens planted with rice, beans, and manioc, keep animals, and hunt in the forest. They also cultivate and manage fruit trees, palms, and other forest species. The rubber tappers' production system "appears to be indefinitely sustainable. Many rainforest areas have been occupied by rubber tappers for over sixty years, and some families have been on the same holdings for forty or fifty years, yet about 98 percent of each holding is in natural forest" (Schwartzman, 1989:156).

The diversity of sources of income is reflected in various aspects of the rubber tappers' culture: their diet is much more varied than that of most urban groups; their average cash income, although not large, is equivalent to twice the Brazilian minimum wage; and their awareness of the links between their livelihood and the maintenance of ecological diversity has enabled them to present their case as a convincing one of sustainable development. Any suspicion that their case has received special attention needs to be set against the fact that most other economic activities in the Amazon receive much higher subsidies, and are usually accompanied by disastrous effects.

In terms of local resource management, the interest in the rubber tappers' activities lies in two important issues. First, unlike much of the conservationist response currently being urged on governments in the South, the extractive reserves advocated by the rubber tappers are not simply another culturally alien "management strategy" urged on unwilling, or oblivious, local people. The idea of extractive reserves is an organized initiative directly undertaken by Amazonian grassroots groups and sympathetic national organizations, designed to change the course of official regional development policy for the benefit of local people. Because the extractive reserve concept was created by a social movement, it does not depend for its effective implementation on government agencies far removed from Amazonian reality. Forest communities have put their own model before the government and multinational lending institutions as a potential strategy for consideration within a wider context of sustainable development. Second, although locally sustainable, the rubber tappers' activities also produce a surplus that finds its way to the larger society: this is a movement that is not only locally initiated, but is also one that generates momentum outside the immediate domain of the *seringueiros* (rubber tappers).

The other two cases presented in Table 1 are less well-known. The tropical colonists referred to in the third case are largely migrants from the Bolivian Andes who migrated to the lowland province of Santa Cruz in the 1960s and 1970s, in search of land. These migrants have concentrated on growing rice for the market, but the difficulties associated with cutting down the forest, and the insecurity of the market for rice have also led some of them to explore (with official encouragement from some quarters) a more mixed farming system, comprising rice, perennial crops, and small-scale animal production. The problems of managing a more sustainable system in an area where conflicts over land are compounded by contraband traffic and the cocaine trade are outlined in Redclift (1987).

The final case is illustrated by the conflict between INDERENA, a Colombian environmental agency, and the people living in the area of the Choco, a reserve situated on the tropical Pacific coast of Colombia. These people were able to receive $10 a cubic meter for hardwoods cut from the forest reserve with chainsaws loaned by a transnational company operating in Colombia, Carton de Colombia. Each load of hardwoods had to be taken by sea, on a homemade raft, out into the Pacific and onto the port of Buenaventura. There was considerable resentment in the area at the attempts, usually futile, of the INDERENA staff to prevent the cutting of wood in this way. For the people involved in illegal cutting, the activity represented an essential livelihood strategy, and there was no shortage of men willing to take the place of those who did not survive the dangerous sea journey.

The tragedy of hardwood logging in the Choco, even on the relatively small scale practiced by "freelance" colonists, is that with sufficient official support, sustainable alternatives for the area could be implemented. It is thought that the Choco possesses "perhaps the most diverse plant communities in the world and extremely high levels of local, as well as regional endemic species" (Budowski, 1989:274). Two sustainable strategies, in particular, have attracted attention, because they would make no serious inroads into the region's ecological diversity but would enable large numbers of people to make a decent livelihood. First, food production could be concentrated on the rich alluvial river banks where, together with agroforestry combinations, larger populations could be supported. Second, if sustainable forestry schemes were promoted, especially in the swamp and secondary forests, numerous opportunities would open up for settlers in the region. The potential for the sustainable yield of freshwater fisheries in the area is even greater (Budowski, 1989:276). Finally, it is clear that the ecological value of the Choco is so great in global terms that international efforts to promote local research activities, and to promote research stations within the region, linked to local communities, would bring about huge advances in our knowledge, especially of better-drained forested areas.

Each of the cases referred to in Table 1 is related, along the horizontal axis, with four dimensions of the conflict: the alternative choices available for resource utilization in the area, the political demand of the participants in the various social movements, the points of tension and forms of resistance employed during the conflict, and the form of outside, state intervention to mediate the situation. In the cases of the Chipko movement and the rubber tappers, the conflict surrounds the defense of an existing, sustainable resource use or livelihood. In the case of the Bolovian colonists, a sustainable alternative to existing resource uses was available, but the incentives to make it attractive top people did not exist. The framework of policy measures and incentives in the Santa Cruz region of Bolivia favored short-term calculations of profit over longer term considerations of sustainability, although the risks carried by involvement with the market also threatened profitability for the colonist farmers. In the case of the Choco, the individuals' logging activities were undertaken independently of any community structure: individual livelihood opportunities were pursued in opposition to the formal, legal framework, but "supported" by a powerful transnational corporation.

The points of tension for each of the conflicts are different, and the interest of outsiders in the conflict vary widely, especially in terms of the commitment of the state to intercede on behalf of one group rather than another. In addition, it is impossible to view these conflicts as divorced from wider patterns of influence on the governments concerned, and in a more general sense in reshaping our awareness of the urgency of ecological issues. Although the local agents seem remote from most people, not only in the North, but also from the population of Indian or Latin American cities, their struggles provide evidence of the interdependence of both economic forces and power relations. Before considering the need to examine these power relations in more detail, it is worth reflecting on the potential value of an approach to resource management that explicitly recognizes the importance of popular participation. It is clear from these and other similar cases that forms of political activity over the environment vary widely: we should not expect popular participation to follow a single trajectory. Second, it needs to be emphasized that in the course of conflicts over natural resources, new priorities and development opportunities are opened up and brought within the compass of popular discourse. The determination of development trajectories is not confined to the offices of experts working for the World Bank or of academic observers; they are worked out in the heads of the subjects themselves. Third, resistance to the "totalizing" effects of incorporation, even at the geographical periphery, into modern society can lead to the formulation of demands that have to be negotiated with governments and international interests.

A commitment to a more democratic discourse on the part of governments or the international development community, however, is only one of several possibilities whose probability depends, critically, on the role of supportive groups and interests, including NGOs, international pressure groups, and classes within the society itself. The mediation of conflicting demands and their peaceful resolution might be the outcome of resource conflicts, but it is unhelpful to assume that general agreement of this kind can be found, and that better environmental management is virtually impossible without it. The discussion of environment and development by international agencies frequently fails to identify the alternatives to consensus, or the role that the recognition of conflicting interests can play in policy formation. The more closely we examine conflicts over resource management in developing countries, the more we need to pay attention to the political and social mechanisms through which interests in the environment are channelled and expressed. It is therefore to this question, for so long ignored in discussions of resource management, that we turn in the final section.

CONTESTED RESOURCES: POWER, RESISTANCE, AND SOCIAL CHANGE

At the beginning of this chapter it was suggested that conflicts over the environment could be analyzed in terms of three dimensions: the economic, the sociopolitical, and the epistemological. It was argued that power and resistance were complementary aspects of the same strategic situation. Further, it was suggested that the way the environment was viewed in different cultures corresponded with distinct epistemological traditions of thought. We should not assume that knowledge, whether "local" or "scientific," could be easily separated from ways of behaving, ways of managing resources, or ways of expressing resistance toward the attempts of others to manage resources.

The current rethinking of mainstream economics, and the greater incorporation of environmental considerations that is highly influential within some development agencies, is helping to fashion a tool for policymakers in the North, but there are limitations to the heuristic possibilities that such techniques provide. Any serious discussion of participation in resource management—and any analysis of the problem—needs to consider the full range of demands that the management of natural resources involves. We should not pursue better resource management within an apolitical, normative conceptual framework of our own making. We need to take seriously the resource politics of people in the South, especially since their own political consciousness is forged through contact with external development agencies, planning institutions, and policymakers.

The articulation of demands governing the use of natural resources inevitably means the exercise of power, and resistance to it. It should come as no surprise, then, to find that environmental demands affect the content of social relationships, as well as the form. They bring new social relationships into being, and with them new power relations, many of them uncomfortably like those they have superseded. In some cases a radical break is achieved, through which existing relations are democratized or opened up, but there is no guarantee that the new relations of power that are established will be more stable. Every strategy of confrontation dreams of becoming a relationship of power, of finding a stable mechanism to replace the free play of antagonistic forces. However, there is no guarantee in history that this will happen. As we have seen, frontier colonists in Brazil and villagers in India do not demand the end of the State or law, but insist instead on respect from the government for rights that are enshrined in tradition, as well as law.

The approach I have outlined to power relations can be used in exploring the contest between human agents over environmental resources. For example, peasant movements may be contained by a chain of State agencies through which power relations are deployed and reformulated (Harvey, 1989). By identifying the weaker and stronger points in this chain, movements can apply pressure to break the former with the goal of eventually breaking the latter. If we begin by identifying the most important points of tension in local society, and the conflicts they generate, we can observe how the specific application of power is resisted and transformed, how new tactics are introduced, and how traditional mechanisms are abandoned.

Bearing these points in mind we can propose a set of questions that can help us establish better methodological guidelines for the comparative analysis of micropolitical change in relation to the environment. We can usefully compare the different ways in which groups seek to control and manage resources, and the concrete implications of these strategies for external agencies whose remit is to help channel and facilitate the expression of local demands. We need to look closely at the way in which different groups establish power relations through their control over resources, and the way in which these power relations change over time. In this respect, the following sets of questions can be posed:

How do legal and institutional changes limit or enable groups to engage in particular forms of political action over the environment? Which groups have most successfully integrated their own microstrategies with wider strategies shared by other members of the society? As it becomes clear that different groups in the wider society acquire different notions of "sustainability," carrying implications for their own political action, it becomes more urgent that local demands are linked to wider social resistance.

How does the recomposition of power relations affect the political priority given to more sustainable resource management? Do new strategies of political mediation, or domination, make certain policy alternatives less feasible, while opening up new ones? How do local agents view the constraints and opportunities that changing resource uses make possible? Are they able to carry their alternative vision of sustainability, their "concrete utopia," into the organs of the state itself?

How do struggles over resources shape the paths of different social groups? Do they channel environmental demands into the institutional arena alone, or do they engage groups in confrontations that highlight basic divisions within the wider society? What are the effects on NGOs and governmental agencies of intervention to secure long-term environmental demands? Is it the case, as the Brundtland Commission hoped, that more contact between development agencies serves to bring forward the urgency of environmental priorities within policymaking circles?

These considerations are offered as a contribution to the resolution of some of the conceptual and methodological issues that surround local resource management. By identifying the points of tension in local systems of power, and comparing their implications for different groups, often possessed of different epistemological systems, we will be able to highlight the changes through which the environment becomes the object of economic, social, and political dispute. The lessons of the past and the of the present are central to any strategy of resistance and liberation, but it is up to us to undertake the necessary analysis, and to place it in the hands of those disempowered by the development process.

REFERENCES

Barbier, E. 1989. *Economics, Natural-Resource Scarcity and Development.* London: Earthscan.

Bartelmus, P. 1987. *Environment and Development.* London: George, Allen and Unwin.

Blauert, J. 1990. Autochthonous Approaches to Rural Environmental Problems: The Mixteca Alta, Oaxaca, Mexico. Ph.D., Wye College, University of London.

Brandt Commission. 1980. *North-South—A Programme for Survival.* London: Pan Books.

Brundtland Commission. 1987. (World Commission on Environment and Development). *Our Common Future.* Oxford: Oxford University Press.

Budowski, G. 1989. Developing the Choco region of Colombia. In *Fragile Lands of Latin America.* J. O. Browder (ed.). Boulder, CO: Westview Press.

Chambers, R. 1988. Sustainable Rural Livelihoods: A Strategy for People, Environ- ment and Development. Institute of Development Studies, University of Sussex.

Conway, G., and Barbier, E. 1988. After the Green Revolution: Sustainable and Equitable Agricultural Development. In D. Pearce and M. Redclift (eds.). *Future* 20(6):651–678.

Devall, B, and Sessions, G. 1985. *Deep Ecology: Living as if Nature Mattered*. Layton, UT: Peregrine Smith.

Ewell, P,. and Poleman, T. 1980. *Uxpanapa: Agricultural Development in the Mexican Tropics*. Oxford: Pergamon.

Feyerabend, P. 1987. *Farewell to Reason*. London: Verso.

Giddens, A. 1984. *The Constitution of Society*. Oxford: Policy Press.

Goonatilake, S. 1984. *Aborted Discovery, Science and Creativity in the Third World*. London: Zed Books.

Guha, R. 1989. *The Unquiet Woods: Ecological Change and Peasant Resistance in the Himalaya*. Delhi: Oxford University Press.

Harvey, N. 1989. Corporations Strategies and Popular Responses in Rural Mexico: State and Opposition in Chiapas, 1970–1988. Ph.D., University of Essex.

Hayter, T. 1989. *Exploited Earth; Britain's Aid and the Environment*. London: Earthscan/Friends of the Earth.

Hecht, S. 1989. Chico Mendes: Chronicle of a Death Foretold. *New Left Review* 173: 47–55.

Hecht, S., and Cockburn, A. 1989. *The Fate of the Forest Developers, Destroyers and Defenders of the Amazon*. London: Verso.

IUCN (International Union for the Conservation of Nature). 1988. *Economics and Biological Diversity: Guidelines for Using Incentives*. Gland, Switzerland.

IUCN (International Union for the Conservation of Nature. 1989. *World Conservation Strategy for the 1990s* (with UNEP and WWF). Gland, Switzerland.

Kunwar, S. S. 1982. Edited, *Hugging the Himalaya: The Chipko Experience*. Gopeshwar.

Latin American Bureau. 1989. *Fight for the Forest*.

McNeely, J., and Pitt, D. 1985. *Culture and Conservation: The Human Dimension in Environmental Planning*. London: Croom Helm.

Meadows, D. H., D. L. Meadows, J. Randers, and W. Behrens. 1972. *The Limits to Growth*. London: Pan Books.

Mill, J. S. 1873. *Principles of Political Economy*. London: Parker, Son and Bourn.

Norgaard, R. 1985. Environmental Economics: An Evolutionary Critique and a Plea for Pluralism. *Journal of Environmental Economics and Management* 12(4):863–879.

O'Riordan, T. 1989. The Challenge for Environmentalism. In *New Models in Geog- raphy*, 77–104, Vol. 1. R. Peet and N. Thrift (eds.). London: Unwin Hyman.

Pearce, D., A. Markandya, and E. Barbier. 1989. *Blueprint for a Green Economy*. London: Earthscan Publications.

Redclift, M. R. 1987. *Sustainable Development: Exploring the Contradictions*. London: Methuen.

Schwartzman, S. 1989. Extractive Reserves: The Rubber Tappers' Strategy for Sustainable Use of the Amazon Rainforest. In *Fragile Lands of Latin America*. J. O. Browder (ed.). Boulder, CO: Westview Press.

Sheridan, A. 1980. *Michel Foucault: The Will to Truth*. London: Tavistock.

Shiva, V. 1989. *Staying Alive: Women, Ecology and Development*. London: 2nd Press.

Shiva, V., and J. Bandyopadhyay. 1986. The Evolution, Structure and Impact of the Chipko Movement. *Mountain Research and Development* 6(2).

Smart, B. 1985. *Michel Foucault*. London: Tavistock/Ellis Horwood.

WCED (World Commission on Environment and Development). 1987. See Brundtland Commission.

World Bank. 1987. Environment, Growth and Development. Development Committee Paper No. 14, Washington, DC.

World Bank. 1988a. Environment and Development Implementing the World Bank's New Policies. Development Committee Paper No. 17, Washington, DC.

World Bank. 1988b. The World Bank and the Environment. Internal discussion paper. Washington, DC.

CHAPTER 8

Sustainability and the Rural Poor: A Latin American Perspective

Miguel A. Altieri

I t has become increasingly apparent that most conventional patterns of agricultural modernization (focusing on high-input agriculture for maximum yield and profit) have been often socially and ecologically harmful and unsustainable (Buttel, 1980). In most countries of the South, despite numerous internationally and state-sponsored rural development projects, poverty, food scarcity, malnutrition, health deterioration and environmental degradation continue to be widespread problems (Altieri and Yurjevic, 1991). As countries of the South are pulled into the existing international order and change their policies in order to serve the unprecedented debt, governments increasingly embrace neoliberal economic models that promote export-led growth. Although in some countries the model may appear successful at the macroeconomic level, deforestation, soil erosion, industrial pollution, pesticide contamination, and loss of biodiversity (including genetic erosion) proceed at alarming rates and these environmental costs are not reflected in the economic indicators. Modernization proceeds in the absence of effective land distribution and benefits favor primarily larger, better-off farmers who control optimum lands. In certain areas, this has resulted in increased concentration of land ownership, peasant differentiation, and more landless peasants (Altieri and Yurjevic, 1991).

Environmental problems and changes associated with modern agriculture

Food for the Future: Conditions and Contradictions of Sustainability, edited by Patricia Allen.
ISBN: 0-471-58082-1 © 1993 John Wiley & Sons, Inc.

have become, in addition to ecological and social processes, inextricably linked with the fluctuation, inequities, and unpredictabilty of the world economic press. As problems of social inequity, rural poverty, and diffusion of inappropriate technologies are not addressed, the dominant style of agricultural development appears less viable with each passing day.

The purpose of this chapter is to point out that development and environmental issues in the South cannot be understood without examining the economic and political linkages with the North. This becomes clear when one examines the effects of the introduction of foreign technologies in the South, which have resulted in abandonment of traditional sustainable agricultural systems and shifts toward crop specialization and technological dependency. The enormous debts of the South must be paid by export cash crops and forest products at the expense of basic grains (Redclift, 1989).

The reason why the new technologies have been distributed so unevenly is because they are biased toward modern, high-input farming. They are also channeled through institutions whose policies perpetrate conditions of land tenure, credit, technical assistance, infrastructure, etc. that favor large-scale farmers. These technologies have continuously proven to be not suited for small-scale farmers.

REDEFINING AGRICULTURAL DEVELOPMENT

The idea of sustainable development emerged after the realization that economic growth alone did not provide an adequate measure of development. Advocates of sustainable development have recognized that while market forces may be excellent allocators of capital and raw materials, they cannot be trusted to safeguard social equity and environmental quality.

At this point, however, sustainable development has been ascribed so many definitions that it has lost significance and clarity (Brown et al., 1987). It is used to serve and suit the interests of a wide diversity of groups from NGOs to the World Bank. Most definitions imply sustainable development as a way of achieving some sort of environmentally benign development (Douglass, 1984). However, the concept is so tainted with an inherent productivist and econmistic character that so far it cannot reconcile the contradictory objectives of economic growth and respect for the environment (Tisdell, 1988). Neoclassical economics has become such a dominant language of politics and development that we must first start by clarifying whether we are for sustainable rural development or the development of sustainability of rural areas. In fact, to even redefine "development" in the midst of a struggle for survival and sovereignty in the South is a paradox. However, it is absolutely necessary to engage in a new development paradigm.

Implementing a sustainable agriculture that favors the rural poor is essential, because peasants play a key role in maintaining the food self-sufficiency of most countries, and if their productivity problems are not met, major social consequences are likely to follow. This is difficult, however, in a region where institutional arrangements, market forces, policies, and research efforts are biased against such type of agriculture (Redclift, 1989). A major challenge is, therefore, to create a new policy framework that enhances sustainable agricultural development and conservation efforts through promotion of agroecological technologies directed at (LACDE, 1990) the following:

- Increasing agricultural small farm productivity to satisfy food needs, increase rural income, and curb the advancement of the agricultural frontier into fragile environments.
- Introducing ecological rationality in agriculture to rationalize use of chemical inputs, complement watershed and soil conservation programs, plan agriculture according to land use capabilities of each region and promote efficient use of water, forests, and other nonrenewable resources.
- Coordinating agricultural and environmental and economic policies related to pricing and taxing policies, land and resource distribution and access, appropriate technical assistance, etc.

Global forces that have shaped the structure of world agriculture include commodification of agriculture, internalization of capital and labor, the increasing dominance of agribusiness and multinational corporations, and the concentration of land (de Janvry, 1982). If these forces are considered along with the conditions penalizing developing countries (LDCs)—debt service obligations, commercial protectionism, declining terms of trade, and an economic order where industrialized countries control and manipulate world markets to their advantages—it becomes obvious that both the macro- and microeconomic contexts are biased against the basic objectives of sustainable agricultural development (SAD) that entail ameliorating poverty, providing food security, and ensuring the conservation of natural resources (Conway and Barbier, 1990).

A basic strategy for achieving sustainable development must solve the principal social problem in the South: poverty. SAD must attack the structural factors that underlie poverty, among them the economic policies that contribute to impoverishment of the population and, in large measure, encourage environmental degradation. Clearly then, for sustainable development to become a reality, it is necessary for the livelihoods of the poor to be given priority. An important question is, however, how can this priority be pursued at the local and national level while the effects of international development systematically marginalize the poor? Yet, improvements in the environment and poverty reduction deserve urgent and simultaneous attention (Barbier, 1987).

A new economic order that assigns overriding importance to satisfying basic needs, while minimizing environmental impacts, must emerge. LDCs must adopt new models of development that distribute the benefits of economic growth in a more equitable manner, that avoid a high level of environmental deterioration, and that truly improve the quality of life (but not only the per-capita income level) of present and future generations (Maihold and Urquidi, 1990).

The future of nations, both industrial and less developed, is interconnected in multiple ways and tied to an inexorably common destiny. It is thus impossible to conceive a strategy for sustainable development in the Third World that is isolated from its international context, and most particularly from events in the industrial countries. The process toward SAD will be viable only to the extent that there is an equally important change within the developed countries themselves, and in the factors underlying their international relation with LDCs (LACDE, 1990).

SEARCHING FOR AN EQUITABLE COMMON FUTURE

International concerns regarding world environmental problems, global sustainability, and the concepts of the common future provide a new dimensional or platform to develop a new North–South partnership to save planet Earth. For this partnership to succeed, both the North and the South must acknowledge their levels of contribution and responsibility to global environmental changes.

The North must admit that its style of development, level of affluence, and patterns of consumption and dissipation exert unsustainable demands on the planet and cause alarming impacts on the environment. The industrial countries' heavy consumption of natural resources is a threat, not only to sustainable development in global terms but also to the natural heritage of the Third World, which is the prime supplier of those commodities. The North has also developed and exported much of the technologies that today we link to erosion, salinization, pesticide pollution, loss of genetic diversity, etc. in LDCs. This practice of using LDCs as a kind of dumping ground by the industrial countries must cease. In addition, the nuclear arsenals and military expenditures created by the great powers are not only immoral and unacceptable, but are the greatest threat to the environment (LACDE, 1990).

The issue of "overpopulation" so commonly perceived by the North at the core of environmental degradation in the South should not be viewed abstractly as a matter of absolute numbers of people. When socioeconomic factors are considered, the population problem in the South becomes clearly more formal than real, given that it is compounded by the lack of people's access to and the other resources (Leff, 1986). In fact, countries considered overpopulated and plagued by hunger problems turn out to be major agricultural exporters to the North. Moreover, family planning policies frequently proposed by the developed countries are not in tune with the reality of LDCs. No matter how difficult it may be, a cutback in population growth can be achieved only by dint of the minimum economic growth indispensable for improving education and the quality of life, raising them to levels compatible with human dignity (LACDE, 1990).

The South, in turn, must recognize the enormous obstacles that failed development policies, bureaucracy, and unequal distribution of wealth within countries exert on regional development. The South is also a repository of biodiversity that provides key ecological services to the biosphere and that are vital for the health of agriculture. Due to its biodiversity and associated ecological services, the South must engage in conservation, but in order to do so, it will require unified and effective support to enhance its capacity to protect natural resources while improving the quality of life of the masses of urban and rural poor.

Within the context of the current international order, with countries in the South burdened by unbearable foreign debts, widespread poverty, etc., fostering conservation in the South without addressing the root causes of underdevelopment is totally unrealistic. In fact, some of the proposed economic arrangements (i.e., debt for nature swaps) to secure environmental preservation in the South take advantage of an inequitable international economic regime that encourages a secondary market in which debt obligations are bought at less than face value. Such debt exchanges, in addition to attempting against the sovereignty of particular countries, result in underpayment for the resources to be conserved, because they do not take into account the real value of the biodiversity and ecological services provided by the protected area. It is important not to forget that a significant portion of the natural resource base and biodiversity currently remaining on the planet is located in Latin America; and that it has been subjected to continuous and intense degradation, largely due to the development policies and exploitation systems created by the industrial countries and imposed by them for so many years (LACDE, 1990).

It is clear that at the present time the South constitutes an impoverished and indebted region, susceptible of being pressured by the countries of the North, which because of the crisis have strengthened their positions. Nevertheless, the invaluable and abundant natural resources constitute the basis of sustainable development for all humankind as well as the potential of its rich biodiversity. These ecological assets are enlarged by the substantial "ecological debt" that the industrialized countries have run up with the region over time and that has not yet been paid. This "ecological debt," is properly accounted for, could be several orders of magnitude greater than the foreign debt so rigidly imposed on the South, despite its social costs.

Some would oppose such concept of an "ecological debt" on the basis that although the South contributed germplasm, carbon sinks, and other resources, the North contributed funds, technology, and development assistance. The problem is that such aid did not become translated into better conservation or improved quality of life as evinced by growing ecological deterioration and persistent levels of poverty, hunger, malnutrition, and disease in the South. In fact, aid has mostly been translated into projects whose ecological toll is unacceptably high and also into increased technological dependency (Maihold and Urquidi, 1990).

It has been suggested that attaining sustainable development will require trade-offs between productivity, stability, and equity (Conway and Barbier, 1990). The worst trade-off is one in which LDCs are urged to protect the global environment at the expense of national development priorities and their ecological sovereignty. The only strategies for averting disastrous global climatic changes likely to find acceptance in LDCs are those that simultaneously address development priorities in the South, and those that curve overconsumption, waste, and emissions in the North.

WORKING TOWARD A SUSTAINABLE AGRICULTURE IN THE NORTH

In the industrialized countries, interest in sustainable agriculture has also emerged, but mostly as a response to the need to deal with the consequences of technology-induced environmental degradation resulting from a sort of "development-oversaturation" (Douglass, 1984). At the core of SAD projects is the desire to develop agricultural systems that sustain production in the long term without degrading the resource base. Thus, the arsenal of options that potentially meets these goals includes a variety of low-input technologies that improve soil fertility and conservation, maximize recycling, enhance biological pest control, and diversify production. Intrinsic to these projects is the conviction that as long as the proposed technologies benefit the environment and are profitable, sustainability will eventually be achieved and all people will benefit. It is assumed that what is good for the environment is good for society at large. This technological determinism has, to a significant extent, prevented environmentalists from understanding the structural roots of environmental degradation linked to modern agriculture (Altieri, 1989).

Sustainable agriculture, as depicted by mainstream institutions, has a relatively benign view of capitalist agriculture. It assumes the persistence of private property in agriculture and seeks to perfect it and/or make it more equitable. This narrow acceptance of the present structure of agriculture as a given condition restricts the real possibility of implementing alternatives that challenge such a structure. Thus, the possibility of a diversified agriculture is inhibited by the present trends in farm size and mechanization. Implementation of a mixed agriculture would be possible only as part of a broader program that includes, among other strategies, land reform and redesign of farm machinery. Otherwise, merely introducing alternative agricultural designs will do little to change the underlying forces that led to monoculture production, farm size expansion, and mechanization in the first place.

Several issues and questions must be addressed if a truly sustainable agriculture is to be achieved in the North.

Capitalist relations of production must be taken into account as a determinant of how and to whom technology is delivered, otherwise, technology (whether sustainable or not) will continue to be delivered preferentially to the class holding the capital and political power.

The focus on low-input technologies as a main component of sustainable agricultural development will cause the development of several biotechnologies (i.e., nitrogen fixing cereals, crops with insecticidal properties, etc.) that clearly are not scale neutral or capital neutral to be proposed as viable options, thereby enhancing farmers' dependence on the private sector.

The issue of economies of scale must be addressed. If it continues to be argued that sustainable production methods should not be restricted to small farmers, corporate and large-scale farmers will rapidly shift to alternative methods because there is a good prospect for monetary gain (i.e., low inputs reduce costs of production, organic produce receives premium price). This shift potentially will displace some small farmers. As competition increases, more small farmers will shift to specialty crops oriented to elite markets as a way to exploit economic windows in order to survive in the capitalist-oriented economy.

If only environmental quality is emphasized, and if the fact that large-scale farmers are at least no longer "polluting" is considered sufficient, other factors that make a sustainable agriculture socially just and humane can be overshadowed. Within this framework, the fact that certain organic farmers do not provide proper wages and living conditions to their farm workers can easily be "overlooked."

The reason why the new technologies have been distributed so unevenly is because they are biased toward modern, high-input farming. They are also channeled through institutions whose policies perpetrate conditions of land tenure, credit, technical assistance, infrastructure, etc. that favor large-scale farmers. These technologies have continuously proven to be not suited for small-scale farmers.

It, therefore, is crucial that scientists involved in the search for sustainable agricultural technologies be concerned about who will ultimately benefit from them. This requires them to recognize that political determinants enter when basic scientific questions are asked, and not only when technologies are delivered to society. Thus, what is produced, how it is produced, and for whom it is produced are key questions that need to be addressed if a socially equitable agriculture is to emerge. When such questions are examined, issues of land tenure, labor, appropriate technology, public health, and research policy unavoidably arise.

THE ROLE OF NONGOVERNMENTAL ORGANIZATIONS IN ACHIEVING SUSTAINABLE AGRICULTURE

Given the current levels of rural poverty, environmental degradation, and inequity in most countries in the South, the top priorities for a sustainable agricultural development strategy should be

- Reduction of poverty;
- Ecological management of productive resources located in fragile ecosystems;
- Food security and self-sufficiency;
- Transforming rural poor communities into social actors capable of determining their own development.

These priorities have yet to be met by most top-down national and international sponsored development approaches that have not reached the poor and have not solved hunger and malnutrition problems. In Latin America, this failure has legitimized the role of nongovernmental organizations (NGOs) as new actors in rural development. In the past 10–15 years, a number of NGOs have become actively engaged in grassroots rural development, focusing their attention on neglected crops, lands, and people (Altieri and Anderson, 1986). Their approach has been to question "conventional wisdom" in technological design through the search for new kinds of agricultural development and resource management that, based on local participation, skills, and resources, enhance productivity while conserving the resource base. By focusing on the root causes of poverty and low land productivity, NGOs, together with the peasants, are also attempting to change the socioeconomic and political environment where their systems operate (Altieri, 1991).

Indigenous knowledge systems and peasant rationale gain unprecedented significance within this new agroecological paradigm. Peasant knowledge about soils, plants, and the environment, which at times has no counterpart in modern agronomy, proves to be the key in the process of technological generation. Some elements of this knowledge are now regarded as crucial to guide sustainable agricultural development. How to systematize this knowledge and apply it in rural development is a major challenge facing researchers and development specialists worldwide (Altieri and Yurjevic, 1991).

A significant number of NGOs have embraced agroecology as a unifying scientific paradigm that approaches agriculture in an integrated manner, by emphasizing the interactions between the biological, technical, cultural, and socioeconomic determinants of sustainability (Altieri, 1987). Agroecology is more sensitive than conventional agricultural development approaches to the complexities of local agricultures, by broadening its performance criteria to include properties of sustainability, food security, biological stability, re-source conservation, and equity along with the goal of increased production (Table 1).

Due to its novelty in viewing the question of peasant agricultural develop-ment, agroecology has heavily influenced the agricultural research and exten-sion work of many Latin American NGOs. Impacts and achievements of NGO-led agroecological projects in various regions have been significant (Table 2). Several characteristics of the agroecological approach to technology development and diffusion make it especially compatible with the discourse of NGOs:

- Agroecology, with its emphasis on reproduction of the household and regeneration of the agricultural resource base, provides an agile frame-work for analyzing and understanding the diverse factors affecting small farms. It also provides methodologies that allow the development of technologies closely tailored to the needs and circumstances of specific peasant communities.
- Low-input and regenerative agricultural techniques and designs pro-posed by agroecology are socially activating since they require a high level of popular participations.
- Agroecological techniques are culturally compatible since they do not question peasants' rationale, but actually build on traditional farming knowledge, combining it with the elements of modern agricultural science.

TABLE 1. A Comparison of Some Features of Green Revolution and Agroecological Technologies in the Context of Latin American Agriculture

Characteristic	Green Revolution	Agroecology
Goals	Increase crop yields	Optimize agroecosystem performance
Crops affected	Wheat, rice, maize, and few others	All crops
Areas affected	Flat lands, irrigated areas	All areas, especially marginal areas (rainfed, steep slopes)
Dominant cropping systems	Monocultures, genetically uniform	Polycultures, genetically heterogenous
Dominant inputs	Agrochemicals, machinery, high dependency on external inputs	Nitrogen fixation, biological pest control, organic amendments, high reliance on on-farm resources
Environmental impacts	Medium-high (chemical pollution, erosion, salinization, pesticide resistance, etc.)	Low-medium (nutrient-leaching from manure, etc.)
Crops displaced	Mostly traditional varieties and land races	None
Technology development and dissemination	Quasi-public sector, private companies	Largely public, NGO involvement, farmers participation
Capital costs of research	Relatively high	Relatively low
Research skills needed	Conventional plant breeding and other disciplinary agricultural sciences	Ecology and multi-disciplinary expertise
Proprietary considerations	Varieties and products patentable and protectable by private interests	Varieties and technologies under farmers' control
Indicators	Crop yield, per area, crop tolerance	Stability and constancy of production

Source: Modified after Kenny and Buttel, 1985.

- Techniques are ecologically sound since they do not attempt to radically modify or transform the peasant ecosystem, but rather identify management elements that, once incorporated, lead to optimization of the production unit.
- Agroecological approaches are economically viable since they minimize costs of production by enhancing the use efficiency of locally available resources.

These groups recognize, however, that ultimate sustainability will be reached as farmers increase their access to land, resources, and a suitable

TABLE 2. Agroecological Projects Led by Non-government Organizations in Latin America

NGO	Characteristics of Intervened Areas	Agroecological and Socioeconomic Constraints	Goals of the Agroecological Strategy	Technical Components of the Strategy	Impacts and/or Achievements
SEMTA (Bolivia)	Pacajes Province, Altiplano (3500–3800 m.a.s.l.). Potato, cereals, andean crops. bovine/ ovine cattle, alpacas.	Frost, low soil fertility, erosion, deforestation, drought. Generalized poverty, low access to credit, public services, and markets.	Slow environmental degradation process and regenerate productive potential.	Organically managed mud-built greenhouses for vegetable production. Terracing, crop rotations for erosion control. Reforestation with native species. Improvement/ management of native pastures.	Early production of vegetables under greenhouses resulted in premium prices in nearby La Paz markets, increasing income of participating farmers.
CIED (Puno-Peru)	Altiplano (3500 m.a.s.l.). Natural pastures (ichu), andean crops. potato, cattle, camelids.	Frost, droughts, flooding, soil and genetic erosion, low productivity. Poverty and marginalization.	Food self-sufficiency, conservation of natural resource base, rescuing of traditional technologies.	Rehabilitation of waru-warus and terraces (andenes). Crop rotations. Reintroduction of alpaca. Improved cattle management and sanitation.	Waru-warus ensure potato production in the midst of frost, therefore reducing risks in food production.
IDEAS (San Marcos-Peru)	Inter-andean valleys of Cajamarca (18°C, 450 mm rainfall). Potato, maize, cereals, cattle	Steep slopes, erosion, seasonal drought. Poverty, low access to good land.	Design of self-sufficient farming systems. Rescuing and enriching traditional technology. Soil and water conservation.	Predial design with rotation and polycultures. Organic soil management. Management of small mammals and poultry.	Organic crop production has proved viable, stabilizing yields without use of toxic chemicals.
PTA/CTAQ (Brazil)	Northeastern Brazil, semi-arid tropics. Eight-11 dry months Perennial cotton, maize, beans.	Rapid organic matter photo-decomposition, low biomass production, low soil fertility, hardpan, and salinity. Poverty, low access to land, marketing problems.	Improve traditional shifting cultivation system (rozado). Offer new productive options for vegetable, fruit, and animal diversification. Water harvesting and conservation. Improved management of animals. in-situ conservation of local germplasm.	Agrosilvopastoral management of catinga (xeric natural vegetation). Design of rotations, agro-forestry schemes and polycultures	Water harvesting techniques and design of drought tolerant cropping systems have enhanced productive potential in semi-arid areas.
NGO	Characteristics of intervened areas	Agroecological and socioeconomic constraints	Goals of the agroecological strategy	Technical components of the strategy	Impacts and/or achievements
CPCC (Paraguay)	Subtropical serrania (600–800 m.a.s.l.). Cassava, maize, peanuts, beans, cotton, sugarcane, and rice.	Seasonal drought (4–6 months). low soil fertility. Low income, small landholdings.	Design of agroforestry systems, soil conservation and diversification of production.	Community tree nursery. Forest enrichment, soil conservation in slopes, organic soil management.	Agroforestry systems have enhanced production of multiple resources and reverted deforestation processes.
CETEC (Colombia)	Southwest of Cauca Valley (1500 mm rainfall). Cassava, tropical fruit trees.	Acid and erosive soils, crop pests and diseases, weed interference. Low income, no access to credit or technical assistance. Low prices of agriculture commodities.	Diversify production with low-input technologies. Natural resources conservation. Alternatives to pesticides.	Improved cassava cropping systems. Soil conservation systems. Home-gardens. Pest control with parasites and botanicals	Soil erosion has been reduced and alternatives to pesticides are proving effective.
INDES (Argentina)	Dry subtropical area (600 mm). Cotton and subsistence crops (maize, squash, cassava).	Drought, high temperatures, wind erosion, low soil fertility. Poverty, unemployment, lack of credit	Food self-sufficiency. Optimize use of local resources	Rationalize cotton based rotations. Improve soil cover avoid erosion. Use of adapted crop varieties.	Diversification schemes have brought new crops into production, challenging dominance of cotton.
CET (Chile)	Chiloe Island Southern Chile (2000–2500 mm rainfall). Potato, wheat, pastures.	Frost, acid soils, phosphorous deficiency, overgrazing of pastures, genetic erosion. Poverty, marketing problems.	Improve and stabilize productive systems through diversification, use of local resources, rescuing of traditional varieties and technologies and soil conservation.	In-situ potato genetic community conservation programs. Pasture-based crop rotations. Rotational grazing systems. Silvopastoral systems.	More than 100 traditional potato varieties rescued. with about 56 families involved in in-situ conservation programs.

technology that allows them to ecologically manage these resources and also become socially organized to secure governance of resources, equity of access to markets for inputs and products, and income derived from harvests.

REQUIREMENTS OF SUSTAINABLE AGRICULTURAL DEVELOPMENT IN THE SOUTH

In view of the above analysis, what are the requirements and criteria for sustainable agricultural development in the South?

Since agricultural development inevitably implies some level of physical transformation of landscapes and artificialization of ecosystems, it is essential to design strategies that emphasize methods and procedures by which environmentally sustainable development can be achieved. Agroecology can serve as a unique guiding paradigm since it defines, classifies, and studies agricultural systems from an ecological and socioeconomic perspective. In addition to providing a methodology to diagnose the "health" of agricultural systems, agroecology delineates the ecological principles necessary to develop sustainable production systems within specific socioeconomic settings. In the past, a lack of holistic understanding contributed to the current environmental and socioeconomic crisis affecting modern agriculture. An agroecological strategy can guide SAD to achieve the following long-term goals (Conway and Barbier, 1990):

- Maintenance of natural resources and agricultural production;
- Minimize environmental impacts;
- Adequate economic returns (viability vs. profitability);
- Satisfaction of human needs and income;
- Provision of social needs of rural households and communities (public health, education, etc.).

There is an urgent need to develop a set of socioeconomic and agroecological performance indicators by which to judge project success, durability, adaptability, stability, equity, etc. Performance indicators must demonstrate a capacity for interdisciplinary evaluation (Conway, 1985). Care should be taken so that the physical and social well-being resulting from agricultural schemes can be quantitatively measured in terms of increased food, real income, quality natural resources, better health, sanitation, water supply, educational services, etc. (Liverman et al., 1988). Whether a system is sustainable or not should be assessed by the local population in relation to the satisfaction of the main goals ascribed to sustainable development. A major measure of sustainability should be the reduction of poverty and its consequences on environmental degradation. Sustainability indices should be derived from an analysis of the way in which patterns of economic growth are harmonized with natural resource conservation at the global as well as local level (Barbier, 1987). Participatory research methods such as Participatory Technology Development, Rapid Rural Appraisal, Agroecosystem Analysis, Farming Systems Research, etc., as well as new economic techniques such as natural resource accounting and ecological economics can all provide valuable instruments for the analysis of sustainability.

Emphasis should be placed not only on understanding when agroecosystems become sustainable, but more importantly on when agroecosystems become unsustainable (Gallopin et al., 1989). An agroecosystem may be considered unsustainable when it cannot maintain the supply of ecological services, economic goals, and social benefits resulting form one or a combination of changes at the following levels:

- Reduction in productive capacity (due to erosion, pesticide pollution, etc.);
- Reduction in homeostatic capacity for adjusting to changes due to destruction of built-in pest control mechanisms or nutrient cycling capabilities;
- Reduction in evolutionary capacity due, for example, to genetic erosion or genetic homogenization through monocultures;
- Reduction in the availability or value of resources necessary to satisfy basic needs (e.g., access to land, water, and other resources);
- Reduction in the capacity to make adequate use of available resources due to inappropriate technology or physical incapacity (illness, malnutrition);
- Reduction in the autonomy of resource use and decision making due to increased foreclosure of options for farmers and consumers.

By defining the thresholds of social and ecological "impoverishment" of a system, it may then be possible to determine a pattern of development that minimizes the degradation of the ecological basis that sustains the quality of human life and the function of ecosystems as providers of services and food. For this to occur, the processes of biological transformation (inherent to agriculture), technological development, and institutional change must all be in harmony so that sustainable development does not impoverish one group as it enriches another and does not degrade the ecological basis that sustains productivity and biodiversity.

There is a need to remove economic policies constraining sustainable development (Conway and Barbier, 1990). This will require the design of appropriate policies and institutional changes at the international, national, and local level (Table 3). The creation of a favorable policy environment will depend on

- The political will of governments in the North and the South;
- Proper economic incentives;
- Methodologies for the economic evaluation of environmental impacts of development;
- Institutional flexibility and capacity;
- Enhanced education and research capabilities in sustainable agricultural development;
- Increasing options for self-reliant, decentralized local development

In addition to North–South cooperation, there will be a need for an enhanced South–South partnership to establish more regional sovereign economic entities. Such institutional arrangements would allow LDCs to

- Exploit interregional comparative advantages;
- Form a block for a unified negotiation for debt relief;
- Push for regional democratization;
- Establish selective relations with northern countries not part of the "group of 7" (The United States, Japan, and major EEC actors)
- Negotiate redistribution of regional resources

CONCLUSIONS

An essential task for the world is the design and promotion of SAD strategies to achieve an environmentally sound agricultural production that satisfies human needs, promote equitable economic growth, and improves the quality of life of rural and urban populations. It is impossible to conceive a SAD strategy in the South isolated from the global context. The process toward SAD will be viable only to the extent that there are similar significant changes in the North. No enduring development and environmental goals will be achieved unless a more just relationship between North and South is assured. Definite actions to alleviate poverty, and foreign debt, to change distorted policies, and to conserve natural resources are immediate steps.

To achieve global sustainability, it will be crucial to engage in a new solidarity pact that will put the world on the path of sustainable development.

TABLE 3. Policy Constraints and Opportunities to Promote Sustainable Rural Development in Latin America

Obstacles	Possible Solutions
Domestic	
1. Land concentration	1. Agrarian reform
2. Government neglect of small-farm sector	2. Government recognition of importance of small-farm sector for national food security and economic development; extension of credit and technical assistance through NGOs
3. Lack of campesino organization and participation in decision-making	3. Government support for rural organizing and for grassroots decision making
4. Lack of infrastructure for basic foods production and marketing	4. Increased government support for national food system that emphasizes food security
5. Lack of appropriate technologies	5. Promote agroecological research, training, and information exchange programs
6. Rural poverty	6. Promote income generation rural activities, provide poor with access to productive resources
International	
7. Low and fluctuating prices for agricultural commodities	7. International commodity agreements and formation of associations of exporting nations
8. Dependence on only a few cash crops for foreign exchange	8. More agroindustrial development and diversification
9. Increasing external debt	9. Renegotiation and cancellation of debt burden
10. Transnational corporations' control of world agribusiness	10. Regulation of TNC investment in agricultural production and local food processing industries, state marketing of exports, commodity agreements, and cartels of exporting nations
11. Protectionist measures of economically powerful larger nations	11. Diversification of trade, regional alliances, and association with other nonaligned nations

Source: After Barry (1987).

This will entail the emergence of a "new ethic" that allows the common interests of nations to prevail over those of individual countries, that affords a more objective appreciation of the great risks faced by humankind unless we resolve situations that, like the rampant poverty or changes in climate, compromise global stability, and that starts with the premise that progress is not viable in the long term if it is not conceived as a process enabling all countries, groups, and individuals—not just a group thereof—to realize their development aspirations in an equitable and egalitarian manner.

REFERENCES

Altieri M A. 1987. *Agroecology: The Scientific Basis of Alternative Agriculture*. Boulder, CO: Westview Press.

Altieri, M. A. 1989. Beyond Agroecology: Making Sustainable Agriculture Part of a Political Agenda. *American Journal of Alternative Agriculture* 3:142–143.

Altieri, M. A. 1991. Traditional Farming in Latin America. *The Ecologist* 21:93–96.

Altieri, M. A., and M. K. Anderson. 1986. An Ecological Basis for the Development of Alternative Agricultural Systems for Small Farmers in the Third World. *American Journal of Alternative Agriculture* 1:30–38.

Altieri, M. A., and A. Yurjevic. 1991. La Agroecologia y el Desarrollo Rural Sostenible en America Latina. *Agroecologia y Desarrollo* 1:25–36.

Barbier, E. B. 1987. The Concept of Sustainable Economic Development. *Environmental Conservation* 14:101–110.

Barry, T. 1987. *Roots of Rebellion: Land and Buyer in Latin America*. San Francisco: South End Press.

Brown, B. J., M. E. Hanson, D. M. Liverman, and R. W. Meredith, Jr. 1987. Global Sustainability: Toward Definition. *Environmental Management* 11:713–714.

Buttel, F. H. 1980. Agricultural Structure and Rural Ecology: Toward a Political Economy of Rural Development. *Sociologia Ruralis* 20:44–62.

Conway, G. R. 1985. Agroecosystem Analysis. *Agricultural Administration* 20:31–55.

Conway, G. R., and E. B. Barbier. 1990. *After the Green Revolution: Sustainable Agriculture for Development*. London: Earthscan.

de Janvry, A. 1982. Historical Forces That Have Shaped World Agriculture: A Structuralist Perspective. Giannini Foundation of Agricultural Economics, University of California, Berkeley.

Douglass, G. K. 1984. *Agricultural Sustainability in a Changing World Order*. Boulder, CO: Westview Press.

Faeth, P., R. Repetto, K. Kroll, Q. F. Dai, and G. Helmers. 1991. *Paying the Farm Bill: U.S. Agricultural Policy and Transition to Sustainable/Agriculture*. Washington, DC: World Resources Institute.

Gallopin, G. C. et al. 1989. Global Impoverishment, Sustainable Development and the Environment: A Conceptual Approach. *International Social Science Journal* 121: 375–397.

Kenny, M., and F. Buttel. 1985. Biotechnology: Prospects and Dilemmas for Third World Development. *Development and Change* 16:61–69.

Latin American Commission on Development and Environment (LACDE). 1990. Our Own Agenda, Inter-American Development Bank—UNEP, New York.

Leff, E. 1986. *Ecologia y Capital.* UNAM, Mexico, D.F. 147. pp.

Liverman, D. M., M. E. Hanson, B. J. Brown, and R. W. Meredith, Jr. 1988. Global Sustainability: Toward Measurement. *Environmental Management* 12:133–143.

Maihold, G., and V. Urquidi (eds.). 1990. *Dialogo con Nuestro Futuro Comun.* Mexico City: Ebert Foundation-Nuestra Sociedad.

Redclift, M. 1989. The Environmental Consequences of Latin America's Agricultural Development: Some Thoughts on the Brundtland Commission Report. *World Development* 17:365–377.

Tisdell, C. 1988. Sustainable Development: Differing Perspectives of Ecologists and Economists, and Relevance to LDCs. *World Development* 16:373–384.

PART III

FOOD FOR THE FUTURE: DEVELOPING STRATEGIES FOR SUSTAINABILITY

CHAPTER 9

After Midas's Feast: Alternative Food Regimes for the Future

Harriet Friedmann

FOOD AND LAND

Food and land are matters of life and death, joy and misery, community and exploitation. From the first domestication of plants and animals, humans posed the problem of creating social relations through which to act in concert upon nature. Successive solutions to this problem posed another problem, that of the possibilities and limits set by nature.[1] About four centuries ago, and intensively for the past hundred years, diets and agriculture began to be organized on a world scale. A specialization of agricultural production stretching across continents opened vast spaces between where people lived and where their foods originated, between the work they did and the objects they used. It severed the direct link between inhabiting the physical world and nourishing the body, between the natural setting of social life and the relations among people. The totality of relations among people and between people and the land became organized on a scale beyond direct observation. This was a watershed in human history.

Access to land and to food is becoming precarious for increasing numbers of people at an accelerating pace. Those with common sense are becoming aware of the fragility of a food system that creates so much distance, both

Food for the Future: Conditions and Contradictions of Sustainability, edited by Patricia Allen.
ISBN: 0-471-58082-1 © 1993 John Wiley & Sons, Inc.

socially and geographically, between an unprecedentedly urban world of consumers and a global farm, linked by the perpetual motion of an oil-fueled transportation network and a shaky international monetary framework. Those with money are seduced by the exotic produce of distant lands—mangoes in Canada and apples in Mexico. Those without money are increasingly forced to enter large networks of buying and selling at the expense of self-provisioning—in order to survive—and even of regional and national markets—in order to earn export revenues to pay national debts.

It is urgent that we understand the changes in food and agriculture that have intensified over the past century. Yet the chains linking crops and diet across the globe fragment individual experiences of food. Complex and often contradictory relationships—which are nonetheless unified and comprehensible—have produced a bewildering array of experiences among social classes, cultures, and regions. These relationships comprise structures that reflect the more or less conscious activities of people responding to their experiences. When these activities become more conscious, they can change the weight between alternative futures.

Food has always been about power and money. As markets for food have deepened and expanded, people, communities, and states have needed to protect themselves from the destructive features of a world governed by the perpetual motion of money. In other words, market expansion has eventually provoked an opposing movement for self-protection. During the past century, this dual movement has taken specific form within two distinct food regimes—that is, two periods of stable relations of power, production, and consumption in the world food economy. Today, as the second food regime is in a prolonged state of disintegration, we hold the opportunity to create a new food regimen based on sustainable, local alternatives to the global regime presently under construction.

FOOD, POWER, AND MONEY: MYTH AND REALITY

Since recorded time, food has been about state-building and about wealth. Rulers have justified the centralization of power through promises of protection, not only against hostile attacks but also against famine. A major way to centralize power is to centralize control over land and labor, and the most direct link between land and labor is food. As for wealth, the link to food is more obvious in ancient stories and myths than in the confusing complexity of our own times. The lessons may help to remind us that, ultimately, wealth is only the useful objects available to its holders. Numbers recorded in accounts and portfolios can evaporate as easily as they were entered. Bluntly, there is no wealth without food.

Food power is as old as our written texts. According to the Biblical story, Joseph became advisor to the king of ancient Egypt, the Pharaoh, after predicting the future from the king's dreams. In the dreams, seven healthy cows were eaten by seven gaunt cows, and seven full ears of grain were swallowed up by seven shriveled ears. Joseph said that both signified seven years of abundance to be followed by seven years of famine. His advice to the king was to empower a wise man to organize the land during the years of plenty. Specifically, Joseph advised him to collect all the food under the king's authority and to store it in the cities, as a reserve for the famine ahead. The king appointed Joseph to do this task. Joseph collected the food in the cities. When the famine came, Joseph rationed out the food. When the famine became very severe, Joseph gathered in all the money as payment for the rations, and gave it to king. When the money was gone, Joseph demanded livestock in payment. Then the Egyptians said,

> "Take us and our land in exchange for bread, and we with our land will be serfs to Pharaoh; provide the seed, that we may live and not die, and that the land may not become a waste" . . . And he removed the population, town by town, from one end of Egypt's border to the other.

Then Joseph gave them seed to sow the land, and

> Joseph made it into a land law in Egypt . . . that a fifth [of the harvest] should be Pharaoh's. (*Genesis* 41–47, in *The Torah*, 1967:74–89).

Since recorded time, rulers have justified the harshest measures through fears of famine, and with those measures have changed relationships among people and between people and the earth.

Food has long been about wealth as well as power. An ancient Greek myth tells of Midas, who, according to the soothsayers, was destined from birth for great wealth. Midas once gave hospitality to a drunken satyr asleep in his rose garden, who turned out to be a former teacher of Dionysus, the god associated with drunken revels. As a reward, Dionysus gave Midas his wish, which was that all he touched would turn to gold. In Graves's (1960) account,

> not only stones, flowers, and the furnishings of his house turned to gold but, when he sat down to table, so did the food he ate and the water he drank. Midas soon begged to be released from his wish, because he was fast dying of hunger and thirst.

As the ancient Greek myth of Midas shows, gold can be more subversive of intentions than power. Where Joseph used food to achieve power for the king, Midas forgot about food in his desire for gold. Where the king of Egypt got both food and power, King Midas got gold but lost his food.

History changed decisively when the magical powers of money became deeply rooted in the real relations among people. Only with industrial capitalism in Western Europe two hundred years ago, and even then never completely or stably, did money come to be the central human need (Soper, 1981). It is not very long, in historical perspective, that it has seemed obvious that each of must sell something in order to buy the food we eat. And in that time, the magic of money has been the justification and the means for continually transforming food itself. Agriculture has changed from a complex culture transparently rooted in the land to an industry whose activity generates wealth and power. Food has changed from an integrated material and symbolic basis of life—breaking bread, in Western culture—into an array of edible products of complex, often global production chains. Some of these products are themselves as close to the durability of gold as modern science can devise: hard tomatoes, frozen disks of industrially minced chicken, reliably standard hamburgers and chips. Others, such as fresh fish and exotic fruits and vegetables, are so fragile that significant social and technical resources are devoted to transporting them quickly and constantly across the globe.

Since recorded time, food has been the contested terrain of power and money. The origins of European states, which eventually became the core of a state system spanning the earth, lay in a variable combination of capital and coercion (Tilly, 1990). In our own time, the different balances between power and money were expressed in the competing blocs of Soviet socialism and Atlantic capitalism. Wendell Berry (1977:41) expresses the parallel in these words:

> I remember, during the fifties, the outrage with which our political leaders spoke of the forced removal of the populations of villages in communist countries. I also remember that at the same time, in Washington, the word on farming was "Get big or get out"—a policy which is still in effect and which has taken an enormous toll. The only difference is that of method: the force used by the communists was military; with us, it has been economic—a "free market" in which the freest were the richest. The attitudes are equally cruel, and I believe that the results will prove equally damaging, not just to the concerns and values of the human spirit, but to the practicalities of survival.

The stories of Joseph and Midas are also about resistance and change. The first shows us that states are built and sustained purposefully, though rarely with the intended consequences. The end of that story was the plagues visited on the land, and the beginning of the story of liberation from bondage and eventually of the Ten Commandments, the first written legal code of Western society.

The second story is about the dangerous magic of money. In a parody of democracy, we have all become King Midas. The two principles of industrial agriculture, practiced with various success in real capitalism and real socialism, are durability and distance. To be good as gold, food must be as durable as gold and travel as far. The kind of food that must be turned to gold in order to be produced at all, we now recognize as dangerous. Even food industries, and certainly their advertising agencies, now assure us that each commodity is natural and healthy. And the use of the land to supply not food, but raw materials for industrial production of edible commodities, is killing the land and many species, even perhaps, the human species.

A SOCIAL CONCEPTION OF SUSTAINABLE ECONOMIES

As most often used, the term sustainable economies means to sustain economic development within the constraints imposed by nature. Yet sustainable economies involve more than connection with the portion of the earth they inhabit and transform. If, to use Karl Polanyi's (1957) word, rules governing production and exchange attempt to disembed economies from other dimensions of social life, the results are eventually disruptive of societies, communities, and households.

Social relations include both what is consciously organized, and what is unconscious, mystified, or simply unrecognized. The consciously organized dimension of social life is in the broad sense political, that is, it consists of negotiation and planning by and for collectivities. For most of us, this has been a small-scale, intermittent, even haphazard experience. But it is that experience that each of us must tap to change the economy in ways we desire, that is, to extend the realm of relations that are negotiated and planned and to shrink the realm of those that are mystified and unrecognized.

Among the latter, what is called "the economy" is the most important. It is so reified that it is difficult to see the human institutions that support and shape it. More and more, the language of the economy has taken on an inevitable, abstract, even magical quality. Far from being a social creation, governed by rules of contrast, property, and entitlements, the economy is talked about in a language that gives it power over society. In the language of adjustment now prevailing, it is not the economy that must be shaped to socially defined needs, but society that must adjust to the supposed "laws" of competition, productivity, and the rest. And society, when we think about it, means people inhabiting the earth, with all their capacities and needs.

Still following Polanyi's usage, sustainable communities are those that embed economic life in society. Obvious as this seems, it is deeply radical. It is diametrically opposed to the widespread insistence on the necessity, even the virtue, of "adjustment" by society to the so-called market. The idea of the market, examined closely, is more complex than rhetoric implies.

The major thrust of capitalism—what Polanyi calls the self-regulating market system—has been to separate economy from society and to force society to adjust to its strange creature, its Frankenstein monster—the disembedded economy. The disembedded economy is mystified when it is treated as a powerful external force rather than a product of human agency, sustained by social institutions.

In *The Great Transformation* (1957), written during World War II, Polanyi argued that since the beginnings of industrial capitalism, two contending movements have dominated history. First, in place of the multitude of markets, fairs, trade routes, and other forms of exchange that existed throughout history, the self-regulating market system consisted of a new set of rules that forced land and labor (that is, the earth and human beings) to behave as if it and we were something we cannot be—commodities. Commodities are goods or services produced for sale, whose very existence presupposes a buyer, and whose failure to find a buyer creates problems rather than utility for their owners. Polanyi called land and labor "fictitious commodities," because they are the very substance of society, not produced for sale (or even "produced" at all in the economic sense), and not able to survive either their unregulated use as commodities, or (in the case of labor) their failure to find a buyer. A third fictitious commodity, which is completely invented and yet the most consequential of all, is money.

Second, and opposed to the rules of the self-regulating market system, is the movement of self-protection. Neither human beings nor the earth can sustain the fiction that they are commodities. Humans not in demand starve and are not available when demand returns later. The earth fails to provide a continuous flow of resources when it is damaged in response to fluctuating prices for its treasures. The communities that link land and labor are disrupted as demand fluctuates. If they retain enough vitality, however, people begin to name specific problems and find ways to limit the effects of the self-regulating market.

The world food economy has shown this tension between expansion of the self-regulating market system and self-protection for more than a century. In the last decades of the nineteenth century, with the rise of a world wheat market, the self-regulating market system commodified basic foods, as opposed to the tropical luxuries of the mercantile era. This was the first food regime. The world wheat market linked new classes of wage workers in Europe to European settlers in the Americas and Australia, on territories made available through expulsion of native peoples. Both workers and settlers soon experienced problems and began movements for self-protection—to demand that governments provide social welfare for workers and price regulation for farmers. These forms of self-protection, after long turbulence, became the basis of a second food regime. In that regime, too, the self-regulating market expanded and movements for self-protection emerged, but in different forms from the first regime. In the second regime, capitalist enterprises grew up on the basis of major state regulation of land, labor, and agricultural products, but sought to expand the self-regulating market system beyond the limits of national economies. Eventually, the problems of disembedding food (and land and labor) from their social contexts led to attacks on the old forms of self-protection, which had become obstacles to further expansion. This is the story of the creation and crisis of two food regimes.

FOOD IN THE SELF-REGULATING MARKET SYSTEM: DISTANCE AND DURABILITY

Food is a special commodity. Unlike land, labor, and money, food is not a fictitious commodity, but it is especially closely tied to land for its production and to labor for its survival. It is therefore an interesting place to begin to trace the effects of the self-regulating market and to find opportunities for self-protection. Reembedding production and consumption in social life involves discovering appropriate ways to institutionalize planning and negotiation over needs and capacities present in communities at ascending scales of size and complexity.

Food is special in that land has bound the products of agriculture more closely than other products to the dictates of time and place. It was only a century ago that large numbers of people began to depend for their sustenance on foods produced in distant places, and that large numbers of farmers producing most of the world's food came to depend on selling it to distant consumers. That tendency has now moved very far and very deep. Most of the world's population now lives in cities, and most of those cities depend completely on the continuous operation of far-flung markets.

The dominant principles of the food sector enforced by the self-regulating market system are distance and durability. When food is subordinated to the market imperative, its only relevant characteristics are its quality and price. Its place of origin is of interest only as a marketing feature. The people whose labor enters into planning, cultivating, harvesting, shipping, processing, packaging, and selling it have no more claim to it than anyone else, and all claims are expressed through money. The distance may therefore be social as much as spatial—the people with money to buy the food may be close or far from its site of production. The principle of distance refers to the attempt to make place irrelevant. The location of production may change as quickly as market conditions change.

For food to be independent of space, it must also be impervious to time. In the period of increasing deregulation of money since the early 1970s, there has also been a differentiation of markets according to the effective demand of rich and poor consumers. For poor consumers (still privileged relative to those without money to enter markets), standard, manufactured foods are produced at lower unit costs, it is hoped. Durable foods must be able to travel long distances and wait for the best time as market conditions fluctuate. In the last half-century, corporations have changed the nature of most agricultural products from final consumer goods to industrial raw materials. Foods, especially in value terms, are overwhelmingly manufactured products that can endure long periods of storage and transport. Delicate as they may seem, for instance, potato chips are far more "durable" in their characteristics of storage and transport than potatoes, and worth the trouble for the value added to a small amount of nutritional substance.

For rich consumers, exotic foods—star fruit in the North, apples in the South—must travel quickly to the most lucrative markets. The foods that remain unprocessed are increasingly reformed (through genetic engineering) and revalued (through proliferation of exotic fruits and vegetables in all locales). The products native to each region become the luxuries of distant regions. Rather than export their surpluses, the export imperative reorients local production to shifting foreign needs, governed by fashion that changes ever more quickly. It is an opportunity for fast money with little investment for anyone with cash or political connections. It is the basis for export agriculture more destabilizing than the old colonial form of plantations locking in the capital of major corporations. Within the framework of the self-regulating market system, local markets and craft preservation are luxuries, and their prices reflect the change.

The subordination of food to the self-regulating market system involves specific types of disembedding of the economic relations of land and people. First, each step of the food chain involves a further separation of the production and consumption of a community. The necessary mediation of production and consumption is done by capital, that is, by enterprises whose profits mostly depend on distance and durability. In the wave of corporate mergers of the 1980s, the historically unprecedented size of agrofood corporations became apparent. Beatrice Foods, Kraftco, and others became major centers of accumulation not only by buying and selling food, but increasingly as temporary expressions of the frenzied motion of money moving in order to grow. The fictitious-commodity money accelerated the depredations of the corporations forcing land to be a fictitious commodity to supply the diets of the fictitious commodity, labor. Now the rules that made agriculture an exception to free trade under the GATT, and to free markets under consistently high state regulation, are under assault, and with them the self-protection achieved by farmers and consumers.

THE GLOBAL FOOD ECONOMY IN THIS CENTURY: TWO FOOD REGIMES

Over the past century, there have been two major periods of stable rules and relations in the world food economy. Each has been followed by a period of crisis, in which the old rules and relations have ceased to work, and new social categories are emerging and contending over the ones that will prevail in the future. We are now two decades into the crisis of the second food regime, and the outcome of contending rules for a third regime is still open. In this section, I shall sketch the historical shape of each of the two stable periods, or food regimes, of the past. In the following section, I shall discuss the changing relation of commodified food to the fictitious commodities of land, labor, and money. Then I shall outline the regional alternatives presently contending with the globalization of the food economy.

The First Food Regime: Creation of a World Market

Between about 1870 and 1914, the first food regime was created. The growth of capitalist production in Europe—industrial and agricultural—included the massive expansion of a class of workers whose income was in the form of money and whose food was obtained through markets. The poverty of that era, plus the possibilities for profiting from burgeoning food markets, produced the elements for a world food market. Many Europeans were unable to survive either on the land or in labor markets. They migrated in unprecedented numbers to the United States, Canada, Australia, Argentina, and a few other "areas of new settlement," which were made available by expanding states through eviction of indigenous peoples by force. Many immigrants became farmers. From the outset they were not peasants, but commercial farmers directly involved in world markets. They produced grain and livestock products more cheaply than those who remained in Europe. The flood of commodified food into Europe further undermined domestic agriculture and created more displaced people potentially entering the flow of migrants.

During the same period, European states revived the colonial expansion that had been dormant for most of the nineteenth century.[2] This was part of the larger rivalry that culminated in World War I in 1914. The new colonialism (called imperialism) was more deeply rooted than before in the systematic production of raw materials for the expanding industries of Europe (Wayne, 1980). Many of the raw materials for textiles, ropes, and manufactured foods—vegetable oils, sugar, tea, coffee—were tropical crops. The European states seeking to consolidate national economies extended their rule to encompass tropical as well as their own temperate climates. In doing so, they reorganized colonial land and labor into specialized exports of these specific tropical crops.

The Second Food Regime: National Agriculture and Transnational Capital

After a long period of instability and experimentation, including two world wars with the Great Depression of the 1930s in between, a second food regime was created in the late 1940s. This regime was centered not on the import countries of Europe, as the first regime was, but on what became—through the implicit rules of the new regime—the leading export country, the United States. Paradoxically, the United States, whose capitalist economy had less significant legacies of noncapitalist production than in Europe,[3] created a link between agriculture and the state that was intensely national, and that was sustained only by mercantile trade restrictions.

When the United States began actively to exercise its power in the world economy and state system, it insisted on rules of agricultural trade that were highly protectionist. This had two results. First, it forced other countries to adopt national regulation of agriculture to a degree well beyond what was permitted in other sectors. Since the U.S. market was protected, they had little choice but to orient themselves elsewhere. Since few countries, had foreign exchange after World War II, effectively this meant a national orientation for Europe. In 1958 protection was extended from national states to the European Economic Community (now the European Community), whose common agricultural policy was even more regulated and protectionist than that of the United States.

Second, the United States invented a new mechanism, foreign aid, that (among other effects) allowed it to get rid of surplus agricultural products abroad through subsidized exports. The surpluses were chronically generated because of the particular form of U.S. domestic agricultural programs. Since the U.S. dollar was world currency under the Bretton Woods monetary system established in 1944, and since the currencies of other countries were not acceptable in world trade for a long time (never outside of Europe and Japan), this trade practice was not available for that period of time to other countries. The motivations for food aid were complex and changed over time, especially from the European Marshall Plan to the programs of food aid created afterward for underdeveloped countries. Yet the effect on world markets was to enhance the vulnerability of other countries to the efficiently produced and subsidized exports of the United States.

The response ranged from even higher levels of protection in Europe, Japan, Taiwan, South Korea, and a few others, to the embrace of cheap imports in many underdeveloped countries. The former ultimately led to trade wars. The latter underdeveloped domestic agriculture and led to dependence on imported food. These are the two major symptoms of the prolonged food crisis that began in 1972 and has yet to be resolved (Friedmann, 1990a,b, 1991a,b, 1992, 1993).

Now that we are two decades into the contentious restructuring of a future food regime, we need to understand how the two past regimes created conditions we face in the present. The legacies of the first regime shaped possibilities for the second. The legacies of the second regime, in turn, open certain possibilities and place certain limits on the shape of a third regime.

FICTITIOUS COMMODITIES AND NATIONAL REGULATION

The two food regimes differed in the ways that land, labor, and money were commodified and organized in relation to each other. These differences were bound up with the number and type of foods that circulated in world markets, and with the ways that national and colonial economies developed specialized export sectors. The emergence and crisis of the first regime set conditions for the possible shapes of the second food regime. And the emergence of the second regime, and its ongoing crisis, set the conditions for alternative future regimes.

The first food regime substituted land for labor by relocating agriculture to the areas of new settlement in the Americas and in Australia and New Zealand. Land use in Europe was intensive. The agricultural revolution of the eighteenth century improved on past practices but did not reduce the fertility of the soil (Duncan, 1989). It was unsustainable in social terms. The transformation of British agriculture created surplus population, by evicting small producers and by shifting to less labor-intensive livestock husbandry. Its exploitive practices in relation to labor were not sufficient to allow it to compete with cheap imports from areas of European settlement.

The settlers, in turn, constituted new social classes of "simple commodity producers." Farmers in areas of new export agriculture were fully integrated into commodity relations, that is, markets for their means of production and articles of consumption, and for their products. They produced the same staples, particularly wheat, as farmers in Europe, and for the same markets, the wage-earning classes of Europe. However, labor itself was not commodified. Wheat farms were based on family labor, and only supplemented by labor hired in labor markets. These simple commodity producers, or farm families, were able to drive out the capitalist farms of Europe, which were highly "efficient" in England and part of Germany, because they could sustain themselves and their enterprises without having to make a profit or pay rent above the cost of maintaining themselves and their farms. They could sustain themselves without rent, however, only because they were "given" land that was seized by states in the "areas of new settlement," and because they "mined" the soil (in the phrase of the time) through extensive cultivation that did not replenish its virgin fertility.

The unsustainable social practices of the old world gave way to the unsustainable ecological practices of the new world. Eventually, the social relations of the new world became unsustainable, too (and in the subsequent food regime eventually established, the ecologically unsustainable practices of the New World were imported into Europe!). When world agricultural markets collapsed in the 1920s, not to recover until a new food regime was established in the late 1940s, there were two combined forms of crisis. The dust bowls of North America resulted from two generations of bad agronomic practices in the fragile prairie soils. The impoverishment of wheat farmers led to migration off the land only decades after immigrants had settled it. Land was now commodified, but could regain even a fictitious value only after a new regime made farming economically viable again.

Money was the key to the first food regime, but only indirectly. The wheat trade was profitable not for wheat farmers, but for merchants and railway and shipping companies. These were possible both because of British sterling as world currency, and because of new forms of incorporation and new forms of state guarantees, grants, and subsidies to the companies. This was to some extent true also for colonial production, for instance of palm and coconut oils for soaps and foods. Although huge fortunes were made through colonial investments, the production was sometimes in plantations organized directly by the companies, but often left to family farmers who, like the wheat farmers of North America, had only one place to sell. Profits were made mainly through the buying and selling and transporting of commodities, not through the direct production of crops or livestock.

The second regime was built on the success of simple commodity producers in gaining their own subsidies and protection from states. These took a mercantile, that is, protectionist, form because simple commodity producers are considerably more attached to place, that is, to the land and the community, than large capitalist firms. It began in the United States, with the New Deal farm programs of the Roosevelt government in the 1930s. These were retained after World War II, and the U.S. blocked proposals for international planning for food and agriculture—a World Food Board located in a strong Food and Agriculture Organization of the United Nations—(Peterson, 1990) and even for trade generally—an International Trade Organization—in order to protect domestic U.S. farm programs. Other states, whose farmers were unable to sell in protected U.S. markets, adopted similar, sometimes even more protectionist, subsidized, agricultural programs (Friedman, 1993).

Beneath the nationalist surface of agriculture, however, they grew up a transnational integration of agrofood sectors. Agriculture became industrialized and agricultural products changed from final consumer goods to industrial raw materials for the manufacture of highly processed, value added foods. Farms became integrated with and subordinated to agrofood industries, which became some of the most dynamic parts of the advanced capitalist economies, linked to key sectors such as chemicals and energy. Farmers in intense competition with each other were on a "technical treadmill." They had to buy industrial inputs (feedstuffs for animals, chemicals and machinery for crops) and sell to food-processing industries, and increasingly they had to borrow, shifting the balance between livelihood for farmers and commercial criteria of farm businesses.

When this happened, each agricultural product became in principle substitutable, not only by the products of farmers in other places, but also by other products entirely. There was a shift from a world market in which Europe imported traditional dietary staples of wheat, meat, and dairy products, to a transnational agrofood sector in which corporations increasingly sought raw materials and markets globally. The durable foods complex was organized by the demand for raw materials by growing food-manufacturing industries. Among the many consequences, perhaps the most significant was the development of temperate and often synthetic substitutes for "traditional," i.e., tropical, export crops of the Third World. The two most important ingredients of manufactured foods, sweeteners and oils, were most significant. Sugar-exporting countries, and sugar workers especially, faced declining demand despite increasing consumption, because food-manufacturing industries invented processes to create sweeteners from domestic grains (cheap because of government sponsored surpluses) and from chemicals. Exporters of palm, coconut, and other tropical oils had to compete with temperate oils. Soy oil was most cost effective because it was a byproduct of the manufacture of soy meal for animals feeds.

Industrialization of agriculture and subordination to agrofood industries had enormous consequences. Farms became increasingly specialized, significantly separating crops and livestock. Socially, grain and soybean producers (the latter a postwar phenomenon on a significant scale) sold to feedstuffs industries, while livestock operations (increasingly industrial, "factory" farming) bought manufactured feedstuffs. Ecologically, animals wastes became a cost to livestock producers, while crop farmers bought industrial chemicals for fertilizers.

Combined with U.S. (and eventually other) farm programs that produced grain surpluses, there was a double effect internationally. For advanced capitalist countries, and some key Third World countries that subsidized agroindustries and national agriculture to support it (e.g., Brazil), this created both an integrated livestock complex, and, eventually, with the breakdown of the regime, international competition verging on trade wars. For most Third World countries, world trade was organized through food aid, directly or indirectly because of the low prices sustained by the scale of subsidized U.S. exports. The wheat complex of the second food regime created a new pattern of trade between the United States as dominant exporter and the Third World (and Japan) as importers increasingly dependent on foreign supplies.

The second food regime more fully incorporated food and agriculture into industrial capital. In the first food regime, both land and labor were outside the main capitalist organization of the economy, though they depend on each other. The main advances were made through using distant locations in areas of new settlement and tropical areas—again, substituting land for labor, and using land for its specific characteristics of fertility or climate. In the second food regime, agrofood corporations brought food and agriculture centrally into the accumulation of capital and into the technical dynamism of industrial corporations. Although they have existed in various forms since the beginning of capitalism, agrofood corporations finally became central to the international capitalist economy. Now they are both very large, from Kraftco to Cargill, and technically and organizationally dynamic. Biotechnologies now rival microelectronics as the revolutionary innovations of capital.

Although land and labor are more fully commodified now, and more directly subordinated to industrial capital, they remain fictitious commodities. There is now a second crisis of the "family farm," which anticipated the present long recession and restructuring, just as the farm crisis of the 1920s anticipated the Great Depression of the 1930s. This time the family farm is different, however. Its industrial transformation changed not only its external relations to the economy, but also its internal relations as a source of livelihood for farmers (Friedmann, 1990b). There is also a more complex legacy of ecological damage due to the prolonged use of chemicals and heavy machinery by industrial agriculture.

Transnational organization of the agrofood sector eventually came into conflict with national farm programs, which, following the lead of the United States, were interventionist and protectionist almost everywhere. This began with the "food crisis" of the early 1970s. Although that name was given to the sudden, unexpected shift from surplus to scarcity, and from low prices to high prices, the crisis was actually deeper than those symptoms. Surpluses returned, because commodity support programs were kept in place. But the old rules were permanently subverted.

The food regime depended on a U.S. monopoly over world trade and the ability to subsidize exports. With high prices and a shift to a deficit in its overall trade balance, the United States sought to expand commercial exports. With the monetary crisis and the oil crisis, other countries gained the economic and political means to enter world markets in ways that undermined the old rules. Europe began to subsidize exports. Brazil and other "new agricultural countries" began to undersell U.S. exports. Importers with hard currency, briefly the Soviet Union, in the longer term Japan, began to gain new clout. And the poor countries borrowed to pay oil and food bills in the 1970s, leading to the debt crisis of the 1980s and 1990s. Now they are forced to export at any cost, and their agricultural exports, particularly what are called "nontraditional" exports, are further transforming the rules and relations of the postwar food regime.

THE SUSTAINABLE ALTERNATIVE

Midas's gift was a joke of the gods. When he requested to be released from his wish, according to Graves (1960:282),

> Dionysus, highly entertained, told him to visit the source of the River Pactolus, near Mount Tmolus, and there wash himself. He obeyed, and was at once freed from the golden touch.

Like Midas, many of us regret the effects of turning food to gold. By emphasizing distance and durability, the present food economy creates specific problems for labor and land as fictitious commodities. Commodified food, which is increasingly global in the sourcing of raw materials and the marketing of food products, is not sustainable socially or ecologically. The simplest sign is the vulnerability of urban populations to a steady supply of food from long distances (Busch and Lacy, 1984). More than that, the nutritional and cultural quality of food is increasingly questioned, from the breeding that favors durability and appearance over nutrition and taste, to the manufacturing that changes the balance of nutrients (too much fats and sugar) and adds chemicals whose effects on health are highly suspect and sometimes known to be dangerous. From the other direction, the ecological effects of factory farming of animals and chemical intensive crops are becoming increasingly potent political issues, from the contamination of soil and water, to deforestation.

What form can self-protection take? Self-protection, to be conscious and effective, must build on the understanding that land and labor are the real substance of society. Rather than subject land and labor to the vicissitudes of the self-regulating market system, the substantive relations between people and the earth must be the center of sustainable agrofood regions.

Piecemeal solutions are not promising. The privileged can create small markets for plants cultivated and animals fed with fewer chemicals. They can even create profitable opportunities for agrofood enterprises, large or small, to bring them across the world. But this private solution ultimately reinforces the distance and durability that are unsustainable features of the present situation. For example, the small health food industries in California in the 1970s were bought up by large food corporations. Although the new owners did not change production practices, they distributed the products through their far-flung networks. This raised the entry cost for similar companies elsewhere. The absence of local manufactures reduced the potential market for local farmers, especially organic farmers. Consumer demand could be satisfied by diversification of corporate production, but at the cost of undermining local food linkage elsewhere.

The promising solution lies in locality and seasonality. In some parts of the world, people are choosing to try to relink the regional components of agrofood relations. A new politics of food, land use, and the environment is based on urban constituencies interested in supporting diverse, local, sustainable farms, which are in turn reorienting themselves from corporate science and markets toward informed experimentation and local consumers. In other parts of the world, desperation at the impoverishment and marginality imposed by the turbulent shifts of the world markets and the export imperative of the debt regime is leading to experiments in self-reliance that are necessarily local. Despite the desperate circumstances of people dispossessed and impoverished by the restructuring of the food regime by capital, the creativity and energy of their solutions offer important lessons for us.

An important aspect of the industrial restructuring now taking place is the surprising success of both small enterprises and regional economies, which are often described under the labels of "postfordism" or "flexible specialization." Agriculture has been very important to the success of the "third Italy," based on a long history of planning by left-wing regional and municipal governments, left to their own devices by both large corporations and the national state. They fostered cooperatives and interesting combinations of public and private agencies to provide technical and marketing services to innovative enterprises (Bagnasco, 1988; Fanfani, 1991).

As a result, traditional food production could be transformed, rather than replaced by durable foods. For instance, Reggio-Emilia is the site of the best Parmagiano cheese. While Parmagiano-Reggiano is still manufactured under the supervision of a Consortio, much like the organizations that enforce quality controls over vintage wines, within that framework the most modern and rational dairying operations are improving the productivity of the milk, and local biotechnical research is engaged in finding a replacement for the one imported component, genetic stock from Canadian dairy cows and bulls.

This is a complex process. On the one hand, like the successful health food companies of the United States, the very success of small enterprises and regional economies makes them interesting to large agrofood corporations. In the Emilia-Romagna case, it is interesting to note that the one major new player in international grain markets to grow up outside the United States—Ferruzzi—is based in Ravenna, the port city of the region. On the other hand, the success of agrofood products, like clothing, ceramics, industrial equipment, and the like, depends on the ability to export from the region.

Yet if this model were generalized, it would mean a world food economy that was regionally based, but included trade among regions of efficiently produced foods specific to each region. And building up through nested regions, in this case, the European Community, Ferruzzi is an example of a transnational agrofood corporation whose interests lie, at least in part, in public support for technical innovations and the markets for them, and therefore, in planned agrofood policies. In particular, like some firms in the United States, this involves potentially important linkages to the energy sector, for instance, in the use of agricultural crops as renewable and potentially nonpolluting alternative fuels.

THE NONSUSTAINABLE ALTERNATIVE

In stark contrast to the prosperity of Emilia-Romagna (which may not survive the larger economic and political changes at play) indebted states, such as Mexico, are vulnerable to the dictates of the IMF to reduce state involvement in the economy and to reorient production toward exports. As national institutions are abandoned, people have been forced to seek alternative local solutions to survive. Even before the postwar food regime, Mexico had a history of deep involvement by the national state in agrofood as in other key sectors. This was encouraged under the rules of the food regime, and like the food regime itself, culminated in the early 1970s. The Echeverria government of that period established a new state organization to administer key aspects of the agrofood system, from agricultural price support programs to local food shops that sold staple foods, mostly processed by state enterprises, at fixed prices. Among its goals were to ensure the physical supply of basic commodities to low income populations at prices that they could afford (Grindle, 1977:Chapter 1).

Since the debt crisis of 1982, and increasingly under the present regime, the state has been withdrawing from these activities. Supermarkets, which had to coexist with a subsidized public sector, are now rapidly expanding; yet they cannot make profits in poor areas where the state established shops. As a result, local communities are extending their long tradition of demanding services for squatter communities into a demand to keep and control local food shops. They are creating new community institutions, such as community kitchens, and new links between consumers and agricultural producers. Yet all this is vulnerable to the dislocation of farmers and the impoverishment of consumers by the official turn to export agriculture, which entails drastic reductions in entitlements to land and to basic foods.

The withdrawal from the national agrofood system by the Mexican state creates, as intended, new spaces for transnational capitalist enterprises governed by principles of distance and durability. Large-scale, private manufacture of staple foods, such as corn flour for tortillas, was not profitable under the system of national regulation. This consisted of two parts. One was a network of state enterprises processing flour, cooking oil, milk, and so on, linked to agricultural marketing agencies and retail marketing agencies. The other was a network of small millers and tortilla makers ideal to the short time frame of traditional manufacture (the traditional meal prepared from dried corn must be used for tortillas while it is still wet). The results of privatization and withdrawal of the state, combined with the growth of supermarkets. is the replacement of the traditional tortilla with a nutritionally inferior, more durable product, which yields a tortilla that is less desirable in culinary and dietary terms. This is roughly the equivalent of industrial bread in North American and Europe, but more rapid and less accessible to the poor. Resistance to distance and durability is built into the temporality and locality of the traditional product, just as it has been for artisanal bread in Europe and now even its revival in North America.[4]

THE CONTESTED ROLE OF NATIONAL STATES

Even though national regulation is in crisis along with the food regime that sustained it, national states are still the locus of political decisions structuring agrofood relations. Their autonomy is now challenged by transnational capital, and by the institutions and ideologies promoting deregulation and the withdrawal of national states from earlier forms of regulation. Like all major changes, the challenges to national states offer both danger and opportunity. The danger is the loss of old forms of self-protection without replacement, and the untrammeled rule of the self-regulating market system—until its damage provokes major movements for protection once again. The opportunity is that local experiments can flower as the state withdraws—or better, changes—its former uniform regulation.

Sweden in 1989 adopted a national policy that may never be implemented because of entry into the European Community, but that shows remarkable creativity and foresight. The conditions were typical of small advanced capitalist countries: national commodity price support programs that generated surpluses, inability of the public finances to sustain the growing gap between high domestic prices and plummeting world prices because of the U.S.–EC competition to subsidize exports, and a growing political awareness of the ecological and health problems of industrial agriculture and the manufacture of durable foods. The ecological party developed an innovative alternative and gained in popularity, leading the ruling Social Democrats to adopt the key points of the policy.

Those points, which had the effect of subordinating agricultural policy to land use, food, and environmental policy (plus a few others, such as animal rights), focused on low-chemical grain production, low-intensity grazing of animals, and public education concerning the virtues of healthy and local foods. This has the merit not only of reframing agricultural policy in relation to land, labor, and food (instead of yields and profits), but also of shifting the political base of agricultural policy from rural to urban areas, where most people now live (Peterson, 1990).

These sketches can only suggest some of the directions and actors for construction of a food regime based on season and locality. Whether from choice or desperation, such movements are occurring widely. By definition each must reflect the specificities of place. This makes diversity, which is a necessary condition of sustainability, a principle of the food regime.

THE CHALLENGE

The Midas story is no more magical than the present belief in the powers of money. The monetarist god is harsh and unforgiving. He demands sacrifice on the altar of the deficit. His church is the International Monetary Fund and He demands allegiance to its prescriptions. His earthly representatives formulate austerity measures and insist on obedience to the letter of the law, no matter what the earthly consequences. His priests are economists, who have the dominant voice in universities and governments. Whether intended or not, the political effects of implementing monetarist ideology is to reorient the states of the world away from their relations to their citizens—their responsibilities for guaranteeing individual and social rights—and toward the institutions—corporations and banks—that turn all they touch into gold. This is Midas's feast.

Midas's feast is undermining state power except that which is dedicated to gold. As an unintended but historically precedented result, Pharaoh's dream of famine is likely to come true after all. The challenge is to face the lean years together, without illusions or magic. The story I have to tell about money, power, and food is my contribution to confronting the past and therefore the future.

NOTES

1. The widespread denial of natural limits has its intellectual roots in the revolutionary conceptions of nature developed in the seventeenth century parallel to the fundamental changes in social relations expressed by political theorists. The scientific method and the social contract expressed relations among people and between society and nature that were both new and newly opaque. See Leiss (1974) and MacPherson (1962). The emotional roots may go much deeper. See Brown (1959) and Becker (1973).

2. Before the 1880s, there had been a shift from the formal empire of the mercantile period to informal "empire of free trade" (Gallagher, 1953). The rights of mercantile companies were abolished in India and Canada, administration was rationalized in British, French, and other colonies, and there was a general, though uneven, movement to informal mechanisms of imperial control and even, for the settler colonies, to self-government. "[T]he European powers remained reluctant to extend their responsibilities and their rivalries" to Asia (Langdon, 1976:641), and Africa was of little interest as an area of formal empire (Robinson and Gallagher, 1976:593).

3. Although the legacy of slavery and the planter class in the South marked U.S. politics for more than a century after the Civil War.

4. In France the successful strategy of artisanal bakers in resisting the encroachment of industrial bread was to convince consumers of the virtues of the baguette, which must be consumed within hours of baking (Bertaux and Bertaux-Wiame, 1981).

REFERENCES

Bagnasco, A. 1988. *La Costruzione Social del Mercato*. Bologna: Mulino.

Becker, E. 1973. *Denial of Death*. New York: Free Press.

Berry, W. 1977. *The Unsettling of America, Culture and Agriculture*. New York: Avon and The Sierra Club.

Bertaux, D. and I. Bertaux-Wiame. 1981. Artisanal Bakery in France: How It Lives and Why It Survives. In *The Petite Bourgeoisie*. pp. 155–182. F. Bechhofer and B. Elliott (eds.). London: Macmillan.

Bienefeld, M. 1989. The Lessons of History. *Monthly Review*, 44:9–41. July/August.

Brown, N. O. 1959. *Life Against Death*. New York: Vintage.

Busch, L., and W. B. Lacy. 1984. *Food Security in the Unites States*. Boulder, CO: Westview Press.

Duncan, C. 1989. The Centrality of Agriculture: Between Humankind and the Rest of Nature. Ph.D. Dissertation, York University, Toronto.

Fanfani, R. 1991. Il Rapporto Agricoltura-Industria tra Passato e Presente, MS.

Friedman, H. 1990a. The Origins of Third World Food Dependence. In *The Food Question*, pp. 13–31. H. Bernstein et al. (eds.). New York: Monthly Review.

Friedman, H. 1990b. Family Wheat Farms and Third World Diets: A Paradoxical Relationship Between Unwaged and Waged Labor. In *Work Without Wages*, pp. 193–213. J. Collins and M. Giminez (eds.). Binghamton, NY: SUNY Press.

Friedmann, H. 1991a. Changes in the International Division of Labor: Agri-food Complexes and Export Agriculture. In *Towards a New Political Economy of Agriculture*, pp. 65–93. W. H. Friedland et al. (eds). Boulder, CO: Westview Press.

Friedman, H. 1991b. New Wines, New Bottles: The Regulation of Capital on a World Scale. *Studies in Political Economy* 36:9–42.

Friedman, H. 1992. Distance and Durability: Shaky Foundations of the World Food Economy. *Third World Quarterly* 13(2):371–383.

Friedman, H. 1993. International Political Economy of Food. In *Food* B. Harriss (ed.). Oxford: Basil Blackwell, in press.

Gallagher, J. 1953. The Imperialism of Free Trade. *Economic History Review* 6(1): 1–15.

Graves, R. 1960. *The Greek Myths*, Vol. 1. London: Penguin.

Grindle, M. S. 1977. *Bureaucrats, Politicians, and Peasants in Mexico: A Case Study in Public Policy*. Berkeley, CA: University of California Press.

Kneen, B. 1989. *From Land to Mouth*. Toronto: NC Press.

Langdon, F. C. 1976. Expansion in the Pacific and Scramble for China. In *Material Progress and World-Wide Problems, 1870–98*, Vol. XI of *The New Cambridge Modern History*, pp. 641–667. F. H. Hisley (ed.). Cambridge: The University Press.

Leiss, W. 1974. *The Domination of Nature*. Boston: Beacon.

Macpherson, C. B. 1962. *The Political Theory of Possessive Individualism, from Hobbes to Locke*. London: Oxford University Press.

Peterson, M. 1990. Paradigmatic Shift in Agriculture: Global Effects and the Swedish Response. In *Rural Restructuring*, pp. 77–100. T. Marsden, P. Lowe, and S. Whatmore (eds.). London: David Fulton.

Polanyi, K. 1957 [1944]. *The Great Transformation*. Boston: Beacon.

Robinson, R. E., and J. Gallagher. 1976. The Partition of Africa. In *Material Progress and World-Wide Problems, 1870–98*, Vol. XI of *The New Cambridge Modern History*, pp. 593–640. F. H. Hisley (ed.). Cambridge: The University Press.

Soper, K. 1981. *On Human Needs*. Atlantic Highlands, NJ: Humanities Press.

Tilly, C. 1990. *Coercion, Capital, and European States, AD 990–1990*. Cambridge, MA: Basil Blackwell.

The Torah, A New Translation of the Scriptures According to the Masoretic Text. (1967). Philadelphia: The Jewish Publication Society of America.

Wayne, J. 1981. Capitalism and Colonialism in Late Nineteenth century Europe. *Studies in Political Economy* 5:79–106.

CHAPTER 10

Scaling Sustainable Agriculture: Agendas, Discourse, Livelihood

David Goodman

It is, by now a commonplace that sustainable development can mean essentially whatever you want it to mean. The phrase has become a slogan, truism, shibboleth, cliché, and, for a considerable body of agnostic opinion, an oxymoron. Sustainability can equally refer to a conceptual framework, an instrumentalist methodology for policymakers, a desirable characteristic of social systems, or a code of ethics. Sustainable development is seen as a binding constraint on economic expansion by "limits of growth" theorists or, alternatively, as the foundation for continuing economic growth. These contrasting positions also are compounded by operational differences between those who believe that sustainable development can be achieved by making running adjustments to the present system and others for whom the system itself must first be transformed if environmental behavior is to be changed.

The level or scale of analysis provides yet a further source of variation. With the caveat that global life-support functions be maintained, the impression frequently conveyed is that sustainable development is an operational strategy intended primarily for Third World countries, rather than the advanced industrial economies as well. In the burgeoning literature on the South, prescriptive approaches or "solutions" are formulated from the level of the field, farm, and village up to national policy (Lélé, 1991). Within the

Food for the Future: Conditions and Contradictions of Sustainability, edited by Patricia Allen.
ISBN: 0-471-58082-1 © 1993 John Wiley & Sons, Inc.

arbitrary boundaries of this discourse, sustainability is presented as a new paradigm of development. This focus on the Third World, I suggest, exemplifies the social construction of sustainable development and its operational agenda. It is readily invoked as a strategy for Amazonia, less easily for the parallel destruction of old-growth forests in the Pacific Northwest and Alaska.

In the 1980s, virtually all international development agencies made some commitment to sustainable development. This consensual strength arguably is again directly related to the focus on Third World issues for there are powerful political economic factors at work in framing the international environmental policy agenda. Northern interest in conserving biodiversity and related biosphere services has brought to the fore the linkages between resource degradation and "overpopulation" in the South, while studiously ignoring class relations, social equity, or reasons for the demise of traditional resource management systems. The population–environment nexus is presented in a neo-Malthusian framework, cloaked in a putative apolitical, value-free scientific discourse. This fails to conceal, however, the reductionism of this view, with its arrogant elitism and scare-mongering defense of existing North–South inequalities. Yet the salience achieved by this formulation completely overshadows the "ecological overpopulation" of the North in terms of its share of global resource consumption, waste generation, and contribution to global climate change (Altieri, 1990). We are invited to believe that Third World population growth is more central to sustainable development than the reproductive behavior of global capitalism.

Questions of how different scales or levels of analysis can be integrated, conceptually and operationally, go to the heart of the problem of identifying "paths" of sustainable development. This chapter explores these issues in the context of formulating operational steps toward sustainable agricultural and rural development (SARD). In the following section, I draw a brief comparison between the discourse of sustainable development as applied to advanced industrial countries and the Third World.

SUSTAINABILITY IN THE NORTH AND THE SOUTH: AGENDAS AND DISCOURSE

The complex hierarchies of interdependence intrinsic to sustainable development, ranging from the microenvironments of field and farm, for example, to local watershed areas, regional agroecosystems, and beyond, are accompanied by widely varying views on operational goals and praxis. Thus mainstream environmental economics, building on neoclassical welfare foundations, puts the onus on individual behavior, which implies that market prices and public policies must incorporate the "proper" *economic* valuation of environmental goods and services (Pearce et al., 1989). While accepting the axiom of intergenerational equity, this orthodoxy rarely ventures into the realm of social and political organization or geopolitical structures and their relationship to sustainable development. The operational challenge rather is to measure the contribution environmental assets or "natural capital" make to social welfare, as revealed by individual preference systems, and to express this valuation in monetary terms so that *trade-offs* between environmental goods and "human-made" assets are explicit. In this view, market failure and policy failure in appropriately valuing environment–economy linkage lead to unsustainable development paths. Within mainstream environmental economics, the practice of sustainable development frequently is reduced to a technical exercise in policymaking to improve the workings of markets by incorporating the economic value of environmental assets into the everyday calculus of individual production and consumption decisions.[1] It is significant that the initial or inherited distribution of income and resource endowments is taken as a datum. This is hardly the paradigm to lead us down sustainable development paths to alleviate poverty among present generations or to entertain other world views and forms of social organization.

The influential report of the Brundtland Commission (WCED, 1987) also advocates proper valuation of the environment to encourage conservation and meet the requirements of intergenerational equity in terms of the constancy or enhancement of "environmental capital." Yet technical criteria and objectives are not allowed to silence a preeminently political question: sustainable development for whom? Advancing the idea of sustainability as both a process and a policy objective, Brundtland introduces the concept of "needs," which explicitly incorporates such issues as "access to resources" and "the distribution of costs and benefits." The report insists that "meeting the basic needs of all" is a condition of sustainable development and identifies the asymmetry in North–South economic relations as an obstacle to the pursuit of sustainability. A prime example is the unilateral "adjustment process" imposed on Third World debtors by the creditors' cartel of multinational banks and the International Monetary Fund (IMF) as a response to the world debt crisis in the 1980s.

The emphasis in Brundtland on questions of access, basic needs, intragenerational equity, and international relations brings the praxis of sustainability to the forefront of discussion. Global to local ecosystems are clearly identified as contested domains, which are structured in ways inimical to the satisfaction of basic needs and sustainable development. "Even the narrow notion of physical sustainability implies a concern for social equity" (WCED, 1987: 43). The approach to sustainability advocated by Brundtland, which gives "overriding priority" to "the essential needs of the world's poor," carries an unequivocally political message in terms of the fundamental changes required at all levels of the global system.

Once basic needs are incorporated into the concept of sustainable development, it is a short but highly charged step to such related concepts as "sustainable livelihood security" (Chambers, 1986) and "resource entitlement" approaches to hunger and deprivation (Sen, 1981, 1987). In the conditions of rural societies in the Third World, the agenda of sustainable development effectively extends to the poor the human rights of citizenship and participation in society. The significance of the Brundtland Report is that it brought home the *social* content of the Third World environmental crisis: the struggle for access to resources essential for day-to-day survival, which frequently, though by no means invariably, leads to unsustainable practices. Livelihood and social reproduction thus are the starting point for the analysis of sustainable development in the Third World. From the perspective of the one-fifth of humanity that forms "the global underclass" (Eckholm, 1982), sustainability signifies basic human rights and empowerment.

When we turn to discussions of sustainable agriculture (SA) in advanced industrial countries we find a radically different discourse and operational agenda.[2] Questions of livelihood and subsistence give way to those of amenity and ecoaesthetic values in disputes concerning alternative land use, matching, for example, farmers against developers or habitat conservation against the encroachment of intensive agriculture, particularly in Western Europe (Lowe et al., 1986; Lowe, 1988; Potter, 1989). In the advanced industrial countries, debates on agriculture and the environment primarily are concerned with the impacts of modern capital-intensive farming practices. These are exemplified by high levels of mechanization and the heavy applications of agrichemicals needed to maintain yields in monocultural cropping systems. Similar intensive, large-scale operations also characterize livestock production, leading to serious water pollution caused by discharges of slurry and silage effluents. So-called "industrial" agriculture, reflecting its dependence on external inputs, is heavily subsidized by commodity-price and income supports, import controls, and other sources of economic "rents," which have underwritten overproduction and artificially fostered technological competition, intensification, and specialization (Goodman and Redclift, 1991).

The damaging environmental impacts of modern agricultural practices are directly related to this "technological treadmill." These include nutrient leaching, particularly the build-up of nitrates in surface and groundwater, the contamination of drinking water supplies by nitrates and pesticides, eutrophication, the depletion of aquifers, the loss of land and aquatic species diversity, and the destruction of wildlife habitat, such as wetlands. Growing recognition of the links between intensive farm practices and environmental degradation also has aroused concern that the quality of the rural countryside as a public amenity resource is diminishing.

This perception has called into question the role of farmers as "stewards" of the land, particularly in Western Europe. As a result, previous widespread reliance on voluntary codes of "good farming practice" is now being superseded by statutory environmental regulations in leading advanced industrial nations. In addition, although still tentative and with varying degrees of commitment, some Organization for Economic Cooperation and Development (OECD) countries, including Denmark, Sweden, United Kingdom, and the United States, have responded to environmental concerns, as well as the fiscal burden of mounting production surpluses, by launching programs to encourage transition to low external input farming systems and to protect fragile rural ecosystems. Agriculture has thus been drawn into environmental politics and, in turn, environmental priorities are being incorporated into the formulation of agricultural policy.

The increasingly explicit weight of environmental goals in OECD agricultural policy does not yet extend to the Third World where the expansion of food and fiber production has highest priority. Although restricted by the fiscal constraints imposed by the continuing debt crisis and structural adjustment policies, the less developed countries are more concerned to accelerate the diffusion of modern capital-intensive agricultural methods, the so-called "Green Revolution" technologies, than to control their environmental impacts. Domestic food security and commodity export surpluses to finance debt service obligations are the primordial goals of agricultural development policy in the Third World. As these introductory comments reveal, the agriculture–environment nexus and the operational implications of SARD thus differ substantively in advanced and Third World countries. Yet at the same time they are inextricably linked through the vectors of international economic relations, the global food system, and technology transfer, as we see in the following sections.

GLOBAL-LOCAL LINKAGES

The operational problems of SARD at different scales of spatial aggregation, ranging from the global biosphere to rural communities, are examined briefly below. This hierarchical organization, of course, is simply a heuristic device since the various levels, in practice, are interdependent in ways that are highly complex and still imperfectly understood. Moreover, "The behavior of higher levels in the hierarchy cannot be reduced to the sum of behavior at lower levels, nor are the latter the simple disaggregate of the former"

(Barbier and Conway, 1990:13). This lacks of uniqueness implies that the transition of SARD will require concerted and coordinated action at all levels.

Global Biospheric Management

The dependence of SARD on on global biospheric sustainability can be exemplified in many ways. In the case of global warming and the loss of biological diversity, the business-as-usual scenario of greenhouse gas emissions elaborated by the Intergovernmental Panel on Climate Change (IPCC) anticipates a rate of global warming that exceeds the known capacity of natural communities, such as forests, to migrate and adapt by several orders of magnitude (IPCC, 1990b). Global warming, possibly accompanied by a greater frequency of extreme climatic events, such as droughts and storms, will severely disrupt ecosystems and result in widespread species loss and the destruction of specific ecotypes. Such violent transitions will be accompanied by global biotic impoverishment and a reduction in the capacity of the Earth to support all forms of life (Woodwell, 1990). In short, global warming threatens the essential foundations of agriculture, irrespective of the characteristics of production systems. Within this sobering global context, the IPCC Working Group II (1990b) suggested that the most severe impacts of climate change would be felt in already-vulnerable and populous regions, including the Sahel, Southeast Asia, and semiarid areas of Central and South America (Parry, 1990). In addition, the northerly relocation of production zones from today's temperate "bread baskets," such as the U.S. Midwest, has unknown implications for world agricultural trade, and hence for food security and basic needs provision at all levels of the international system.

Sustainable Global Relations

The threat to the biosphere caused by the high, and in global terms, unequal consumption levels of the North has been widely seized on by groups arrayed across the political spectrum as evidence of the unsustainability of fossil fuel-based industrial development. One corollary of this perception—that Northern consumption levels place unsustainable demands on environments and economies in the South—has met determined resistance in international fora, however. As Redclift (1991) notes, some developing countries, notably Brazil and Mexico, recently attempted, unsuccessfully, to place the IPCC Report on global climate change on the wider agenda of negotiations on the New International Economic Order, embracing trade, aid, and development. Yet, as the Brundtland Report made clear, it is imperative that such interdependence and its consequences be recognized if the global biosphere and, by extension, agroecosystems, are to be managed sustainably. These goals are plainly and simply unrealistic in operational terms unless structures of international governance and global relations are informed explicitly by the principles and normative axioms of sustainable development.

The asymmetries of North–South political and economic relations that militate against sustainable development are exemplified by the continuing Third World debt crisis. Intragenerational, North–South inequality has continued to worsen in the 1980s, with profound, immediate consequences for the one billion absolute poor in the Third World (World Bank, 1990). Concomitantly, pressures to intensify agricultural export production, including nontraditionals, threaten long-term sustainable resource management and household food security. A second related example concerns the distortions of international trade, created principally by protectionist policies in the advanced industrial countries, which foster unsustainable agricultural development both in the North *and* South. In the industrial countries, notably the United States and EEC, agricultural protectionism, subsidies, and unfair export practices have contributed to overproduction, unsustainable use of resources, agricultural pollution, and the dumping of agricultural surpluses on world markets. The heavily subsidized competition between the two agricultural "superpowers" has exacerbated agricultural trade distortions and deprived specialized "third" country agricultural exporters, such as the 14-strong Cairns group, of valuable exchange earnings (Goodman and Redclift, 1989). "Sustainable trade relationships" therefore constitute an important element of a global strategy for SARD.

As these examples suggest, a framework of international relations and governance that actively responds to global biospheric "imperatives" is the essential foundation of SARD at the global level. This framework, in short, must be conductive to the adoption of sustainable resource management practices that, in turn, create sustainable rural livelihoods and household food security.

Macroregional Perspectives

Since agroecosystems and environmental degradation do not respect administrative boundaries, SARD may require policies and programs that operate at the intermediate level between global relations and individual national contexts. Efforts to restrict the loss of biological diversity associated with the clearance of tropical moist forests or the degradation of drylands come to mind here. A recent OXFAM (UK) report estimates that more than 6 million hectares of land turn to desert each year, and over 20 million hectares of tropical forest are cleared for timer and agriculture. At this rate almost one-fifth of the land now under crops in the South will have disappeared by the year 2000 (*New Scientist,* 5 October, 1991).

Dryland degradation occurs in virtually all arid and semiarid regions, including the Middle East, Sahelian Africa, Central Asia, Northeast Brazil, the western United States, and Australia. The human impact of desertification is most significant in Africa and Asia, where the livelihoods of an estimated 230 million people are affected (Mabbutt, 1984), and these cases accordingly have attracted most attention in discussions of rural poverty and the environment. Rural poverty undoubtedly is the main proximate cause of the overexploitation of drylands, characterized by shorter fallow periods, overgrazing, rising fuelwood needs, and more intensive use of highly erodable areas. However, such behavior is shaped by social forces operating in politicoeconomic structures that encourage land concentrations, undermine traditional resource management systems, condone the privatization of common property resources, and subsidize unsustainable technologies. Rising numbers of rural households are excluded from access to productive land and other sources of entitlements necessary to meet basic needs.

There is little purpose in assigning blame for resource degradation, as often occurs in neo-Malthusian formulations, to "overpopulation," "population pressure," or other euphemisms for the poor farmers and herders who bear the brunt of declining production and more vulnerable, unstable livelihoods. For example in Northeast Brazil, the highly unequal distribution of property rights and land-use changes associated with heavily subsidized agricultural modernization programs are the principal causes of rural poverty, as well as outmigration to fragile, unfamiliar environments in Amazonia. By ignoring societal dynamics, such formulations obscure rather than elucidate the critical operational questions of how to reverse processes of resource degradation and rural pauperization.

Within this macroregional perspective, agricultural settlement pressures on regional ecosystems may exacerbate global environmental problems. It is estimated, for example, that the biomass carbon, methane, and nitrous oxide released by tropical deforestation constitute 20% of current global emissions of greenhouse gasses. In this case, global biospheric imperatives and SARD intersect since there is convincing evidence that many tropical soils are unsuitable for permanent settlement based on intensive cropping systems (Furley, 1990).

Conversely, urban–industrial activities can contribute directly to resource degradation and the loss of ecosystem services on a regional scale, as exemplified by emissions of sulfur dioxide and nitrogen oxides that are precipitated as "acid rain." Soils without alkalines to buffer rain lose fertility and production potential through the leaching of nutrients. In addition, acid rain is widely identified as the prime cause of the degradation of terrestrial and aquatic ecosystems, particularly forests, lakes, and wetlands in North American and Northwest Europe, including Scandinavia.

SARD clearly will be one among many constituent elements of a transition to global sustainable development. The global-local politics of this transition and its institutional forms are only dimly discernible, if at all. Nevertheless, there is a growing realization that sustainable development requires fundamental, root-and-branch transformations in the ethos, cultural values, and "styles" of material life characteristic of postindustrial and industrializing societies in the late twentieth century. With the United Nations Conference on Environment and Development (UNCED) meetings in Rio de Janeiro, we will all be able to see how wide the gap remains between this perception and its translation into the effective actions by the global community and its hegemonic powers. In the remaining pages, I concentrate on the operational priorities and requirements for SARD in national and local contexts, particularly in the Third World.

OPERATIONALIZING SARD IN THE SOUTH: RETHINKING MEANS AND ENDS

As a broad generalization, we can characterize Third World agricultural policies as being productionist in orientation, following a commodity-by-commodity approach imitative of the advanced industrial countries. This one-dimensional, typically "top-down" perspective must be superseded by more holistic, integrated strategies, which combine productivity criteria with equity and sustainability considerations. If this is to be accomplished, the watchword in articulating national policies is sensitivity to local and regional conditions. These must be constructed on the basis of "bottom-up" approaches that incorporate the best of local traditions and local knowledge of sustainable resource management. This attention to local ecosystems and social organization implies also that SARD will integrate high and low external input farm systems, large- and small-scale production, private property, and community-held land. The diversity of agroecological systems and associated social institutions opens up different yet complementary trajectories towards SARD. The common denominator is that these pathways would be selected for their capacity to support sustainable production systems and, therefore, to generate sustainable rural livelihoods and household food security.

This multidimensional approach will prompt reconsideration of current investment priorities and the roles assigned to well-endowed agricultural environments and fragile, resource-poor systems. In the former, SARD may lead to the less intensive utilization of irrigation systems, for example, in order to rectify salinization, water logging, and other threats to long-term production potential. In other cases, SARD may depend more on improvements in the efficiency and equity of water use, involving revisions of user charges and reforms in local water markets. The critical challenge in these better endowed agricultural environments is to maintain high production levels sustainably and equitably; that is, to protect future productive capacity and hence ensure the continuing contribution of these areas to food supplies and rural livelihoods.

Implementation of SARD strategies similarly will prompt reassessment of production priorities in so-called "marginal lands" or fragile environments. Due to past neglect, even relatively modest investment programs, including research and training, may offer high incremental returns in terms of production and sustainability goals. With their concentration of poverty and resource degradation, marginal lands represent key targets of SARD strategies. This emerges from recent research at the International Food Policy Research Institute (IFPRI) on the distribution of the rural poor, defined by calorie consumption, in the six major agroecological zones of the Third World, classified according to the Food and Agriculture Organization of the United Nations (FAO) criterion of growing-season days (GSD). This study reveals that "to reach the greatest number of poor, technological improvements in agriculture in the seasonally dry (210–269 GSD) and semi-arid zones (90–209 GSD) would have the greatest effect. . . . In aggregate the seasonally dry and semi-arid 1 zones (150–269 GSD) represent 45–50 percent of total calorie production in all geographic regions except South America (22 percent)" (IFPRI, 1990:28).

In addition to the investment resource reallocation and institutional changes implied by the passage cited above, SARD must give immediate priority to improving the material conditions of small farmers and the rural poor. This involves maintenance of the natural resource systems that provide their sustenance in ways that reinforce positive feedback mechanisms between livelihood, material welfare, and sustainability. Without an equitable stake in SARD, resource-degrading practices imposed by the immediate struggle for survival—deforestation, overgrazing, soil mining—are likely to continue. Because it is imperative to engage and reward the energies of the poor, SARD in the Third World calls for fundamental changes in political and economic participation, beginning with the principle of individual equality and the fundamental rights of citizenship widely recognized in the advanced industrial countries. If it is restricted to "getting prices right" and administrative reform, significant as these are, SARD will become yet another empty slogan of planning ministries and the "development community."

By contrast with the industrial countries of the North, SARD in the Third World involves political issues of societal, rather than mainly sectoral, significance. Given the magnitude of rural poverty, its priorities strike directly against unequal structures of income, wealth, and power. The distribution of life-chances and the social origins of "winners" and "losers" would be transformed. A moment's reflection on the equity and basic needs priorities of SARD suggests that it would inevitably be accompanied by reforms of private landownership, common property systems, tenure relations, and other social arrangements that determine land use and rights of access to resources and livelihoods. SARD here speaks directly to issues of social reproduction and cultural identity: small farmers, pastoralists, squatters, rubbertappers, indigenous peoples. The rural environment is the arena of these struggles, faithfully mirroring social structure and the ways in which its contradictions are embedded in resource access, property rights, surplus extraction, gender relations, ideology, and culture. To anticipate these structural changes is an heroic assumption, indeed.

LOCAL CONTEXT: COMMUNITY AND HOUSEHOLD

I have suggested that SARD strategies need to be carefully differentiated and adapted to local agroecological systems while, at the same time, addressing basic human needs, such as household food security. It is worth emphasizing that in low-income developing countries, where agriculture and related activities absorb 70–80% of the total labor force, resource degradation immediately reduces productive capacity, incomes, and the food security of households and communities. Of equal concern to those at the margin of subsistence, resource degradation is accompanied by the greater instability of production, raising the frequency of reduced harvests and prospects of malnutrition and famine. Nutrient mining, overgrazing, and other resource-degrading practices typically are survival strategems adopted by poor rural households in social systems that deny them access to natural resources and other entitlements necessary for sustainable livelihoods. For this reason, in principle and practice, sustainable resource management and basic needs provision are interdependent, mutually reinforcing dimensions of SARD. In the local context, this interdependence implies that SARD strategies must be highly responsive to the complexities of agricultural and livelihood systems in order to influence behavior at the farm, community, and grassroots levels.

This point can be illustrated by the limitations of the prevailing top-down, transfer-of-technology model of agricultural research and extension in responding to the needs and constraints of poor farmers. This model, diffused by the international agricultural research centers and their national counterparts, has reinforced unsustainable, high external input, single commodity systems by developing crop innovations suited to well-endowed, highly managed biophysical environments. This agroecological and commodity bias is compounded by the conspicuous neglect of local knowledge acquired by farmers experimenting with traditional and new technologies, often in highly sophisticated ways (Richards, 1985). The conventional agricultural research agenda also has given low priority to soil and water conservation, integrated pest management, integrated nutrient conservation, and other practices of sustainable agriculture. Reform of this agenda to eliminate this bias against resource-poor, usually dryland, systems and to move the locus of research on to farmers' fields in an important operational component of SARD. Local community groups, grassroots movements, and nongovernmental organizations (NGOs) are developing innovative ways of bringing farmers, extension agents, and research workers together to share local knowledge and expertise, including farmer-to-farmer extension and farmer innovator workshops.

Attention to local context also would reveal that rural livelihood systems incorporate a multiplicity of sources, ranging from tree crops, wild products, "weeds," and common property resources (CPRs) to kinship sharing arrangements and migrant remittances. for example, research by Jodha (1991) in 82 villages in the dry tropical zone of India indicates that 84–100% of poor households depend on CPRs for food, fuel, and fodder, and derive 14–23% of total income from this source. Similarly, small-scale producers in Africa (Rochelau, 1987), Central America (Wilken, 1977), India (Chambers et al., 1989), and elsewhere in the Third World have for generations incorporated trees in polyculture farming systems, "long before formally-trained scientists popularized the term 'agroforestry'" (Thrupp, 1989:41). SARD strategies must recognize the complexity of rural livelihood systems and strengthen them in sustainable ways.

These measures would include support for collective action to manage local resources sustainably and equitably. Sustainable agricultural practices, such as soil and water conservation, cut across individual holdings, and require community organization, in this case at the catchment level, to be effective. Community mobilization and empowerment are vital to the implementation of SARD, and consistent with its commitment to equity and participatory politics. Such developments would be facilitated by reforms to enhance the local accountability of state agencies and encourage closer working relationships with local grassroots organizations and NGOs.

CONCLUSION

This chapter has explored some operational requirements of a SARD strategy whose central aims are to meet the basic needs of the rural poor while maintaining the productivity and biological diversity of agroecosystems. The enabling conditions for such a strategy are many and complex, beginning with effective international action to sustain the global biosphere. Such action necessarily would address the interdependence between global ecology and global economy in order to establish a framework of international relations consistent with the practice of sustainable development. As the Brundtland Report observes, "Ecology and economy are becoming ever more inter-woven—locally, regionally, nationally, and globally—into a seamless web of causes and effects" (WCED, 1987:5). Yet the Northern discourse that dominated the treaty negotiations on global warming and biodiversity pre-ceding the 1992 UNCED meetings continues to ignore this interdependence, focusing conveniently and myopically on ecology to the virtual exclusion of economics.

Nevertheless, such interdependence does not hold efforts to promote SARD in thrall. The transition to SARD is a process of social change fought out simultaneously in many arenas, global and local. In the Third World, this struggle is a question of survival and sociocultural reproduction for millions of rural poor; in the First World, one of habitat conservation, clean water, and food free of chemical residues. Yet it is this multidimensional, cross-cultural, and global-local process of social struggle that forms the link between, say, rubbertappers' extractive reserves in Brazil, farmers' markets in the United States, and NGO campaigns for a global climate treaty. Linking these different scales into effective systems of global-local environmental resource management demands new forms of empowerment and governance. As one Third World environmentalist argues, "the existing nation-states must give up some of their sovereignty to the village republic in the first place, and the global republic in the second" (Anil Agarwal in Pearce, 1992:41). Finally, an instrumental, *ex ante* operational approach provides a useful spec-ulative exercise but it is no substitute for the praxis of SARD. Here, our experience is just beginning.

ACKNOWLEDGMENTS

I would like to acknowledge my debt to collaborative work in 1990 with David Baldock, Edward Barbier, and Julian Pretty under the auspices of the FAO/Netherlands International Conference on Agriculture and the Environment. The usual disclaimers apply.

NOTES

1. For a revealing critique of neoclassical environmental economics, see Redclift (1991).
2. In advanced industrial countries, on-farm workers, including farm families, typically repre-sent below 5% of the total labor force, and in some cases considerably less.

REFERENCES

Altieri, M. 1990. How Common Is Our Future. *Conservation Biology* 4(1):102–103, March.

Barbier E. B., and G. R. Conway. 1990. *After the Green Revolution: Sustainable Agriculture for Development.* London: Earthscan.

Chambers, R. 1986. Sustainable Livelihoods. Brighton, Sussex: Institute of Development Studies, mimeo.

Chambers, R., N. C. Saxena, and T. Shah. 1989. *To the Hands of the Poor: Water and Trees.* London: Intermediate Technology Publications.

Eckholm, E. P. 1982. *Down to Earth.* London: Pluto.

Furley, P.A. 1990. The Nature and Sustainability of Brazilian Amazon Soils. In *The Future of Amazonia: Destruction or Sustainable Development.* D. Goodman and A. Hall (eds.). London: Macmillan, 309–359.

Goodman, D. E., and M. R. Redclift (eds.) 1989. *The International Farm Crisis.* London: Macmillan.

Goodman, D. E., and M. R. Redclift. 1991. *Refashioning Nature: Food, Ecology and Culture.* London: Routledge.

Intergovernmental Panel on Climate Change. 1990a. *Climate Change: The IPCC Scientific Assessment.* Report prepared for IPCC by Working Group I. J. T. Houghton, G. J. Jenkins, and J. J. Ephraums (eds.). Cambridge: Cambridge University Press.

Intergovernmental Panel on Climate Change. 1990b. *The Potential Impacts of Climate Change: Impacts on Agriculture and Forestry.* Geneva and Nairobi: World Meteorological Organization and the United Nations Environment Programme.

IFPRI: International Food Policy Research Institute. 1990. *IFPRI Report 1990,* Washington, D.C.

Jodha, N.S. 1991. *Rural Common Property Resources: A Growing Crisis Gatekeeper Series, No. 24.* London: International Institute for Environment and Development.

Lélé, S. M. 1991. Sustainable Development: A Critical Review. *World Development* 19(6):607–621.

Lowe, P. 1988. Environmental Politics and Agriculture in Western Europe. *Agriculture et Environmement.* Arlon, Belgium: Fondation Universitaire Luxembourgeoise.

Lowe, P., G. Cox, M. MacEwan, T. O'Riordan, and M. Winter. 1986. *Countryside Conflicts.* Aldershop: Gower.

Mabbutt, J. A. 1984. A New Global Assessment of the Status and Trends of Desertification." *Environmental Conservation* 11.

Parry, M. 1990. *Climate Change and World Agriculture.* London: Earthscan.

Pearce, D. W., A. Markandya, and E. Barbier. 1989. *Blueprint for a Green Economy.* London: Earthscan.

Pearce, F. 1992. No Southern Comfort at Rio? *New Scientist* 1821, 16 May, 38–41.

Potter, D. 1989. Approaching Limits: Farming Contraction and Environmental Conservation in the UK. In *The International Farm Crisis*. D. E. Goodman and M. R. Redclift (eds.). London: Macmillan.

Redclift, M. R. 1991. Environmental Economics: Policy Consensus and Political Empowerment. In *Sustainable Environmental Economics and Management: Principles and Practices*. R. Kerry Turner (ed.). London: Belhaven Press.

Richards, P. 1985. *Indigenous Agricultural Revolution: Ecology and Food Production in West Africa*. London: Hutchinson.

Rochelau, D. 1987. Women, Trees and Tenure: Implications for Agro-Forestry Research. In *Trees and Tenure. Proceedings of an International Workshop on Tenure Issues in Agroforestry*. J. R. Raintree (ed.). Nairobi: ICRAF.

Sen, A. K. 1981. *Poverty and Famines: An Essay on Entitlement and Deprivation*. Oxford: Clarendon Press.

Sen, A. K. 1987. *Hunger and Entitlements*. Helsinki: WIDER/United Nations University.

Thrupp, L. A. 1989. Legitimizing Local Knowledge: From Displacement to Empowerment for Third World People. *Agriculture and Human Values* 6(3):13–24.

Wilken, G. C. 1977. Integrating Forest and Small Scale Farm Systems in Middle America. *Agroecosystems* 3:291–302.

Woodwell, G. M. 1990. The Effects of Global Warming. In *Global Warming: The Greenpeace Report*. J. Leggett (ed.). London: Oxford University Press.

World Bank. 1990. *World Development Report*. New York: Oxford University Press.

WCED—World Commission on Environment and Development. 1987. *Our Common Future*. Oxford: Oxford University Press.

CHAPTER 11

Sustainable Agriculture and Domestic Hunger: Rethinking a Link between Production and Consumption

Katherine L. Clancy

W e are the best-fed nation in the history of the world, though I would add the caveat 'not that there isn't hunger'. However, it has never been the fault of the farmer.

Kika de la Garza, Chair of the House Agriculture Committee. May 1991 (CNI, 1991c)

In a sense de la Garza is correct. In a market economy people's lack of food is not the fault of food producers, but is the result of poor peoples' "inability to establish entitlement to enough food" (Sen, 1982:8). Clearly farmers cannot be blamed for the low incomes of a fifth of the U.S. population that is the proximate cause of hunger. Still, for many reasons, farmers and the agriculture establishment should not expect to be let off easily from any real or perceived connection to the problem of U.S. hunger. Despite the insistence of many agriculturalists that farming is an industry producing inputs for other industries, the public perception is that farmers produce food. Support for farm programs has been forthcoming for decades both because of farmers'

Food for the Future: Conditions and Contradictions of Sustainability, edited by Patricia Allen.
ISBN: 0-471-58082-1 © 1993 John Wiley & Sons, Inc.

promises to feed the world, and of their willingness to trade off support for farm programs with support for domestic food programs. The present disinterest of the most powerful agricultural organizations in the problems of hunger and poverty can at least be seen as a contradiction to this promise. Their inattention also stems from a myriad of complex and controversial ideologies, political decisions, perceptions and misperceptions, and sometimes totally incompatible goals within two of the major spheres of U.S. policy making—agriculture and welfare.

The purpose of this chapter is not to resolve any of the many complex issues, but to describe the different problems and needs within the sectors, to suggest that those working on the improvement of the food system look more carefully at some of the contradictions in their aims, and to search for places where some reconciliation of agendas can occur. The first section examines the linkages between agriculture and hunger and why I think these connections should be strengthened. The second section sets the context for the remaining analysis describing the extent of poverty and hunger in the United States. The third section is a brief history of the interaction and interconnections of hunger and welfare policy with agriculture and environmental policy in this century. This will be followed by a more in-depth discussion of the reasons why it may be difficult to bring poverty and hunger policy into the purview of farm and environmental organizations. The final sections look at the reasons why these groups should, at this time, shift some of their attention to consumption issues, including hunger, for both self-serving and humanitarian reasons, and how dialogue might begin on addressing the problems.

WHY THERE SHOULD BE STRONGER LINKAGES BETWEEN AGRICULTURE, FOOD, AND HUNGER

Some would argue that because farm prices are only one of many factors influencing food prices (and a dwindling one at that) that agriculture's role in improving access to adequate food is a trivial one. The myriad of factors that provide the explanation of why I think this is erroneous are discussed in the text. Among them are (1) the symbolic and political factors involved are quite consequential, (2) the possibility that the enactment of environmental provisions in farm policy may affect the poor adversely needs consideration, and (3) the growing efforts to increase the access of local farmers to low-income areas are a sign of recognized needs on both sides.

One of the several links between hunger and agriculture is the rhetoric found in the preambles of farm bills, which promises abundant and affordable food in exchange for public tax support for farm programs. There are others. Since the Depression of the 1930s the existence of surplus foods simultaneously with hunger among the poor has been seen as immoral, and the symbol of irrationality in the economic system (Poppendieck, 1986). No change has occurred over the intervening decades, and the farm and food assistance programs still fall under the jurisdiction of the same federal agency and Congressional committees (except child nutrition programs that are under the jurisdiction of the Education and Labor committee in the House). This has had two often interrelated effects: (1) a consciousness of the paradox of "want in the midst of plenty" sometimes informs deliberations about authorizations and appropriations; (2) trade-offs in funding are made between programs serving the farmers and the poor.

The connections between farm and nutrition programs used to be stronger than they are now, and some might argue that the weakening of the link is welcomed. I want to suggest that the opposite is true; that precisely because support for various components of agriculture policy is waning in the United States, recapturing that support will require the farm lobby to look carefully at a number of food-related issues. One of these is hunger, which has risen again in the present recession, is high in both rural and urban communities, and is contributing to a general uneasiness throughout the country about the entire domestic agenda.

Historically those farm and commodity groups that have been most successful in controlling the agriculture agenda have not been supportive of welfare and food stamp programs, and have only tolerated the log-rolling in the Congressional agriculture committees between urban-supported food stamp legislation and rural farm price support programs. While direct trade-offs between food stamps and price supports occur infrequently now (Feingold, 1988), new arguments have been formulated in Congress against an agricultural support program because of its impact on the costs of nutrition programs (CNI, 1991f). And worsening rural poverty should be a concern to rural residents of all incomes because of the effects of the entire rural economy.

For these reasons, and given the present prevalence of hunger in the United States, the many actual and proposed changes in agricultural policy that are likely to raise the cost of food, and questions about equity in the food system, I believe that it is valid and relevant to explore the question of whether it is politically savvy for farmers, farm organizations, and scholars, especially those involved in sustainable agriculture, to continue to ignore U.S. hunger and the policies and programs that could help alleviate it.

The chapter addresses two phenomena. One is the lack of attention paid by American agriculture, in general, to poverty and hunger issues. The second is the lack of specific attention to this issue by those espousing concerns about the long-term sustainability of the agriculture enterprise. These include both those whose interests are most aligned with "saving the family farm" as well as those most concerned about "saving the environment." The discussion of these phenomena takes place in the context of an "exploding universe of interests" on the agriculture agenda, which includes what Don Hadwiger first described as the externalities/alternatives (ex/al) sector, i.e., those who represent the interests and values of consumers, the poor, environmentalists, and other groups (Browne, 1988a). Browne concludes that the presence of such a wide array of different interests is responsible for several problems including the collision of agendas and the impossibility of engendering a comprehensive U.S. agriculture policy. This chapter will illustrate both of these problems, the contradictions in positions held by different interest groups, and the propensity of these groups to define their demands narrowly and pursue these demands with the specific subcommittees in Congress and agencies where they are likely to get a more favorable reception (Browne, 1988b).

As a departure from this path, the chapter urges sustainable agriculture supporters to stop disregarding the poverty and hunger issue, to recognize the conflicts between farm welfare and the public welfare in present and proposed farm policies, and to be prepared to support the proposals of the antihunger lobby that could lower the incidence of hunger. Advocates for alternative farming, recognizing the problems outlined by Browne, complain that it is unfair to ask them to take on issues that the traditional agriculture establishment has not addressed, and feel that it is both unfair and risky, given the uncertainty of their successes in the policy arena, to add social justice to the agenda (Gershuny and Forster, 1992; CNI, 1991a). But I want to argue again (Clancy, 1984, 1991) that exactly because conventional agricultural policy has neglected the linkages through the food system, up to and including hunger, it is critical that sustainable agriculture join them together. Thompson (1988:43) states that "the philosophical aim of farm policy is to derive a unified and sensitive portrayal of agriculture's contribution to national interests and to the common good." Widespread hunger can hardly be considered in the national interest, and the sustainable agriculture community has a role to play in eliminating it. "Breadlines knee-deep in wheat" are no more rational in the 1990s than they were in the 1930s when they were criticized as "obviously the handiwork of foolish men" (Poppendieck, 1986:xi). People proposing adopting of a saner, more thoughtful, environmentally beneficent production system have much to gain from bringing this sensitivity to the hunger issue as well.

POVERTY AND HUNGER IN THE UNITED STATES

Although agricultural economists often ignore such issues, Wharton, in his 1990 Fellows address at the American Agricultural Economics Association meeting, said it looked like the United States had spent the last four decades gradually "importing" poverty from the Third World, and that some of this poverty was looking "intractable" (Wharton, 1990:1131). About 33.5 million people or 13.5% of the U.S. population is officially poor (U.S. Department of Commerce, 1992), defined as an annual income of less than $13,400 for a family of four (U.S. Department of Health and Human Services, 1991). Another 5% are near poor with incomes between 100 and 125% of poverty. About 40% of the poverty households in the United States ("the poorest of the poor") have incomes that are under 50% of the poverty level (U.S. Department of Commerce, 1992), a percentage double what it was in the 1960s and 1970s, and one of the most obvious signs of the widening disparity between the rich and the poor (Porter, 1992).

The United States has a much higher poverty rate than the other developed countries to which it is often compared—Canada, Australia, France, and similar European countries. One of the indicators of this is the infant mortality rate on the scale of which the United States ranks twentieth among developed countries (Children's Defense Fund, 1991). Smeeding (1992) has recently calculated that in the years between 1983 and 1990 many of these countries experienced the same amount of market-induced poverty, but that more generous social policies kept the numbers of those defined as poor much lower than in the United States. The incidence of hunger in these countries is also low because welfare benefits provide adequate allowances for food (U.S. GAO, 1988; Kramer and Elliott, 1989).

Children are the age group with the highest poverty rate in the United States. Between 1979 and 1989 child poverty increased 21%; today 23% of children under the age of six are poor (Children's Defense Fund, 1991). Besides children, a disproportionate percentage of the African-American and Hispanic population are poor (33 and 27%, respectively), although by far the largest numbers of poor people are white (Beegle, 1991). The poverty rate in rural areas, at 16.9%, is higher than the average rate (Beegle, 1991). But rural poverty is not farm poverty. Only 8.5% of the rural poor were farm residents in 1986 (Parliament, 1990).

Poverty is increasing in the United States due to falling wages [about 20% of full-time workers are paid wages too low to raise a family of four to the poverty level (Center on Budget and Policy Priorities, 1992)], an increase in the number of female-headed households, the weakening of programs such as Aid to Families with Dependent Children, the benefits from which have fallen 40% in real dollars between 1970 and 1990, and growing income disparities between the rich and the poor (Greenstein, 1991). Still, much poverty is not persistent—many families move out of poverty within a 5-year period (Beegle, 1991).

The poverty level was originally calculated from the percentage of income spent on food by families participating in the USDA food consumption survey of 1955. The figure at the time was 30% and the poverty level still is equal to three times the value of the least costly of the food plans developed by USDA in the 1940s. By definition, therefore, incomes below the poverty line place households at risk of hunger. The USDA estimated cost of the lowest cost food plan (the Thrifty Food Plan) for a family of four was $358 in March 1992 (USDA, 1992), which represents 30% of the poverty-level income. Low-income households spend this much and more. An estimate from 1989 data of the amount spent on food by households in the lowest income quintile is 42% of their after-tax income (Manchester, 1991). Other studies show that very-low-income families are spending 50% or more (Senauer et al., 1991). This contrasts with the U.S. average of 11.8% (Dunham, 1991).

The higher percentages are similar to numbers from Third World countries (Korb and Cochrane, 1989). Unlike Third World countries, however, which can provide little food assistance to their famished populations, hunger in the United States is not manifested as obvious starvation. The clinical definition of hunger as malnutrition and serious health problems is the end result of a long-term shortage of food. To identify risk factors before starvation and hunger occur, and for purposes of policy making, an operational definition is now used defining hunger as the "involuntary going without food due to a lack of resources" (Randolph, 1989). Because the government has refused to survey its incidence, apparently because of an official belief that there is no hunger problem (or perhaps a concern that it might be proved wrong), in the last 10 years hunger has been measured in two alternative ways. One is through private studies that suggest that millions of children in low-income households are hungry and millions more are at risk of hunger (FRAC, 1991b). The other measure is the still growing requests for food in the Emergency Food System (Murock, 1991). Researchers completing a series of studies across the country of families with children under the age of 12 recently projected that one out of 12 children (5.5 million) is hungry, based on a scaled set of responses to questions regarding the availability of food in the house over a year's time (FRAC, 1991b). Another 30% of households were at risk of hunger. The children identified as hungry experienced three times more illness, and twice the amount of school absenteeism of those not so identified. Their families were significantly poorer, more likely to be unemployed, and had average per capita food expenditures that were significantly lower compared to nonhungry families (Samets, 1991).

The other sign of hunger is the growth in emergency food assistance. Private food assistance has grown from a few soup kitchens and food pantries before 1980 to a national network of more than 180 food banks, 23,000 food pantries, and 3,300 soup kitchens (Murock, 1991). In 1991, food banks distributed 516 million pounds of food (Second Harvest, 1992), but over 60% of the food assistance sites have had to turn people away because of lack of food (Murock, 1991). Due to high demand the majority of food pantries can provide people with only enough food for 3 days time, and only every 3 to 6 months. Until recently much of the food given out was surplus commodities. However, surplus holdings in 1992 are almost depleted (Becker, 1992).

Hunger persists, despite the availability of emergency food. The reasons are numerous. The most obvious are the lack of any serious welfare reform or job programs in the last 15 years. Given the unlikelihood of change in these and related areas, the burden for food assistance falls mainly on the food stamp program, which is both inadequate to meet the needs and under-utilized. In March 1992 a record number of people were beneficiaries of the Food Stamp program: 10% of the population or 25.7 million people (*Dallas Morning News*, 1992). But for many reasons food stamps do not meet people's needs. About 50% of the persons coming to food pantries are already food stamp recipients (Kelly et al., 1989). People using them are at risk of hunger because many other expenditures take precedence [e.g., housing, for which 45% of low-income renters pay more than 70% of their incomes (Leonard et al., 1989)]. Another problem is the inadequacies in the Thrifty Food Plan (TFP), which is the basis for the food stamp allotment (U.S. House of Representatives, 1985). Also, the nationwide benefit level does not reflect the significant variations in food costs by region or location (Crockett et al., 1992; Morris et al., 1990). Finally, a large percentage of households eligible for Food Stamps do not receive them for a number of reasons, including lack of information, access problems, and personal feelings about food stamp receipt (U.S. GAO, 1990a).

The goal of the numerous antihunger advocacy groups such as the Food Research and Action Center, Bread for the World, the Center on Budget and Policy Priorities, and the Children's Defense Fund is an end to what has been defined as food insecurity, or "any lack of access by people at any time through normal channels to enough nutritionally adequate food for an active healthy life" (U.S. House of Representatives Select Committee on Hunger, 1990). Food security exists when families receive adequate cash or food stamps to purchase food in retail markets, and end their reliance on surplus foods or emergency food systems. Operationalizing the goal can occur through avenues that provide adequate incomes: jobs that pay at above the poverty level, sufficient numbers of low-income housing units, and welfare reform to increase benefits. Barring any improvements in these areas the task can be accomplished through food stamp reform that would remove the cap on shelter deductions and at a minimum provide benefits that would (1) recognize the inability of poor households to spend 30% of their incomes on food and still meet other expenses, (2) account for regional differences in food costs, (3) eliminate harassment of recipients, and (4) increase the numbers of eligible households. Either cash or food stamps would then eliminate the need for poor households to rely on inadequate charity donations of food from food pantries, soup kitchens, and other sources.

HISTORY OF THE INTERFACE OF AGRICULTURE AND FOOD POLICIES IN THE UNITED STATES

The Great Depression spawned major welfare programs, including farm price supports, food stamps, and Social Security, and minor programs such as commodity distribution. This was the decade when the interactions of farm problems and hunger were most obvious to the citizenry, and was the beginning of a cycle seen several times in the next 50 years that affected farmers' livelihoods, the environment, and food sources for the poor. The cycle was defined by prices below the cost of production to farmers and other policies that led to food surpluses. These in turn brought calls for giving surplus food to the poor and proposals for taking land out of production.

The Agricultural Adjustment Acts (AAA) in the 1930s appeared to be a perfect compromise, simultaneously helping farmers and promoting cheap food for cities, balancing rural and urban interests early on (Constance et al., 1990). However, a careful review of the history of the AAA and the antihunger programs started at the same time shows that only some rural and urban interests were being served. In her thoughtful treatment of this history, Poppendieck carefully traces the origins and the development of AAA and the Federal Surplus Relief Corporation (FSRC), the first ongoing federal food assistance program. The essence of this history is that the program of distribution of surplus commodities to poor people became within 2 years "a convenient outlet for products acquired in efforts to increase the income of commercial farmers" (Poppendieck, 1986:xv).

This transition occurred through the interaction of a number of factors. Farmers, whose average incomes were well below that of the rest of the populace, had endured a depression in 1921 and then again starting in 1929. They felt betrayed by the government and believed that consumers had benefited unfairly, if inadvertently, from the economy's discrimination against agriculture. Soon after the election of Franklin Delano Roosevelt, who believed that the resurgence of agriculture was critical to the recovery of the general economy (Feingold, 1988), an omnibus farm measure titled the Agriculture Adjustment Act (AAA) was passed. It authorized a domestic allotment, parity prices, marketing agreements, and a processing tax (declared illegal not long after). The entire Act would clearly have a regressive effect on consumers, but little attention was paid to this as it was debated in Congress. In 1933, when millions of people were hungry and farmers slaughtered millions of little pigs because there was no market, the citizenry was outraged. "The juxtaposition of hunger and abundance became a central symbol of the irrationality of the economic system" (Poppendieck, 1986:xi) and pressure was strong to find a solution. In haste the New Dealers set up the FSRC as a nonprofit corporation to buy food supplies for relief.

The FSRC was run by people from the Federal Emergency Relief Corporation, the Public Works Administration, and the Department of Agriculture. The latter administrators understood quickly how the Corporation could be a flexible tool for protecting the interests of farmers. At the beginning of the food relief program a "balance was struck between the needs of farmers to survive, the needs of the urban to eat, and the needs of the government to provide food to the cities to avoid unrest" (Constance et al., 1990:29). But soon the complaints of the Farm Bureau and the National Association of Manufacturers against direct relief and its purported interference with commercial sales became very strong. Passage of the Social Security Act in 1935, which authorized states to provide assistance, provided the way out. The federal government could stop direct relief and turn over the FSRC to the USDA. The corporation was renamed the Federal Surplus Commodity Corporation, and a process was begun by which food assistance was increasingly divorced from relief and integrated into the price support program. Furthermore, money from tariff receipts that had been earmarked under legislation passed in 1935 (Section 32) for the purchase of commodities for relief was defined by USDA as "farmers" money and not spent to feed the hungry. It is not surprising, then, that after the transfer to USDA "fewer people got food and they got less of it." Poppendieck calculates that in 1936 individuals received about five pounds of food per month. This would have been about 5% of what they needed.

Despite the ability of USDA's commodity specialists to ignore the problem, even as a committee of nutritionists and economists from other agencies within the department argued strongly for enhanced food assistance, hunger did not go away and a plan for a Food Stamp program was devised to help subsidize food purchases by the poor. It was adopted in 1939 and discontinued in 1943, during the Second World War. After the war, new pesticide and fertilizer technologies were adopted, farmers produced larger crops, a sizable surplus developed, and changes in federal farm policies lowered prices paid to farmers. Commodity food distribution to the poor continued but the Southern Democrat dominated agriculture committees refused to consider food stamp proposals (Andrews and Clancy, 1985).

Continued heavy food surpluses in the 1950s led to the development of the Soil Bank to idle land (Terry, 1990). In 1962 the Committee of Economic Development (composed of the presidents of a number of large corporations and economists from the University of Chicago) and officials from USDA took up an idea that had been around for awhile (Busch and Lacy, 1983) and argued that farmers were not moving out of agriculture fast enough and that price supports were the major problem (Ritchie, 1979). The Committee urged adoption of a two-tier farm system. The bulk of support payments would go from the farm programs to the 25% of large farms that produced 80% of the commodity sales. The remaining farmers would take off-farm jobs or receive a minimum annual income as was then being proposed in Congress for the entire low-income population. The workings of the price support legislation and a host of new oil-based technologies enabled the first part of the scenario to succeed, and millions of farmers left the land. A minimum annual income was not enacted, however, and the continued migration of farmers who were not trained for employment outside of farming [a situation described as "the central failure of U.S. agriculture institutions and leaders" (Thurow, 1991:1–56)], put a heavy burden on welfare programs in the cities, which was to be a major contributing factor to the urban riots later in the decade (Piven and Cloward, 1971; Poppendieck, 1986).

Because of USDA policies that prevented commodity surplus disposal from rising above a low level, and the inability of urban representatives to pass food stamp legislation, many poor people went without food over the decades of the 1940s and 1950s, but their plight was not visible. A campaign promise by J.F. Kennedy led to the reestablishment of a modest Food Stamp Program (FSP) in 1961 and an improved commodity surplus effort called the Needy Family Program (Lipsky and Thibodeau, 1985b). Food stamps were not expanded significantly until later in the 1960s with the "discovery" of hunger by Robert Kennedy and Joseph Clark in Mississippi (Kotz, 1971). Over those years logrolling of food stamps with farm policy was the order of the day (Andrews and Clancy, 1985), and the numbers receiving food stamps gradually expanded. In 1973 the oil embargo, the devaluation of the dollar, the purchase of large amounts of U.S. grain by the Soviets, and crop shortfalls in various parts of the world led to increased farm and food prices, and the elimination of surpluses of most domestic commodities (Becker, 1988). Congress authorized the nationwide implementation of food stamps into all counties that had refused to offer them, and participation increased enormously due to this action, high unemployment, and rapid food inflation (Andrews and Clancy, 1985; Buttel, 1980). In the same year many farm conservation projects were stopped to plant as much land as possible to crops for export.

All-out production for the expanded export market did not bring price stability to farmers and in 1973 also a target price system was adopted to provide some income protection through deficiency payments and to keep food prices from rising (Constance et al., 1990). However, while the high commodity prices of the mid-1970s led to increased net farm incomes, inflation prompted consumers to lobby for decreased support for agriculture (Constance et al., 1990). This period saw the entry of many new participants onto the agricultural agenda (Browne, 1988a).

All problems intensified in the 1980s. The new Republican administration was intent on cutting the budget and moving to more "free-market" agricultural programs. Due to lower commodity prices, farm debt was the highest in history, farm prices fell, and many bankruptcies occurred (Constance et al., 1990). Special asset rules were adopted to allow farmers to receive food stamps (Iowa Department of Human Services, 1985). Simultaneously, an increase in environmental problems was documented, due to increased sensitivity to the environment as an issue, the effects of intense production on too-fragile land, and the lobbying of a number of conservation groups (Browne, 1988a). Also in the 1980s unemployment rose, budget cuts were made in food assistance programs, and homelessness became a visible problem in both urban and rural areas (Cohen and Burt, 1989; Wilkerson, 1989). The Emergency Feeding System came into existence to provide food outside of normal commercial channels (grocery stores and supermarkets) to feed the hungry. Pressure was again put on USDA to utilize surplus food, which was being stored in record breaking amounts, and in 1983 a new commodity distribution effort called the Temporary Emergency Food Assistance Program was authorized (Becker, 1992).

The 1985 Farm Bill contained a set of new conservation titles due to the need to find solutions to the problems of surpluses and strong lobbying by environmental groups. Unfortunately, the Food Stamp Program did not fare well, even though in 1985 the demand for food from food pantries and soup kitchens was higher than at any time since the Depression (Lelyveld, 1985). The 1990 Farm Bill contained changed environmental policies in flexibility provisions, wetland conservation, integrated farm management, and water quality (Hassebrook, n.d.) but food stamp benefit expansion was defeated (FRAC, 1991a) in the face of the federal deficit and legislation that prohibited the movement of dollars from the military budget to domestic programs.

WHY AGRICULTURE HAS IGNORED HUNGER AND FAILED THE POOR

There are a number of theories and explanations for why many farmers, farm lobbies, and domestic agricultural researchers have been missing from the national dialogue on poverty and hunger. In this section those reasons have been divided into the oft-used categories of ideological, political, and economic and will be discussed in that order. The section will conclude with a discussion of the inattention paid to these topics by agricultural scientists.

Ideological

Many of the social scientists writing about farm policy in the last decades have looked at various agrarian myths as one of the factors holding back reform of U.S. farm policy (Browne et al., 1992; DeLind, 1987; Buttel, 1980). A hard working, independent, and virtuous farmer, tied to the land and acting for the common good, has been the model of the optimal citizen since the founding of the Republic (Browne et al., 1992). Farmers were perceived as being free of the moral depravity of urban life, and rural areas considered superior to cities which bred economic concentration and corruption (Bonnen and Browne, 1989). Farmers values were better and should be cherished and emulated. But as so well described in the following passage, the myth has stood in the way of reality:

> when agricultural myths prevail, it is easy to perceive farmers as an undifferen-tiated collection of remarkably independent and unique individuals. Simul-taneously, it is difficult to recognize small commercial producers as an easily manipulated source of cheap labor and cheap farm commodities for agribusi-ness. It is even more difficult to recognize that other populations share similar characteristics arising from their dependency on those who control capital and power (e.g., labor, small business, minorities, the poor). (De Lind, 1987:7)

The historical roots and massive structural and economic changes in agri-culture over time have fostered and moved to the forefront, especially at the national levels, farmers' innate conservatism as defined by the agrarian model. But this has not been salutary to the general response of agriculture to pov-erty. The little empirical evidence available at present shows that support by farm groups and agribusinesses for food assistance for low-income families is very weak. In a 1985 study of national leaders' views of agricultural and food policy, seven major farm organizations whose political ideology ranged from conservative to progressive were surveyed. Six of these groups (84%) were not in favor of expanding benefits for people receiving food stamps or expanding the population eligible for food stamps. Opposition by agribusi-ness groups and commodity promotion groups was even stronger (Spitze, 1985). When farmers themselves were asked about the programs, in 13 states the most frequent response was that the amount spent on food stamps should be decreased. In three states the choice was to keep expenditures about the same. In four states the second most popular response was to eliminate food stamps completely (Guither et al., 1984). Some farmers may simply be responding as their working-class counterparts in other industries would. But this is less clear when compared to a recent survey that reported that those most willing to earmark $100 extra in taxes to end hunger were those with incomes between $15,000 and 25,000—the lower middle class, blue-collar range (Breglio, 1992).

In the 1980s more attention may have been paid by farmers to poverty issues than in the past, especially in the Midwest. Yet despite work like that done in Iowa in the mid-1980s to sensitize farmers to welfare needs and programs, most farmers' views of the poor seem to reflect a basic conservative ideology with overtones of Calvinism that blames hunger on laziness, or "a defect in personality or behavior" (Katz, 1989:236), and not economic and class structures.

Some farmers have supported leftist movements similar to Populism, and a significant percentage of small- and medium-sized farmers presently belong to the National Farmers Union. Researchers have found that farmers with smaller holdings and incomes hold more positive views on issues like job creation and progressive tax structures (Buttel et al., 1980). Farmers with larger farms and incomes, however, tend to be more conservative, members of the Farm Bureau, and/or aligned with the positions of commodity groups. The Farm Bureau position in the 1930s was strongly antirelief, particularly the federal government's role in exerting upward pressure on farm wages (Poppendieck, 1986).

The conservatism in agriculture is also illustrated by the injustice built into the agricultural system itself in farmworker relations. Those agroindustrial sectors that rely on farmworkers, especially migrants, have accepted social injustice and hunger as part of their day-to-day world. Circumstances of inadequate wages, housing, food, water, health care, education, and other necessities are widely observed in migrant areas throughout the country, especially the south (Delfico, 1991; U.S. GAO, 1990b). MacCannell's studies in California also show that where industrial-type farm structures are dominant, there will be depressed standards of living in the surrounding communities (MacCannell, 1988).

Another clear example of the conservative bias in agriculture is the attitude toward and the lack of attention to rural and farmer minorities. It has been shown that rural minorities are particularly disadvantaged compared to rural white poor due to institutional racism, human capital deficits in education, employment, etc., and geographic concentration (Allen and Thompson, 1990). Also, between 1954 and 1987 the total number of non-Caucasian-owed farms declined at a rate double that of Caucasian-owned farms (Christy, 1991). Thus the structural changes in agriculture have had a disproportionately negative impact on African-Americans "because of their inaccessibility to capital markets, educational opportunities, and technical assistance from public and private sources" (Christy, 1991:III–103). And if a finding from the only study directed specifically at racial discrimination in agriculture still obtains, the consequences of discrimination worsen as economic growth proceeds (Tang, 1959 in Bryant et al., 1981). Although no research was found on the topic, one might infer that the racism in agriculture (also seen in the lack of support by general farm groups for new proposals for support for minority farmers) extends to the urban poor.

Sexism may also play a role in farmer's negative bias toward the poor. More than 80% of the recipients of Aid to Families with Dependent Children (AFDC) and food stamps are women and their children; 80% of the elderly who receive food stamp benefits are women (U.S. House of Representatives Select Committee on Hunger, 1990). Sexism in agriculture has been observed for decades, but described empirically only in the last 15 years (see Friedland, 1991b and Haney, 1991 for recent reviews of this literature). The conclusions reached by scholars in this field are that the significant contribution of women to agriculture has been minimized to the point of being invisible, as women's labor has been treated as irrelevant and their subsistence and domestic activities devalued (Sachs, 1991; Friedland, 1991b). Women have had limited access to agricultural land and capital, and have found it very difficult to function autonomously in farming. This phenomenon has a parallel in national welfare policy where women and their children are the principal clients of that part of the welfare system that is oriented to households and designed to compensate for family failures, as opposed to the welfare system oriented to individuals and tied to participation in the work force with programs such as Social Security and unemployment insurance (Fraser, 1989). Note that price support programs would fall in this category. Beneficiaries of the latter programs, predominantly men or wives receiving their husband's benefits, are considered to be receiving their due, in the form of cash payments, and with minimal personal harassment. Beneficiaries of AFDC and Food Stamps are considered charity cases, receive in-kind or tightly designated payments, and are constantly harassed (Fraser, 1989).

Racism, sexism, and class differences probably account for most of the ideological explanation of why many farmers and farm organizations have either ignored or taken punitive positions with regard to poverty and hunger. Although there are progressive farm organizations that have recognized the problem and made useful responses to it, altering these prejudices presents a real challenge. One reason is the strong propaganda directed to their members by the Farm Bureau and the commodity organizations that reinforces the stereotype of the model citizen farmer and the reprobate welfare mother. The "welfare bashing" occurring in the media and states in the early 1990s further abets conservative attitudes and positions. The second reason is that the traditional predictability of support for more liberal positions from small farmers mentioned earlier may not carry over to small part-time farmers. The results of a survey of full and part-time New York farmers found only minor differences in political beliefs between them, leading Buttel to suggest that part-time farmers, because of their changed and more privileged income position (the result of holding good off-farm jobs), could become "anchors of conservative politics in rural communities" (Buttel et al., 1982:10). As stated above, "family farmers are petty bourgeois in interest and ideology and may not be sympathetic to a more progressive agenda" (Buttel, 1981: 2). As Strange (1988:4) points out, "no tradition is more glorious in its acclamation of egalitarian values than the agrarian tradition, yet none tolerates and even admires the accumulation of wealth more."

There may be reason for thinking that despite these factors, farmers complaints about big business and free enterprise "can be shaped in progressive directions" (Buttel, 1980:477). But to build a bridge between farmers and the poor there has to be some modicum of recognition that "poverty is not an unfortunate accident, it has always been a necessary result of America's distinctive political economy" (Katz, 1989:237).

Political

Echoing the preceding statements is Heffernan's (1989:156) assertion that "it is possible to provide a viable explanation of the rural crisis from the point of view of power not efficiency." This standard for agricultural policy-making [economic efficiency rather than equity (Parliament, 1990)] has been in place for decades (Bonnen, 1984), and while it has been questioned at times, those making the queries have apparently had little success in changing policy maker's minds about its usefulness and validity. The one place where there has been some softening is on the issue of environmental damage done by farming as a result of not calculating those costs in the measure of efficiency.

It is interesting to note that when questioned 53% of a national sample of the public disagreed that "obtaining greater efficiency in food production is more important than preserving the family farm" (Variyam et al., 1990). But despite public sentiment, the political barriers to incorporating equity standards into farm and food policy remain high. There are several key reasons for the present situation: (1) the general conservatism in government over the past decade, which has supported continuing structural change in agriculture toward a more concentrated elite form, with little attention paid to those leaving farming, (2) an apparent trade-off in agriculture policy making between environmental concerns and social justice issues, and (3) what some scholars have called the "farmers' non-instinct for preservation" (Constance et al., 1990) evidenced by the lack of attentiveness of most farm groups to the role consumer groups must play in support of agricultural policy.

Structural change in agriculture has concentrated resources in the sector. The decline in farm numbers, due to constant increases in productivity from new technologies and cheap fuel over the past 40 years, leaves the United States with 2.2 million farms and 2% of the population living on farms (Senauer et al., 1991). From half to all of the income earned by the smallest 71% of farms is earned off the farm, but this income, along with the prodigious incomes of the largest farms, accounts for the fact that the average income of farm families is now one-fifth higher than the average income of non-farm households (Kalbacher and Brooks, 1990). The largest farmers have experienced windfall profits from price supports (Bonnen, 1984) and, in conjunction with the agriculture chemical and pharmaceutical sectors, have been able to accumulate capital and profits (Bonanno and Calasanti, 1988). Mid-sized farms, with lower assets and more debt, have been in a more precarious position, and their situation has been exacerbated by the rise of global production and consumption of foodstuffs (Bonnen, 1984). The demand for U.S. food is more responsive to price, as well as changes in foreign markets and capital. Remaining competitive with international producers means lowering commodity prices even though production costs increase (Bonanno, 1991). In a vicious cycle export enhancement subsidizes foreigners and agribusinesses (Paarlberg, 1990), lowers prices, and hastens the exodus of farmers (Bonnen, 1984).

Rural communities have borne much of the brunt of these actions. Farm towns have experienced long-term deterioration of community and capital resources, labor displacement, loss of viability, health care deficiencies, and a decline in services and facilities (Johnson and Bonnen, 1991). These losses are reflected in rural resident's lower educational backgrounds, lower job and residential security, and lower levels of per capita income (Beegle, 1991). Many of these communities are no longer dependent on farming, and economic recovery has been difficult because of high unemployment and stagnating or declining real wages.

The decline in farm numbers has not been stanched [although it has been slowed (Buttel and LaRamee, 1991)]. Government programs may have delayed the demise of some farms but cuts in these supports could increase the rate of movement out of farming again. The three types of support-supply management, marketing agreements, and price supports have been criticized for their effects on markets, taxes, and food prices. The dairy, peanut, tobacco, and sugar programs increase farm and retail prices, marketing agreements raise retail prices of fruits and vegetables, and deficiency payments are reflected in higher taxes (Senauer et al., 1991). The latter programs are one of the clearest examples of inequity in the food system. Seventy-three percent of deficiency payments go to 15% of the farmers (Browne et al., 1992), because payments are based on the volume of production. This linkage has been blamed for contributing to the problems of overproduction, and the continuing need to find outlets for the food produced, that for various economic and biological reasons, cannot be absorbed by domestic markets.

Overproduction of food in turn has led to many of the environmental problems catalogued over the past decade, including pollution of surface and groundwater, salinization and erosion of soils, decreased genetic diversity, and human health hazards (National Research Council, 1989). Alternative production methods include crop rotations and biological pest control, and many other practices have been promoted, but government policies have discouraged environmentally desirable farming practices as farmers protected their base acreage for price support programs. Conservation is the new agriculture policy area that has gotten support in the last 6 to 7 years, spurred by early and continuing public concerns about pesticides and moving more recently to a focus on the nexus of commodity programs and environmental problems (Reichelderfer and Hinkle, 1989). Some relaxation of the farm program requirements occurred in the 1990 Farm Bill, but the policies may not be strong enough to attract large numbers of farmers to sustainable practices (Hassebrook, n.d.).

How conservation proposals will fare as agricultural programs continue under scrutiny is unknown. One possible point of concern is Browne's observation that in the 1985 Farm Bill social policy concerns were replaced by conservation groups, "not necessarily with that purpose in mind but certainly (in) producing that effect" (Browne, 1988a:231). Consumer groups such as Public Voice and Consumer Federation of America at that time fought in opposition to the price increasing dairy, sugar, and peanut programs (in a losing battle), and food stamp advocates like the Food Research and Action Center were "devoting shrinking organizational resources to protect programs already reduced in 1981" (Browne, 1988a:233). However, these advocates had recognized that passage of supply management legislation (the Harkin Bill) would increase food prices and would not support the proposal without provisions for increasing food stamp benefits.

Economic

The effect of the declining effective demand for food of the U.S. poverty population on farmer's income is not known. Farmers may not harbor much interest in doing such a calculation for several reasons. As mentioned earlier, they on average receive a small percentage of the food dollar and are minor contributors to the Gross National Product. Because such a large percentage of the food supply is processed, food price increases for most differentiated products are often not accompanied by increased farm returns. In fact many farmers argue that the country's cheap food policy means that "the American consumer [is] subsidized on the back of the producer" (de la Garza in CNI,

1991c:5), because food prices stay relatively low even while producers of some products like grains and dairy cannot recover their costs of production.

Possibly adding to farmers disinterest in poverty programs is the fact that until 20 years ago eligible farmers and rural residents in general received much less aid from welfare assistance programs than did the urban poor (Bryant et al., 1981), and there is still a bias against rural areas in various welfare policies such as school aid, medicare, and income testing for program eligibility that precludes property ownership (Flora and Flora, 1988). Taken together these facts might support an argument that farmers have little economic reason to care about poverty or hunger because they benefit only marginally from programs that attempt to assuage it.

Comparing farm program and food stamp benefits over the past decade, farmers and the poor have benefited about equally. From 1983 to 1991, the average federal payout for food stamps has been just a little higher than all government costs for farm programs [$11.9 billion for food stamps and $11.3 billion dollars in price supports (Constance et al., 1990; Levedahl, 1992)]. There are, of course, differences in the payments and procedures in the programs. All the food stamp benefits (an average of $710 per year per person) go the the 25 million poorest individuals in the population; 60% of the loan payments go to the richest 323,000 farmers. Also the levels of food stamp appropriations have steadily increased each year over the time period, while the government payments were quite variable. In 1991 the total outlays for food assistance were three times higher than for federal payments to farmers (CNI, 1991f). However, it should be noted that real outlays for food stamps had declined from the 1970s, and that between 1980 and 1987 there was a marked decline in the number of food stamp recipients as a percentage of the poverty population (Center on Budget and Policy Priorities, 1988).

What is the return to farmers of food stamps? The coupons do raise the value of food expenditures to a higher level than people could otherwise afford, and the marginal propensity to consume is higher with food stamps than with cash (Andrews and Clancy, 1985; U.S. Senate Committee on Agriculture, Nutrition, and Forestry, 1985). Two reports place the return at between 1 and 2% of farm income (U.S. Senate Committee on Agriculture, Nutrition, and Forestry, 1985; Browne et al., 1992). Using 1991 data for total food stamp benefits ($17.3 billion), total farm receipts (including government payments ($179 billion), and a 24% share of the retail cost of food, the percentage is 2.6. The return to some farmers is greater depending on what the farm-retail spread is for their particular commodity.

The effect of commodity donations on farmer income is less discernible. The dollar value of commodities given to the poor in 1990 was small—$152 million (Becker, 1992), far lower than the billions worth of commodities given out between 1982 and 1985 (Lipsky and Thibodeau, 1985b). Surplus commodities are donated overseas as well, in large amounts, and the government would own the commodities anyway, so the direct economic benefit to farmers is unclear. The "benefit" to the Treasury is clearer. Surpluses cost taxpayers money in loans and storage costs, and their downward impact on commodity prices and costs was, of course, the original argument, made in 1937–1938 and again in 1972–1973 in favor of the Food Stamp Program. Commodities were undependable and commercial markets were losing business to the government. It was strongly argued that it was both good business, and better farm and food policy, to let low-income households purchase food through regular channels than to rely on government surplus (U.S. Senate Committee on Agriculture, Nutrition, and Forestry, 1985). But deliberate policy decisions keep food stamp benefits inadequate, and the cycle of surplus build-up has never had any relationship to need.

The benefits to farmers and the public from food assistance programs are contradictory and inequitable. As poverty and the number of food stamp recipients rise while farmers' incomes remain stable, the return to farmers from food stamps increases, although these gains could be offset some by the reduction of sales due to loss of income [between 1981 and 1987 food purchases by the poor decreased 13 percent (O'Neill, 1992)]. In those times when surplus farm commodities coincide with rising hunger, as in the 1930s and the mid-1980s, taxpayers benefit from decreased storage cost as the food is distributed to the poor. A more rational system—one that anticipated the food demands of the poor, and removed the onus of relying on surpluses that may or may not exist—would seem to be in order. But until this ideal situation presents itself, and as food stamps claim clear title to the largest item in USDA's budget, farmers may find it in their best interests to support the program.

The Role of Research

The intellectual wing of agricultural institutions has provided little assistance in understanding the phenomena just addressed. For 25 years the humanistic critics of what Johnson calls the Agricultural Research Establishment have decried the lack of attention to many issues including malnutrition, rural poverty, rural structures, and other overlooked areas (Johnson, 1989). But despite acknowledgment by social scientists themselves that there is so much work to be done that there are not enough social scientists to carry it out (Hite, 1991), the critics complaints have hardly engendered a social science renaissance in the land grant colleges. One reason is that budgets were reduced in the 1980s for social science research in general, a move intended, according to Ruttan (1991:1–71), "to reduce the challenge to ideology on policy design." This may be the case, and could continue for some time until the political leadership of the country changes.

Still, internal reasons do a better job of explaining the aversiveness than funding curbs. The reasons include the fact that the philosophical underpinnings of much agricultural (and other) research discourages such inquiry. Logical positivism, which maintains that it is not possible to include knowledge about values as characteristic of the real world (Johnson, 1989), has allowed its adherents to avoid research on the controversial problems in agriculture, even though "the most significant of which in the social sciences are of this type" (Schultz, 1941 in Busch and Lacy, 1983). Classical economic models, which turn on narrow utilitarian theories that fail to capture much of reality (Madden, 1986), are still considered normative. Both philosophical positions allow researchers to adhere to the posture of "value neutrality" in their work, within as Harding (1940:1099) colorfully put it, "a rigid compartmentalization that has done so much to sterilize scientific knowledge by depriving scientific specialists of a broad social vision."

Some research on rural poverty has been conducted over the years, but the effort has been inadequate. One explanation for the neglect is in Busch and Lacy's (1983) retelling of how the Farm Bureau succeeded in having social scientists in USDA dismissed and the Bureau of Agricultural Economics dismantled in the 1940s. The message sent to researchers was a powerful one and apparently is still reverberating. Bryant and his co-editors do hold that the one instance that represents an involvement by agricultural economists in research on poverty policies and programs is the food programs (Bryant et al., 1981). Their interest originated in the effects of programs on the expansion of demand, and moved to some work on the nutritive value of the bonus and on participation. At no time in this literature, however, has there been any emphasis on hunger "per se" (Schertz, 1991).

Several sociologists of agriculture (e.g., Friedland, 1991a; Buttel, 1991) recently raised important questions about structural changes and their effects on rural poverty, but these writers are silent, again, on the paradox of hunger. Perhaps the best but frustrating example of the inattention of social scientists to these issues is the invisibility of rural or urban hunger problems in the new plan for social science research in agriculture developed by a distinguished group in agricultural social scientists to "help empower rural America." The Social Sciences Agricultural Agenda Project (SSAAP) is proposing needed, varied, and ambitious agenda of social science research, but even now in the approximately 650 pages of the report titled Social Science Agricultural Agenda and Strategies (Johnson and Bonnen, 1991) the topic of domestic hunger is nearly imperceptible, being mentioned maybe two or three times near the end of lists of domestic issues. Poverty is addressed but not its consequences—even though the people developing the agenda must share the notion that there is some theoretical relationship between agriculture and food. The Food Stamp Program is mentioned just once as part of the history of past attempts by the federal government to expand domestic markets (Kirkendall, 1991).

With little empirical data it is difficult to assign reasons to this oversight, but some theories seem plausible. It may be that agricultural social scientists are not convinced that hunger is a problem in the United States or that if it is a problem it has no relationship to agriculture and therefore can be ignored. To be fair, they are not alone in their biases. Katz (1989) has acknowledged that American social science in general remains largely silent on the politics or the political economy of poverty in the United States, and its consequences. Another partial explanation may be in the backgrounds of agricultural scientists. Many researchers in the study done by Busch and Lacy 10 years ago came from farm families, and may share the same bias toward the poor as the families, farmers, and commodity organizations with which they work.

At another level, perhaps classical economics models prevent today's social scientists from looking for alternate explanations of hunger and poverty that incorporate different judgments and symbolic meanings. As Harding opined 50 years ago, "reductionism has the real danger that it leads scientific workers to assume that all scientific questions are independent of ethics" (Harding 1940:1100). Another related and possibly contributory factor is that fewer than 2% of agricultural scientists are female (Sachs, 1991), and perhaps the overwhelmingly male cadre of agricultural researchers is wary of discussing hunger because of an inaccurate perception that it is not their area, but "belongs" to nutrition or home economics. Fifty years ago, their predecessors were writing in the 1940 Yearbook of Agriculture about the desperate need for merging the natural and social sciences (Hambidge, 1940). Contemporary agricultural scientists have weak cross-disciplinary training and find it difficult to think in an interdisciplinary way (Busch and Lacy, 1983). Earlier agricultural scientists had strong backgrounds and training in the humanities and were not so wedded to rigid positivistic models. They were acutely aware of the issues of hunger and poverty among rural people, of the inextricable relationship of nutrition and agriculture, and of the two-sided nature of nutrition as both a natural and social science. Despite the force of their intellects and their best intentions, however, their pleas have been largely ignored.

SUSTAINABLE AGRICULTURE AND FOOD POLICY

In contrast to the hunger crisis, for which the groups advocating its alleviation have, for the most part, complementary goals, the problems in agriculture, both conventional and sustainable, are defined with much less consensus. The positions held by those groups lobbying hardest on sustainable agriculture, which include progressive farm groups such as the National Family Farm Coalition and the National Farmer's Union; farm advocacy groups like the Center for Rural Affairs, the Sustainable Agriculture Working Group, and the Land Stewardship Project; and environmental organizations like the National Wildlife Federation, Center for Resource Economics, American Farmland Trust, and the National Audubon Society, are not always in agreement. Visions of saving the family farm, targeting price support payments, reducing economic concentration in the farming and food processing industries, promoting rural development, supporting research on alternative farming systems, etc., are not necessarily shared goals (CNI, 1991a; Browne et al., 1992). In most cases, however, the positions are in strong opposition to those of conventional farm lobby organizations.

Sustainable agriculture supporters were successful in getting some environmental legislation passed in 1985 and renewed their efforts in 1990. Proposals for conservation, integrated farm management options, and flexibility provisions again were on the table, as well as the National Organic Production Act (NOPA). The consumer groups that in 1985 put their attention to reining in price supports were lobbying for NOPA, but only after their traditional stance of not supporting changes that might increase food costs had been adjusted. Organic farm representatives had educated them about the trade-offs between environmental protection, more sound farming practices, and the cost of production, which translates into higher costs of food. The antihunger food stamp advocates were not involved in the legislative and lobbying efforts surrounding the passage of the Act, but were in general not supportive. According to an active participant, except for the conversation with a few consumer representatives, alternative farmers and environmentalists themselves never discussed the effects of their proposals on food prices or their effects on the poor (Blobaum, 1991).

Although the poverty/food cost problem can be considered peripheral to the main point of getting national standards for organic food, ignoring it did display a lack of attention to constituency building and legitimation. There are two issues here. One is the role of consumers on the agricultural agenda; the other is the specific problem of consumer support for sustainable/organic agricultural practices and the policies/legislation needed to enable them.

There seems to be some consensus, at least among scholars studying agricultural policy, that consumers will become a critical component of the food and farm lobby (Constance et al., 1990:65). Traditional farm organizations may accept this or not, although it is difficult to see how they could ignore the fact that the relevance of the farm sector has changed and its economic importance decreased (Bonnano, 1991), which calls for a new look at the positioning of farm programs to taxpayers. As Leman and Paarlberg (1988) have suggested, farmers get support when they are in crisis, like in the mid-1980s. In the 1990s the perception, whether accurate or not, is that many are prospering again (Robbins, 1990) and perhaps not as favored in the taxpayers eye.

Organizations espousing the tenets of sustainable agriculture have strongly taken consumer concerns about food safety into account and have entered coalitions with a few consumer organizations. But they have not given inadequate consideration to food cost and accessibility issues and will probably have to increase their interaction with a much broader array of groups because of the complex legislative agenda they have taken on will need broad based support for passage. At least a good part of the sustainable agriculture lobby is working for three major simultaneous changes (1) mandated environmentally conserving agricultural practices, (2) either across-the-board increases in price supports, supply management, or targeted price supports, claimed by their respective proponents as the way to improve the chances that mid-sized farms can remain viable, and (3) higher food prices, one of the anticipated results of the first two.

The cost question is integral to understanding and explicating the central issue here—the acceptance of sustainable agricultural policy by middle income and poor consumers. Of course, if food costs were not to rise above normal inflation levels with the implementation of the changes listed, the issue is moot. But this has not been the prediction of economists arguing either for or against more widespread adoption of alternative practices and policies (Faeth et al., 1991; Knutson et al., 1990). Accepting the assumption that costs would rise we turn to look very briefly at four key food cost issues: the cheap food phenomenon, the components of food costs, the perceptions of consumers of the costs of alternatively produced foods, and the food costs of the poor.

Whether food is "cheap" depends on the economic condition of a household as discussed earlier. Americans spend a smaller portion of their incomes on food because we have higher average incomes than people anywhere else. Food is "cheap" relative to these incomes, and declining real incomes in a large enough proportion of the population could cause a rise in the percentage spent on food. The prices of some foods produced here are lower than in other countries because of rising gains in productivity, and because U.S. farmers are supported through tax policy more than consumer prices (Browne et al., 1992). The net effect, however, is that if you are poor your income is too low to purchase adequate types or amounts of food to provide a healthy diet.

The poor are not the only segment of the population concerned about food costs, especially foods produced in a more environmentally beneficial way. It needs to be noted that the only identifiable foods in the market grown in an alternative system are organic, so this is the benchmark for perceptions of the "costs" of saving the environment. One estimate posits that the substitution of available organic food in the diet will increase food costs by about $1000 per year (CNI, 1989). Support for organic agriculture is fragile even among the most affluent, who are most likely, of course, to purchase organic food, but who will still not pay prices that some growers would consider fair (Jolly et al., 1989; Goldman and Clancy, 1991). For many people their only knowledge of organic agriculture may be the statement of Earl Butz a decade ago to the effect that if the United States adopts an organic production system 50 million Americans would go hungry. This is not a comforting thought, nor is it accurate, but perceptions will be maintained until education or experience intervene.

The final area to be considered on the question of food costs is the fact that the low-income consumer in the United States has been always considered marginal to the mass market and is quickly becoming more so. The flight of supermarkets to the suburbs and the growing dependence on the Emergency Food System (U.S. House of Representatives Select Committee on Hunger, 1987) highlight the fact that this segment of the population has been written off by many of the institutions within the food system. Whether it is in the best interests of alternative agriculture to abandon this market as well is the question being addressed here.

This lack of constituency building is another example of farmers acting against their own best interests. Most sustainable agriculture interest groups working on domestic issues at the national level have not incorporated a social dimension—except as related to the impact of policies on moderate-sized family farms (Hassebrook, n.d.). This is consonant with having separated themselves from the major social movements of the last decades like civil rights and women's equality (Strange, 1988). There is no research at present on whether consumer interests converge regarding support for family farms and support for organic or alternatively produced food stuffs. Nor do we have any information on the attitudes of the sustainable agriculture leadership toward the issues of poverty and hunger either urban or rural. Such intelligence could be most helpful in charting political strategy.

MAINTAINING PUBLIC SUPPORT FOR SUSTAINABLE AGRICULTURE: A PRAGMATIC RESPONSE

While farmers, in proportion to their numbers, have maintained political strength (Leman and Paarlberg, 1988), in the 1990s some of this support seems to be eroding. According to Doering (1991:5), "current agricultural and rural policies appear to have reached the end of the line in terms of public support . . . and a politically supportable agricultural structure and income distribution." Two studies, one conducted in 1986 with a national sample (Variyam et al., 1990) and the other in 1989 with a sample from a few selected states around the country (Guither, 1990), both suggest that public sentiment on the continuation of family farming and preferences for various agriculture policy vehicles is variable and not strong on many issues. In the smaller, more recent study 45% or more of the respondents would reduce farm costs, phase out present support programs, and phase out target prices. Forty percent would end the dairy program, and 75% would phase out the sugar and peanut programs. The agreement on the need for conservation was much higher. Between 80 and 90% agree with both the regulation of farming practices to reduce water pollution, and continued conservation compliance (Guither, 1990). In the earlier study 75% of the sample agreed that the family farm should be preserved and 57% felt that the government should have a special policy to ensure this. However, queries about support for food price increases found less agreement—35% agreed that family farms should be supported even if it meant higher food prices.

The income of the respondent was a significant determinant of a preference of policy. As income increased, support for the family farm and government intervention declined. Less support was also present in those with higher education levels, residence in town or country, higher age, and residence in the West (Variyam et al., 1990). Women, Democrats, and urban dwellers tended to be the groups who favored support for farm policies, but no distinction was made in the questions of farm size. Over decades the public has failed to differentiate among truly poor farmers, besieged mid-sized farmers, and well-off farmers (Poppendieck, 1986). According to Browne and his coauthors (1992:42), this "stereotypical homogeneity is the prime generator of public support." The attitudes of higher income and educated consumers may presage a change in the ability to distinguish more from less well-off farmers, a talent, of course held by members of Congress for some time, but with no obvious effects on policy.

In the 1990 federal budget agricultural programs took a 25% cut, greater than any other major program (Doering, 1991); even some environmental proposals fared poorly (Hassebrook, n.d.). These actions reflect the fact that the government is in a bind as it is called on to support the continued accumulation of profits by large farms and businesses at the same time that demands for welfare assistance increase (Bonnano, 1991).

Since the 1930s farm programs have been protected by confusing them with food programs (Browne et al., 1992), but this may be starting to unravel. There is a greater visibility now of poverty and hunger, in homelessness, soup lines, and urban riots. This is simultaneous with the disappearance of the discrepancy between farmers incomes and the incomes of the rest of the population. Consumers are certainly willing to provide some protection to farmers in return for a stable and secure food supply (Bonanno, 1991), and to recognize that farmers face unique difficulties. But in the face of the federal deficit trade-offs seem more inevitable.

Other signs are around. In early 1992, a national survey found that a larger percentage of respondents (61 vs. 48%) believed that hunger was a very serious issue compared to the environment (Breglio, 1992). In an earlier section the lack of support by antihunger groups of a supply management proposal was noted. This section will present several other illustrations of how the farm and environmental agendas are being affected by the crisis in distributive justice. It briefly discusses the more or less useful options available to address the situation. It concludes with an outline of some of the components of a pragmatic response that could be made by the sustainable agriculture community that acknowledges the conflicts and looks for a way to resolve them.

One of the more interesting and perhaps foretelling debates took place in the spring of 1991 in both the House and Senate Agriculture committees, in the context of new proposed amendments to the dairy price support program. The Senate acted first, and in mid-March passed a bill proposed by Senator Leahy of Vermont to increase milk prices because prices received by dairy farmers had fallen to the lowest levels in 15 years (U.S. House of Representatives Committee on Agriculture, 1991). In June the House Agriculture Committee passed a different piece of legislation that would have, among other things, raised the support price of milk (U.S. House of Representatives Committee on Agriculture, 1991). USDA strongly opposed both efforts, the President threatened to veto them (Madigan, 1991), and neither bill made it to conference. But before these events occurred a linkage between food assistance to farm programs was made in both committees. To safeguard domestic feeding programs, a significant portion of which costs are dairy costs (Bertini, 1991), it was proposed that dairy producers be required to pay an assessment on their milk marketings to reimburse the Government for the additional costs to the feeding programs associated with the increase in the support price.

With some debatable assumptions, a prediction was made by the CRS that 112,000 fewer WIC participants would be served in an average month (Jones, 1991). In opposition to the proposed increase Senator Domenici (R-NM) offered an amendment that stated that "if the Secretary found that 50,000 WIC participants or more are going to be eliminated because of increased dairy costs then he cancels the new Dairy Program" (Congressional Record, 1991). Leahy countered with an amendment to require the Secretary to distribute funds from the price support to states to prevent a decline in

WIC. The conference dropped the provision because the House had not considered it, but the floor debate was fascinating. Poverty and welfare per se were never mentioned, but the WIC program and other feeding programs (School Lunch and Food Stamps) were cited as programs that would bear heavy costs from a 10-cent milk price increase. In the House committee it was made clear that no dairy bill would pass without the support of urban legislators who would not allow a harmful cost impact on food programs (CNI, 1991f). On the other side some committee members vigorously opposed the idea of compensation, saying it was not fair to ask dairy farmers to fund the poor, and stating that if the farmers went out of business they would become poor themselves.

An issue that did not arise but that is relevant is that decreased participation in food programs is not the only effect on poor people of farm price supports. In an ironic, perverse way food availability is affected as well. The poor have a connection to farmers that most consumers do not have—surplus commodities, i.e., the cheese, nonfat dry milk, honey, cornmeal, flour, and other products handed out until this year through food pantries, and 25 years ago through government warehouses and community action agencies. Low-income families have been put in the position of relying on the CCC programs, and interested in their continuation. A recent report verifies that the depletion of commodities in 1991–1992 put some food pantries out of business (Ballenger and Mabbs-Zeno, 1992). Farm policy proposals that would cut surpluses risk being criticized not just for their impact on food prices, but for their contribution to the demise of the surpluses themselves. The same argument could be made for the conservation reserve, which has functioned as a supply control mechanism and helped contribute to a reduction in surpluses as well as environmental damage.

Of the possible conflicts among farm support, environmental legislation, and the poor, the surplus food problem is, however, a minor one. The potentially more significant issue is cost, one of the key points on which public support for agriculture turns. Consumers want farmers to protect the environment, and many want to save the family farm. However, it is the higher income population, less likely to support the maintenance of a small or medium sized family farm structure, that is the market for organic food. As mentioned earlier, the willingness of many consumers to pay more for organic food is already low, and several commentators have recently predicted it will stay that way because of either (1) the higher cost of organic produce and consumer preference for perfect appearance (Huang, 1991), or (2) the fact that most consumers appear to accept the relative safety of conventionally grown produce (Lane and Bruhn, 1992). What import this has for support for sustainable agriculture is entirely unknown because we do not know if consumers analogize their attitudes toward organic food to all forms of sustainable agriculture, or from produce to other food groups. We also do not know what the effect of the adoption of sustainable practices by a significant number of farmers will have on the cost of food. If research establishes that there is true effect on price, or if there is a public perception that an effect will be significant there are several options. One is to keep food costs low by adoption of sustainable agriculture methods on a sufficient scale for supply and demand to comes into balance (Merrill, 1976). It is unclear how successful this option can be. Buttel (1990) suggests that for some time into the future it is small farmers, who constitute a small proportion of the agriculture economy, who are more likely to retain a commitment to resource conservation. In the meantime, though, the success of sustainable agriculture will depend on the largest, highly capitalized family farms shifting toward better environmental management. These farms will continue to grow and "dominate most on-farm commodity sectors . . . under conditions of rapid technological change" (Buttel, 1990:7). Unless targeting of price supports occurs it will continue to be difficult for medium sized farms, let alone small ones, to compete with the new technologies. Another angle on this dilemma has been acknowledged by Gershuny and Forster (1992:7) when they write "we [organic agriculture proponents] may have built ourselves a Procrustean bed by advancing large-scale organic farming as evidence of our legitimacy, and by implication disavowing the centralist vision of small part-time farming and regional food self-reliance."

The second option is to hide the costs. If identifiable markets for products produced in alternative systems are not developed, so that much of the food (mainly grains) grown in such systems continues to be commingled with that produced conventionally, consumers will have no way of knowing the production method, nor the source of any increased cost, at least at the supermarket level. Under this scenario it might be easier for sustainable farmers to maintain unwitting public support for their legislative proposals, overcoming the problem of having the transparency of policy costs lead to significant opposition from the cost bearers (Rausser and Irwin, 1987 quoted in Variyam et al., 1990).

What is lost, of course, is the opportunity to educate consumers about the "true" cost of food and the reasons why prices should reflect the cost of environmental conservation (IAA, 1991). This is the third option, but it really is not elective. It is unlikely that any but a few consumers understand the bases for the phenomenon of moderate food prices, and the structural upheaval in agriculture, the tax policies, and the environmental damage that have made it possible. Some attempt at explaining this history will probably need to be made if support for more economically and environmentally rational policies is to be forthcoming. Agricultural organizations, from beef producers to organic produce growers, have done fairly little to educate consumers about production practices and food costs, especially with regard to where many food price problems lie: in the massive economic concentration in the food industry in the past 50 years (Connor, 1988; Constance and Heffernan, 1989), the accompanying increased advertising expenditures for the most intensively processed foods, and growing industry profits that have increased 90% between 1975 and 1990 (calculated from Elitzak, 1991), while the CPI for food went up only 73% (U.S. Department of Labor, 1992).

Based on the material reviewed so far, there appear to be several areas where consideration of the situation of the poor can add to the legitimation of policy reforms in both agriculture and social welfare. They are (1) markets, (2) environmental interests, (3) rural development, and (4) research. Due to the inelasticity of food demand, and the low average percent of disposable income spent on food, growth in domestic demand will come only from "improvement of the status of poor Americans" (Tutwiler and Carr, 1988: 224). Undoubtedly the emphasis on export markets has lessened interest in this market, and the increasing hunger makes it appear less promising. But the development of increased domestic demand can be seen as one way to improve farmers' incomes and farming's balance sheet as price supports falter. Farmers benefit by remembering that "welfare programs are highly effective from the accumulative point of view, as their economic impact is almost entirely translated into consumption" (Bonanno and Calasanti, 1988:253).

Organic growers may not need to totally dismiss low-income households as potential markets. The Hartford Food System project found that low-income mothers recognized that organic foods, because they are probably safer, are better, and that they would prefer to purchase them if resources permit (Winne, 1990). Low-income households do not have the food dollars to purchase fruits and vegetables in the quantities that better off households can, but some of the alternative marketing schemes mentioned in the next section respond to this problem.

Nor are the benefits of sensitizing sustainable agriculture to welfare needs confined to urban areas. There is a growing understanding of the mutuality between farm and rural well-being, including the need to maintain the viability of rural community's nonfarm economies so that (1) small farms can be maintained (Swanson, 1988), and (2) larger farms looking for off-farm income can find it (Stevenson, 1992). Rural development cannot succeed without strong food assistance and welfare components, along with education, jobs training, assistance to small business, etc.—the same programs that must be instituted in urban areas.

And it looks like there could be help on the way from the research community. Friedland (1991a), Buttel (1991), and Doering (1991) have all recently described already instituted as well as needed changes in rural sociology and agricultural economics that would legitimate study of the linkages between agricultural policy and other social phenomena. Friedland, in his review, chronicles major changes within rural sociology over the past 15 years or so that have seen the emergence of "the sociology of agriculture," and the beginnings of a critical examination of the social relations within agriculture. Doerings' critique questions the present usefulness of agricultural economics research staying at the margins of agriculture policy, and being unable to provide a broader analysis or prescriptions for change. Both see political economy as a model that allows the disciplines to study problems in a more holistic and realistic way. This approach "emphasizes the impossibility of disaggregating the social from the economic, political, cultural, or, indeed, the scientific" (Friedland 1991a:17).

Another source of optimism on this topic is the emerging role of women as subjects and authors of research on women in agriculture. This phenomenon is the source of two benefits to agricultural social science research. One is the challenging of assumptions about farm structure, agriculture production, and agriculture policy-making that has occurred through the study of women in farming. This research has also renewed attention to social class differentiation in agriculture, and generally opened up agricultural studies to broader and better explanations (Haney, 1991). The second benefit is the different style women scholars bring to research: a greater ability to make connections and a willingness to work in interdisciplinary modes (Sachs, 1991). Women may also not feel so strongly that their scientific findings have to be stripped of compassion before being presented to their peers and the public.

Both Friedland, and the SSAAP in great depth, lay out the areas to which greater attention should be given. One of these is the linkages between natural-ecological processes and agriculture production systems, including the role of environmental concerns in shaping food and agriculture policies, and the "articulation of agriculture and food related movements with larger 'social movements'" (Friedland 1991a:25). I hope it is obvious that hunger and the other biological and social consequences of poverty that are the subject of this chapter fit well in this agenda, and could play a significant role in expanding its interpretation. There are a myriad of other questions identified in the chapter that warrant study. What is the attitude of alternative farmers to the poor? To the paradox of hunger? To domestic food demand? What political, demographic, cultural characteristics affect these attitudes? What trade-offs do consumers see between the environmental and structural characteristics of new farm policy? What are the trade-offs between rural development and sustainable agriculture?

RECOMMENDATIONS FOR MOVING TOWARDS DIALOGUE ON AGRICULTURE, POVERTY, AND HUNGER

It is unclear whether the traditional agricultural establishment, having shown its disinterest in the paradox of hunger over six decades, will decide to weigh in on the issue. Now would be the time, as the hunger problem grows toward levels seen in the 1960s, and farm policy support looks more fragile. Conservative politics, racism, inertia, and an eroding but still strong power base may prevent it from making any attempt to confront the issue and propose humane solutions. A recent proposal by the New York State Farm Bureau to cut welfare, medicaid, and taxes is not exactly going in the direction suggested.

The sustainable agriculture establishment, on the other hand, is working from a smaller and weaker constituency base. There are pluses and minuses in this situation, particularly with regard to the issues of food and the poor. On the negative side are several difficult issues: (1) those promoting more environmentally sound and sustainable policies may be tarred with the same brush as conventional agriculture if they are perceived as wanting to maintain the status quo on price supports and a continuation of the inequity in farm incomes, which also results in less money for food programs; (2) the long held preference for conservation as a means to reduce overproduction has positive environmental effects, but may reduce supplies of surplus food. Despite the fact that virtually all antihunger advocates think that reliance on this source as a principal means for feeding the poor is irresponsible and unjust, until more appropriate food assistance avenues are opened up the need will remain; and (3) as illustrated by the last item, competing interests pose another problem for the sustainable agriculture lobby. Farm and rural interests compete with environmental interests, and both compete against the interests of the poor, as described in earlier sections. Although traditional farm interests have always held the upper hand and fought against food assistance, accommodations have occurred through log-rolling and fragile alliances around surplus commodities. Sustainable agriculture does not have the clout to log-roll on anything yet, and sustainable proposals work against surpluses. Supplanting food assistance with environmental issues on the farm agenda (CNI, 1991a), or event the perception that this has occurred, may very well turn out to be viewed as politically short-sighted, as well as unethical if thought to have been done consciously.

On the positive side are several factors: (1) sustainable farm and environmental groups, none of whom is part of the mainline traditional farm lobby, do not have to maintain the old alliances. Some of the key personalities in these groups came out of antipoverty groups and know the issues and players (Strange, 1988). They can bring a broad group of externalities/alternatives representatives to the table, and look for congruencies in agendas without appearing to be consorting with the enemy of forgetting their roots; (2) the sustainability lobby also does not have to do everything the same old way. Many of the solutions to the problems described here lie in health and welfare initiatives and committees that are not part of agriculture. As suggested earlier, they are proposals, like jobs and improved real wages, as much needed in rural and farm as in urban areas. Support for these proposals, through testimony, supporting letters, etc. would not be difficult to offer once some of the ideological problems are overcome. Things are not so easy in the agriculture committees, where, without a peace dividend, the alternative agriculture lobby will have to honestly face the trade-offs between funding for farm and food assistance programs. As suggested throughout the chapter the arguments for these trade-offs hinge on two key facts: (1) the country, including the agricultural sector, is not going to prosper if a growing portion of the population, especially children, remains unskilled and hungry; and (2) poverty and food accessibility issues may be used as arguments against improved environmental policy, just as they are now in the Third World.

Moving toward an understanding of and resolution of some of these conflicts requires at least three steps: demythologizing poor people and farmers, acknowledging the need for education and attitude changes on both sides, and agreeing on alternative policies and actions. Dialogue has to begin with the recognition that the 200-year-old stereotypes of virtuous farmers, depraved urban residents, hard working yeomen, and lazy poor people are inaccurate. Some people are undoubtedly as described, but the exceptions far outnumber the true types within each category. Farmers work hard, but so do many poor people. Few farmers are wealthy, but poor farmers are now a small fraction of the total poverty population. Farmers have been defined as the repositories of family values (Browne et al., 1992), and the latter has become a code word used by conservatives to denigrate the poor, especially minorities. But most poor parents want the best for their children, even if they feel powerless to provide it. Most farmers do not consciously act to increase the food insecurity of the poor, but they have paid little attention to the effects of their programs.

One serious problem standing in the way of reconciliation of the two groups is the accurate perception by low-income communities that environmental groups have ignored them and the environmental problems in urban and poor areas in preference to wilderness and natural resource issues (Steinhart, 1991; Pena, 1992). Steinhart writes that environmental organizations are attempting to readjust their foci to recognize that problems in low-income areas of cities (and rural areas too) like hazardous waste sites, heavy pollution from industry, industrial water contamination, etc., are health and social justice issues, not just environmental ones. Although food costs and surplus commodities are in no way as obvious or compelling as toxic waste problems, the possibly discriminatory effects of environmental proposals have to be taken into account, as a first step toward moving beyond the class and racial antagonisms between farmers, environmentalists and the poor. They may produce strong tensions, but the worth of working toward a rapprochement seems evident.

The second step is education about each others issues—the causes and effects of food insecurity, the role of food and agricultural policies in perpetuating both social inequities and environmentally destructive practices, and the arguments for identifying all components of the costs of food. Paul Thompson (1991:71) thinks that the latter is one topic about which "a recognition [can occur] that the food system binds producers and consumers into a community of interests" not just with regard to environmental costs, but the economic and health costs of the food system as well that are imposed on all consumers, including farmers, by the other sectors of the food industry. The poor, because of their distance from it, and because they are not well served by the food system, cannot be expected to have much commitment to change in agriculture (after Buttel, 1980). But organized grass roots and national groups probably are ready for good faith efforts, both at the policy and local action levels. The urban and rural dialogues instituted in Minnesota are one model for activity of this sort (Taylor 1992).

The third step is to lay out various proposed solutions to the respective problems and sort out where there is the possibility of reconciliation of the agendas. The policy changes needed to reduce domestic hunger include most of the items already mentioned: education, skills development, and jobs programs; serious welfare reform (Bane and Ellwood, 1991); many improvements in the Food Stamp Program to increase the numbers of eligible recipients and to provide adequate benefits so that the use of emergency food is unnecessary; and changes in the assumptions in the calculation of the poverty level and food stamp benefits that penalize households with average housing costs. The chances for legislative approval of any of these reforms is slim—similar to the likelihood of passage of many farm and environmental proposals. But finding common ground is a significant benefit in itself and might provide the impetus for real reform of the food system.

It is not necessary, nor in most cases productive, for agriculture or environmental groups to take the lead on any of the hunger and poverty issues either at the national or local level; they do not have the expertise, credibility, or resources to do so (Browne, 1988a). The political clout of sustainable agriculture on these issues comes from having even considered them, acknowledging that there are contradictions, seeing where compromise or win–win situations exist, and being ready to support the antipoverty, antihunger proposals that are identified in the process as being compatible with the other agendas.

Simultaneous with the long-term process of legislative change is the opportunity to work on local development activities to build stronger links between farmers and low-income consumers. These include the WIC/Farmer's market coupon program initiated in state departments of agriculture, through which WIC participants are given $10 or $20 in coupons to purchase food in local farmer's markets. Ten states have been part of a USDA demonstration project and all except one have reported success and satisfaction with most of the results (CNI, 1991d). There is even greater support for encouraging the use of food stamps in these markets (CNI, 1992). Along the same lines a consumer organization has recently organized a farmer's market in downtown Hartford, Connecticut to serve a mixed clientele of low-income and middle class shoppers. This extends and complements another program in the same city, called Community Farm Stands, that brings thousands of pounds of local produce (much of it organic) into inner-city neighborhoods to be sold at reasonable prices (Hartford Food System, n.d.).

The final set of promising activities, ones that move in the direction of changing the production relationship between farmers and consumers (Buttel 1980) are the development of consumer-producer cooperatives. Community Supported Agriculture (CSA) is a newer example of this (Kane, 1990), serving, as Buttel suggested such entities might, as a vehicle for community social and economic development (1980).

CONCLUSIONS

This chapter describes many facets of both the farm and poverty "problems." In the end, however, the key question being addressed is from whence support for sustainable agriculture is going to come. For a large segment of its present supporters that question is coterminous with another one, "Who and what do we want to sustain?" (Allen and Van Dusen, 1990). If the answer to the question is only soil and water, upper-income environmental constituencies can be probably relied on for continued support. If the answer is the environment, mid-sized farmers, and rural communities, broader groups of environmentalists, progressive farm groups, rural communities themselves, and some of the general public will lend support. If the answer is the above, plus farmworkers and low- and middle-income consumers, the coalition expands to include many progressive antihunger public health and consumer groups, minority political groups, and low-income consumers themselves.

Farm policy in the next decades will face many challenges from all the forces outlined here, and probably many new ones. Agricultural and environmental leaders of many different political persuasions will have to work with these interests, despite the myths and class and racial prejudices that make this difficult. The mutual interests of farm and antihunger lobbies in improving the food security of the U.S. population should become more evident, and the links between the farm good and the social good more clear. By strengthening the connections between production and consumption we may be able to sustain agriculture, the environment, and the health and well-being of the population.

ACKNOWLEDGMENTS

I am most appreciative of Patricia Allen's unwavering belief in the worth of this project and her editorial assistance. I am grateful to Geoff Becker for his assistance with materials and for reading an early, needy draft; and to Mark Ritchie for helpful comments on a later version. Thanks also to Kathy Porter for patiently answering many questions and supplying data; to Otto Doering for his encouragement and suggestions; and to all of the people who attended the roundtable at the Michigan State diversity meeting and provided helpful insights and support. Mary Lou Orr and Rita Lowery typed most of the bibliography, for which I am most thankful, and Jean Bowering gave untiring and uncomplaining technical assistance; they were assisted by Meghan Clancy-Hepburn, who receives her mother's gratitude for helping and heartening.

REFERENCES

Allen, J., and A. Thompson. 1990. Rural Poverty among Racial and Ethnic Minorities. *American Journal of Agricultural Economics* 72(5):1161–1168.

Allen, P., and D. Van Dusen. 1990. *Sustainability in the Balance: Raising Fundamental Issues.* Santa Cruz, CA: Agroecology Program.

Andrews, M., and K. Clancy. 1985. *The Political Economy of the Food Stamp Program. National Center for Food and Agricultural Policy.* Disc Paper Ser. No. RR85-08. Washington, DC: Resources for the Future.

Ballenger, N., and C. Mabbs-Zeno. 1992. Feeding urban poor in the nation's capital—an interview with Rev. Charles Parker. *Choices* Second Quarter:10–13.

Bane, M. J., and D. Ellwood. 1991. Is American Business Working for the Poor? *Harvard Business Review* 58–66.

Becker, G. 1988. *Farm Support Programs: Their Purpose and Evolution.* Washington, DC: Congressional Research Service.

Becker, G. 1992. Commodity Credit Corporation Surpluses: Food for the Hungry? *CRS Report* 92–104ENR, January 23.

Beegle, J. A. 1991. Poverty and Caucasians in Nonmetro America. In *Social Science Agricultural Agendas and Strategies,* pp. III-78–85. G. Johnson and J. Bonnen (eds.). East Lansing: Michigan State University Press.

Bertini, C. 1991. Letter to Charles Stenholm, Committee on Agriculture. U.S. House of Representatives. June 17. Office of the Secretary, USDA.

Blobaum, R. 1991. Personal communication.

Bonanno, A. 1991. The Restructuring of the Agricultural and Food System: Social and Economic Equity in the Reshaping of the Agrarian Question and the Food Question. *Agriculture and Human Values* 8(4):72–82.

Bonanno, A., and T. Calastani. 1988. Laissez-faire Strategies and the Crisis of the Welfare State: A Comparative Analysis of the Status of the Elderly in Italy and in the United States. *Sociology Focus* 21(3):245–263.

Bonnen, J. 1984. Distributional Issues in Food and Agricultural Policy: Concepts and Issues. In *Increasing Understanding of Public Problems and Policies,* Farm Foundation 34th National Public Policy Educational Conference, Illinois.

Bonnen, J., and W. P. Browne. 1989. Why Is Agricultural Policy so Difficult to Reform? In *The Political Economy of U.S. Agriculture—Challenges for the 1990s,* pp. 7–33. C. Kramer (ed.). Washington, DC: National Center for Food and Agricultural Policy—Resources for the Future.

Breglio, V. J. 1992. *Hunger in America: The Voter's Perspective.* Lanham, MD: Research/Strategy/Management, Inc.

Browne, W. 1988a. *Private Interests, Public Policy, and American Agriculture.* Lawrence, KS: University Press of Kansas.

Browne, W. 1988b. The Fragmented and Meandering Politics of Agriculture. In *U.S. Agriculture in a Global Setting: An Agenda for the Future,* pp. 136–153. M. A. Tutwiler (ed.). Washington, DC: National Center for Food and Agricultural Policy—Resources for the Future.

Browne, W., J. R. Skees, L. E. Swanson, P. B. Thompson, and L. J. Unnevehr. 1992. *Sacred Cows and Hot Potatoes: Agrarian Myths in Agriculture Policy.* Boulder, CO: Westview Press.

Bryant, W. K., D. L. Bawden, and W. E. Saupe. 19891. The Economics of Rural Poverty—A Review of the Post World War II United States and Canadian Literature. In *A Survey of Agricultural Economics Literature,* Vol 3. *Economics of Welfare, Rural Development, and Natural Resources in Agriculture, 1940s to 1970s.* L. R. Martin (ed.). Minneapolis: University of Minnesota Press.

Busch, L., and W. B. Lacy. 1983. *Science, Agriculture and the Politics of Research.* Boulder, CO: Westview Press.

Buttel, F. H. 1980. Agriculture, Environment, and Social Change: Some Emergent Issues. In *The Rural Sociology of the Advanced Societies: Critical Perspectives,* 453–488. F. Buttel and H. Newby (eds.). Montclair, NJ: Allanheld, Osmum.

Buttel, F. H. 1981. American Agriculture and Rural America: Challenges for Progressive Politics. Cornell University Rural Sociology Bulletin No. 120.

Buttel, F. H. 1990. The Political Agenda for Sustainable Agriculture Short-, Medium-, and Long-Term Considerations. Paper presented at conference, Sustainable Agriculture: Balancing Social, Environmental and Economic Concerns, 28–30 June, University of California, Santa Cruz.

Buttel, F. H. 1991. Social Science Institutions, Knowledge, and Tools to Address

Problems and Issues. In *Social Science Agricultural Agendas and Strategies*, pp. I-26–43. G. Johnson, and J. Bonnen (eds.). East Lansing: Michigan State University Press.

Buttel, F. H., and I. La Ramee. 1991. The "Disappearing Middle: A Sociological Perspective. In *Towards a New Political Economy of Agriculture*, pp. 151–169. W. Friedland, L. Busch, F. Buttel, and A. Rudy (eds.). Boulder, CO: Westview Press.

Buttel, F. H., D. Larson, C. Harris, and S. Powers. 1980. Social Class and Agrarian Political Ideology. Revised version of paper presented at annual meeting of the American Sociological Association, August 1979, Boston.

Center on Budget and Policy Priorities. 1988. *Analysis of Changes in Food Stamp Participation*. Washington DC.

Center on Budget and Policy Priorities. 1992. *New Census Report Shows Dramatic Rise Since 1979 in Workers with Low Earnings*. Washington, DC.

Children's Defense Fund. 1991. *Child Poverty in America*. Washington, DC.

Christy, R. 1991. The African-American, Farming and Rural Society. In *Social Science Agricultural Agendas and Strategies*, pp. III-102–107. G. Johnson, and J. Bonnen (eds.). East Lansing: Michigan State University Press.

Clancy, K. L. 1984. Can Sustainable Agriculture Engender a Sustainable Diet? Paper presented at Second Conference on Sustainable Agriculture, April 16–17, Pomona College.

Clancy, K. L. 1991. A New Vision for Agriculture in the 1990s. Paper presented at National Conference on Organic/Sustainable Agriculture Policies, February 15, Washington, DC.

Clancy, Katherine L., and J. Bowering. 1992. The Need for Emergency Food: Poverty Problems and Policy Responses. *Journal of Nutrition Education* (Supplement). 24(1):12–17.

Cohen, B., and M. Burt. 1989. *Eliminating Hunger: Food Security Policy for the 1990s*. Washington, DC: The Urban Institute.

Community Nutrition Institute. 1989. Organic Life Insurance? *Nutrition Week* 19(28):1.

Community Nutrition Institute. 1991a. Organic Farming in the 1990s: Learning to Play Hardball. *Nutrition Week* 21(8):4–5.

Community Nutrition Institute. 1991b. Congress Almost Makes WIC More Like Entitlement. *Nutrition Week* 21(13):3.

Community Nutrition Institute. 1991c. House Puts U.S. Hunger in Gulf War Perspective. *Nutrition Week* 21(19):2–3.

Community Nutrition Institute. 1991d. Experiments Link Farmers to Food Program Clients. *Nutrition Week* 21(22):4–5.

Community Nutrition Institute. 1991e. Food Assistance Programs Serve Tens of Millions. *Nutrition Week* 21(24):4–5.

Community Nutrition Institute. 1991f. Dairy Overhaul Stumbles over Irreconcilable Goals. *Nutrition Week* 21(27):4–5.

Community Nutrition Institute. 1992. WIC and Farmer's Markets Don't Mix, Says USDA. *Nutrition Week* 22(19):6.

Congressional Record. March 20, 1991. S 3671–3673.

Connor, J. M. 1988. *Food Processing—An Industrial Powerhouse in Transition.* Lexington, MA: Lexington Books.

Constance, D., and Heffernan, W. 1989. One Rise of Oligopoly in Agricultural Markets: The Demise of the Family Farm. Paper presented at Agriculture, Food, and Human Values Society meeting, November 10–12, Little Rock, AK.

Constance, D., J. Gilles, and W. Heffernan. 1990. Agrarian Policies and Agricultural Systems in the United States. In *Agrarian Policies and Agricultural Systems,* pp. 9–75. A. Bonnano (ed.). Boulder, CO: Westview Press.

Crockett, E., K. Clancy, and J. Bowering. 1992. Comparing the Cost of a Thrifty Food Plan Market Basket in Three Areas of New York State. *Journal of Nutrition Education* (Supplement) 24(1):71–78.

Dallas Morning News. 1992. Number of Food Stamp Recipients Hits Record. May 30, p. 11-A.

Delfico, J. 1991. Farmworkers Face Gaps in Protection and Barriers to Benefits. USGAO Testimony Before the Select Committee on Aging, House of Representatives. GAO/T-HRD-91-40.

DeLind, L. 1987. For Whose Benefit? A Second Look at Fund Raisers and Other Charitable Responses to the U.S. Farm Crisis. *Agriculture and Human Values* 4(2–3):4–10.

DeLind, L. 1992. Cheap Food: A Case of Mind over Matter. Paper presented at Conference on Diversity in Food, Agriculture, Nutrition and Environment, June 4–7, Michigan State University, East Lansing.

Doering, O. 1991. Looking Back While Going Forward: An Essential for Policy Economists. *Journal of Agricultural Economics Research* 43(1):3–6.

Dunham, D. 1991. Food Costs—From Farm to Retail in 1990. *USDA-ERS Agricultural Information Bulletin.* No. 619. Washington, DC.

Elitzak, H. 1991. Food Costs Beyond the Farm Gate. *Food Review* 14(3):34–37.

Faeth, P., R. Repetto, K. Kroll, O. Dai, and G. Helmers. 1991. Paying the Farm Bill: U.S. Agricultural Policy and the Transition to Sustainable Agriculture. Washington, DC: World Resources Institute.

Feingold, K. 1988. Agriculture and the Politics of U.S. Social Provision: Social Insurance and Food Stamps. In *The Politics of Social Policy in the United States,* pp. 199–234. M. Weir, A. S. Orloff, and T. Skopcol (eds.). Princeton: Princeton University Press.

Flora, C. B., and J. Flora. 1988. Public Policy, Farm Size, and Community Well-Being in Farming—Dependent Counties of the Plains. In *Agriculture and Community Changes in the U.S.,* pp. 76–129. L. Swanson (ed.). Boulder, CO: Westview Press.

Food Research and Action Center. 1991a. *Food Stamp Facts.* Washington, DC.

Food Research and Action Center. 1991b. *Community Childhood Hunger Identification Project—A Survey of Childhood Hunger in the United States.* Washington, DC.

Fraser, N. 1989. *Unruly Practices: Power, Discourse and Gender in Contemporary Social Theory.* Minneapolis: University of Minnesota Press.

Friedland, W. 1991a. Shaping the New Political Economy of Advanced Capitalist

Agriculture. In *Towards a New Political Economy of Agriculture*, pp. 1–34. W. Friedland, L. Busch, F. Buttel, and A. Rudy (eds.). Boulder, CO: Westview Press.

Friedland, W. 1991b. Women and Agriculture in the U.S. In *Toward a New Political Economy of Agriculture*, pp. 315–338. W. Friedland, L. Busch, F. Buttel, and A. Rudy (eds.). Boulder, CO: Westview Press.

Gershuny, G., and T. Forster. 1992. Should 'Organic' Mean 'Socially Responsible'? *Organic Farmer* 3(1):5–6.

Goldman, B., and K. Clancy. 1991. A Survey of Organic Produce and Related Attitudes of Food Cooperative Shoppers. *Journal of Alternative Agriculture* 6(2):89–96.

Greenstein, R. 1991. Testimony before the House Committee on Ways and Means. March 13.

Guither, H. 1990. Trouble Ahead for Farm Commodity Programs. *Choices* Third Quarter: 34–35.

Guither, H., B. Jones, M. Martin, and R. Spitze. 1984. U.S. Farmers' Views on Agricultural and Food Policy: A Seventeen State Composite Report. North Central Regional Research Publication 300.

Hambridge, G. 1940. Farmers in a Changing World—A Summary. *Farmers in a Changing World—1940 Yearbook of Agriculture*. Washington, DC: USDA.

Haney, W. 1991. Theoretical Advances Arising from Studies of Women and Farming. In *Social Science Agricultural Agendas and Strategies*, pp. II-38–43. G. Johnson, and J. Bonnen (eds.). East Lansing: Michigan State University Press.

Harding, T. 1940. Science and Agricultural Policy. In *Farmers in a Changing World—1940 Yearbook of Agriculture*. Washington, DC: USDA

Hartford Food System. n.d. Summary of Farm-to-Family Services. Hartford, CT.

Hassebrook, C. n.d. *The 1990 Farm Bill: Its meaning for Sustainable Agriculture and Future Directions for the Sustainable Agriculture Movement*. Walthill, NE: Center for Rural Affairs.

Heffernan, W. D. 1989. Confidence and Courage in the Next Fifty Years. *Rural Sociology* 54(2):149–168.

Hite, J. 1991. Rural People, Resources, and Communities: An Assessment of the Capabilities of the Social Sciences in Agriculture. In *Social Science Agricultural Agendas and Strategies*, pp. III-21–30. G. Johnson and J. Bonner (eds.). East Lansing: Michigan State University Press.

Huang, C. L. 1991. Organic Foods Attract Consumers for the Wrong Reasons. *Choices* Third Quarter: 18–21.

Institute for Alternative Agriculture. 1991. *Understanding the True Cost of Food: Considerations for a Sustainable Food System*. Greenbelt, MD.

Iowa Department of Human Services 1985. *Food Stamps for Farmers and Self-Employed Iowans*. Ames, IA.

Johnson, G. 1989. Ethical Dilemmas Posed by Recent and Prospective Developments with Respect to Agricultural Research. Paper presented at Conference on Agriculture, Food, and Human Values, November 2. Little Rock, AK.

Johnson, G., and J. Bonnen (Eds.). 1991. *Social Science Agricultural Agendas and Strategies*. East Lansing: Michigan State University Press.

Jolly, D., H. Schutz, K. Diaz-Knauf, and J. Johal. 1989. Organic Foods: Consumer Attitudes and Use. *Food Technology* 43(11):60.

Jones, J. 1991. Memo to Hon. Richard Lugar on Estimated Effect on WIC Program Costs and Participation of Formula Price Increase for Fluid Milk as Proposed under S 671. *Congressional Record* March 20:S 3670–3671.

Kalbacher, J., and N. Brooks. 1990. Farmers Are Part of American Mainstream. *Choices* First Quarter: 22–23.

Kane, M. 1990. Community Supported Agriculture: Linking Farms to Communities. *NOFA-NY News* 8(5):1,3.

Katz, M. 1989. *The Undeserving Poor: From the War on Poverty to the War on Welfare.* New York: Pantheon Books.

Kelly, G., B. Rauschenbach, and C. Campbell. 1989. *Private Food Assistance in NY State: Challenges for the 1990s.* Ithaca, NY: Division of Nutritional Science, Cornell University.

Kirkendall, R. 1991. A History of American Agriculture from Jefferson to Revolution to Crisis. In *Social Science Agricultural Agendas and Strategies*, pp. I-14–25. G. Johnson and J. Bonnen (eds.). East Lansing: Michigan State University Press.

Knutson, R., C. R. Taylor, J. Penson, and E. Smith. 1990. Economic Impacts of Reduced Chemical Use. *Choices* Fourth Quarter: 25–31.

Korb, P., and N. Cochrane. 1989. World Food Expenditures. *National Food Review* 12(4):26–29.

Kotz, N. 1971. *Let Them Eat Promises: The Politics of Hunger in America.* New York: Doubleday-Anchor.

Kramer, C., and B. Elliott. 1989. The Consumer's Stake in Food Poilcy: The United States and the European Community. National Center for Food and Agricultural Policy. Disc. Paper Ser. No. FAP89- 04. Washington, DC: Resources for the Future.

Lane, S., and C. Bruhn. 1992. Organic Foods: Their Demand Will Remain Low. *Choices* First Quarter: 3.

Lelyveld, J. 1985. Hunger in America: The Safety Net Has Shrunk But It's Still in Place. *New York Times Magazine* June 16:20–24, 52–53, 59, 68–69.

Leman, C., and R. Paarlberg. 1988. The Continued Political Power of Agricultural Interests. In *Agricultural and Rural Areas Approaching the 21st Century: Challenges for Agricultural Economics*, pp. 91–119. R. J. Hildreth et al. (eds.). Ames, IA: Iowa State University Press.

Leonard, P., C. Dolbeare, and E. Lazare. 1989. *A Place to Call Home: The Crisis in Housing for the Poor.* Washington, DC: Low Income Housing Information Service.

Levedahl, W. 1992. Personal communication.

Lipsky, M., and M. Thibodeau. 1985a. *Food in the Warehouses, Hunger in the Streets. A Report on the Temporary Emergency Food Assistance Program.* Cambridge, MA.

Lipsky, M., and M. Thibodeau. 1985b. Delivery of Food Assistance. Paper presented at the National Neighborhood Coalitions Conference, November 24–26.

MacCannell, D. 1988. Industrial Agriculture and Rural Community Degradation. In *Agriculture and Community Change in the U.S.*, pp. 15–75. L. E. Swanson (ed.). Boulder, CO: Westview Press.

Madden, J. P. 1986. Beyond Conventional Economics. In *New Directions for Agriculture and Agricultural Research: Neglected Dimensions and Emerging Alternatives*, pp. 221–258. K. Dahlberg (ed.). Totowa, NJ: Rowman and Allenheld.

Madigan, E. 1991. Letter to Kika de la Garza, Committee on Agriculture. July 10. Office of the Secretary, USDA.

Manchester, A. 1991. Food Spending. *Food Review* 14(3):24–27.

Merrill, R. 1976. Toward a Self-Sustaining Agriculture. In *Radical Agriculture*, pp. 284–327. R. Merrill (ed.). New York: Harper and Row.

Morris, P., M. Bellinger, and E. Haas. 1990. *Higher Prices, Fewer Choices: Shopping for Food in Rural America*. Washington, DC: Public Voice for Food and Health Policy.

Murock, B. 1991. Food Bank Networks Respond to U.S. Hunger Crisis. In *Hunger 1992: Second Annual Report on the State of World Hunger*, pp. 12–21. M. Cohen and R. Hoehn (eds.). Washington, DC: Bread for the World Institute on Hunger and Development.

National Research Council. 1989. *Alternative Agriculture*. Washington, DC: National Academy Press.

O'Neill, M. 1992. As the Rich Get Leaner, the Poor Get French Fries. *New York Times* March 18:C-1.

Paarlberg, R. L. 1990. The Mysterious Popularity of EEP. *Choices* Second Quarter: 14–17.

Parliament, C. 1990. Human Capital, Economic Development, and the Rural Poor: Discussion. *American Journal of Agricultural Economics* 72(5):1182–1183.

Pena, D. 1992. The 'Brown' and the 'Green': Chicanos and Environmental Politics in the Upper Rio Grande. *Capitalism, Nature and Socialism* 3(1):79–103.

Piven, F., and R. Cloward. 1971. *Regulating the Poor: The Functions of Public Welfare*. New York: Pantheon Books.

Poppendieck, J. 1986. *Breadlines Knee Deep in Wheat: Food Assistance in the Great Depression*. New Brunswick, NJ: Rutgers University Press.

Porter, K. 1992. Personal communication.

Randolph, L. 1989. Testimony before the House Select Committee on Hunger. Ser. 101-2, March 23, p. 15.

Rasmussen, W. 1991. Personal communication, October 21.

Reichelderfer, K., and M. K. Hinkle. 1989. The Evolution of Pesticide Policy: Environmental Interests and Agriculture. In *The Political Economy of U.S. Agriculture—Challenges for the 1990s*, pp. 147–173. C. Kramer (ed.). Washington. DC: Resources for the Future.

Ritchie, M. 1979. *The Loss of Our Family Farms: Inevitable Results or Conscious Policies?* San Francisco, CA: Center for Rural Studies.

Robbins, W. 1990. Down on the Farm, Things Are Looking Up. *New York Times* October 23:A–12.

Ruttan, V. 1991. The Role of the Social Sciences in Rural Development and Natural Resource Management. In *Social Science Agricultural Policy Agendas and Strategies*, pp. 1-69–77. G. Johnson and J. Bonner (eds.). East Lansing: Michigan State University Press.

Sachs, C. 1991. Women and the Sustainable Agriculture Movement. Paper presented at Rural Sociology Society Meetings, April 17–24, Columbus, Ohio.

Samets, I. 1991. *Running Out of Food: The Results of the Community Childhood Hunger Identification Project in Suffolk County*. Albany: Nutrition Consortium of New York State.

Schertz, L. 1991. Personal communication.

Second Harvest. 1992. *1991 Annual Report*. Chicago: Second Harvest.

Sen, A. 1982. *Poverty and Famines*. Oxford: Clarendon Press.

Senauer, B., E. Asp, and J. Kinsey. 1991. *Food Trends and the Changing Consumer*. St. Paul: Eagan Press.

Smeeding, T. M. 1992. Why the U.S. Antipoverty System Doesn't Work Very Well. *Challenge*:30–35.

Spitze, R. 1985. Policy Leaders' Views on 1985 Agricultural and Food Policy. *Policy Research Notes 19a USDA-ERS*. Washington, DC.

Steinhart, P. 1991. What Can We Do about Environmental Racism? *Audubun* 93(3): 18–21.

Stevenson, G. W. 1992. Interfacing Agriculture and Rural Development. Paper presented at Conference on Diversity in Food, Agriculture, Nutrition, and Environment, June 4–7, Michigan State University, East Lansing.

Strange, M. 1988. *Family Farming: A New Economic Vision*. Lincoln; San Francisco: University of Nebraska Press; Institute for Food and Development Policy.

Swanson, L. 1988. Farm and Community Change: A Brief Introduction to the Regional Studies. In *Agricultural Community Change in the U.S.* L. Swanson (ed.). Boulder, CO: Westview Press.

Taylor, K. 1992. The Minnesota Food Association: A Brief History of an Organization Whose Time Has Come. *Culture and Agriculture* 43:14–15.

Terry, D. 1990. Fifty Years of American Agricultural Policies: What Stakes for Environment, Development, and World Trade? *Rongead Infos* Number 1:5–7, Spring.

Thompson, P. 1988. The Philosophical Rationale for U.S. Agricultural Policy. In *U.S. Agriculture in a Global Setting*, pp. 34–45. M. A. Tutwiler (ed.). Washington, DC: Resources for the Future.

Thompson, P. 1991. Constitutional Values and the Costs of American Food. In *The True Cost of Food: Considerations for a Sustainable Food System*, pp. 64–74. G. Youngberg and O. Doering (eds.). Greenbelt, MD: Institute for Alternative Agriculture.

Thurow, L. 1991. Agricultural Institutions under Fire. In *Social Science Agricultural*

Agendas and Strategies, pp. 1-55–64. G. Johnson, and J. Bonnen (eds.). East Lansing: Michigan State University Press.

Tutwiler, M. A., and A. B. Carr. 1988. An Agenda for the Future. In *U.S. Agriculture in a Global Setting*, pp. 49–71. M. A. Tutwiler (ed.). Washington, DC: Resources for the Future.

U.S. Department of Agriculture. 1992. *Cost of Food at Home Estimated for Food Plans at Four Cost Levels, March 1992, U.S. Average*. Hyattsville, MD: Human Nutrition Information Service.

U.S. Department of Commerce. 1992. *Poverty in the U.S. 1990*. Current Population Repts. Ser. P 60 No. 1850. Washington, DC: Bureau of Census.

U.S. Department of Health and Human Services. 1991. 56 *Federal Register* 6859. Poverty Guidelines. February 20, 1991.

U.S. Department of Labor. 1992. *Consumer Price Index—Food*. Washington, DC: Bureau of Labor Statistics.

U.S. General Accounting Office. 1988. *Food Assistance: Practices of the European Community*. GAO/RCED 88–100. Washington, DC.

U.S. General Accounting Office. 1990a. *Food Stamp Program: A Demographic Analysis of Participation and Nonparticipation*. GAO/REMD-90-8. Washington, DC.

U.S. General Accounting Office. 1990b. *Rural Development: Problems and Progress of Colonia Subdivisions Near Mexico Border*. GAO/RCED-91-37.

U.S. House of Representatives Committee on Agriculture. 1985. *A Review of the Thrifty Food Plan and Its Use in the Food Stamp Program*. Washington, DC: Government Prinitng Office.

U.S. House of Representatives Committee on Agriculture. 1990. *Food and Agricultural Resource Act of 1990*. Report on HR 3950. Report 101–569.

U.S. House of Representatives Committee on Agriculture. 1991. *Milk Inventory Management Act of 1991*. Report No. 2-173, Washington, DC.

U.S. House of Representatives Select Committee on Hunger. 1987. *Obtaining Food: Shopping Constraints on the Poor*. Washington, DC: U.S. GPO.

U.S. House of Representatives Select Committee on Hunger. 1990. *Food Security in the United States*. Committee Report. Washington, DC: U.S. GPO.

U.S. Senate Committee on Agriculture, Nutrition, and Forestry. 1985. *The Food Stamp Program: History, Description, Issues, and Options*. Washington, DC: Government Printing Office.

Variyam, J., I. Jordan, and J. Epperson. 1990. Preferences of Citizens for Agricultural Policies: Evidence from a National Survey. *American Journal of Agricultural Economics* 72:257–267.

Wharton, C. 1990. Reflection on Poverty. *American Journal of Agricultural Economics* 72(5):1131–1138.

Wilkerson, I. 1989. As farms falter, rural homelessness grows. *New York Times* May 2:A-1,A-14.

Winne, M. 1990. *Project Sustain*. Hartford: Hartford Food System.

CHAPTER 12

The Sustainable Agriculture Policy Agenda in the United States: Politics and Prospects

Garth Youngberg, Neill Schaller, and Kathleen Merrigan

F or U.S. agriculture, the 1980s was a decade of awareness-building and reexamination. Important new groups and ways of thinking influenced the politics of American agriculture. To an unprecedented degree, the technologies, goals, and consequences of conventional, chemical-intense agriculture were scrutinized and reevaluated. Many farmers and agricultural scientists, respectively, began to reassess their production practices and research missions. In some cases, whole institutions began to rethink the purposes and consequences of production agriculture. The critical need to balance agricultural production with environmental quality and economic viability became a widely recognized and unavoidable priority. Issues surrounding social justice, farm structure, animal rights, and food safety and quality also gained heightened attention during this period. Out of this intellectual, economic, and political turmoil, there emerged the notion of agricultural sustainability.

At least on the surface, by the end of the decade the concept of sustainability had been embraced by virtually every constituency with an interest in agriculture: environmentalists, consumers, researchers, farmers, rural communities, policymakers, and farm input manufacturers and distributors. As

Food for the Future: Conditions and Contradictions of Sustainability, edited by Patricia Allen.
ISBN: 0-471-58082-1 © 1993 John Wiley & Sons, Inc.

an abstract symbol, sustainability had become agriculture's central goal and rallying cry—its predominant ideology. The symbolism of sustainability had profoundly altered the politics of American agriculture. And yet, the outcome of these developments is uncertain.

In the pages that follow, we examine the impact and implications of sustainability for agricultural policymaking. Sustainability, after all, is an enormously powerful symbol. In terms of its emotional and evocative meanings, it probably ranks alongside such concepts as freedom, liberty, and democracy. Indeed, while these latter American political icons may or may not be essential to the continuation of life itself, the very thought of an unsustainable agriculture immediately conjures up images of massive human deprivation and suffering and, ultimately, mass starvation. What could be more important than sustainability? It is difficult to imagine a more powerful symbol.

While the boundaries of this chapter preclude an in depth analysis of the symbolism of sustainability, even a cursory examination of these phenomena will, we believe, help to explain the present character of, and future prospects for, sustainable agriculture policy in the United States. It will also help to illuminate our discussion of selected recent policy decisions, including certain major features of the 1985 and 1990 farm bills. Here, we show not only the symbolic impact of sustainability on the U.S. agricultural policy process and its policies, but also how the process itself influences the notion of sustainability. The goals of sustainability, however laudable, easily can be lost in the rhetoric of the policy process. Finally, we outline a possible strategy for the future, offering some thoughts about how sustainability can and should be utilized as a future guide for agricultural policymaking. We begin with a look back.

SUSTAINABLE AGRICULTURE: THE SEARCH FOR SCIENTIFIC AND POLITICAL LEGITIMACY

During the 1960s and 1970s, the major critics of conventional agriculture coined and proffered such terms as organic, biological, and ecoagriculture to represent those philosophies and technologies deemed preferable to the chemical and energy intensive production systems that had come to dominate U.S. agriculture (Youngberg, 1984). Spokespersons for these alternative schools of thought rarely, if ever, used the word sustainable as a label for their diversified, low-chemical systems. Instead of arguing, as they later would, that such technologies and cultural practices as crop rotations, biological pest controls, integrated pest management (IPM), green manure crops, and shallow and reduced tillage were, indeed, sustainable, proponents of organic farming (the most generic agricultural alternative of the 1960s and

1970s) favored such approaches simply because of their alleged environmental, conservation, and social benefits to society.

Such technologies, it was claimed, could help address a wide range of concerns that had come to be associated with conventional agriculture. According to USDA's organic farming report (1980), these concerns were as follows:

- Increased cost and uncertain availability of energy and chemicals;
- Increased resistance of weeds and insects of pesticides;
- Decline in soil productivity from erosion and accompanying loss of organic matter and plant nutrients;
- Pollution of surface waters with agricultural chemicals and sediment;
- Destruction of wildlife, bees, and beneficial insects by pesticides;
- Hazards to human and animal health from pesticides and additives;
- Detrimental effects of agricultural chemicals on food quality;
- Depletion of finite reserves of concentrated plant nutrients, for example, phosphate rock;
- Decrease in numbers of farms, particularly family-type farms, and disappearance of localized and direct marketing systems.

Despite these and other alleged problems and concerns of conventional agriculture and the claim that organic approaches could address them in many positive ways, the attribute of sustainability per se was not explicitly and routinely associated with organic farming and other low-chemical approaches until the early 1980s (Harwood, 1990).

From Organic Farming to Sustainable Agriculture

Prior to the 1980s, organic farming and other low-chemical production systems had been summarily dismissed by most of conventional agriculture. Nonetheless, in 1979, Secretary of Agriculture Bob Bergland asked USDA's Science and Education Administration (SEA) to conduct a study of organic farming in the United States. The Secretary felt the U.S. Department of Agriculture should know more about the potential of such systems for addressing economic, energy, environmental, health, and farm structure issues in American agriculture. A multidisciplinary team of scientists was assembled to carry out this investigation. A year later, USDA published its *Report and Recommendations on Organic Farming* (1980).

Despite its official status, the report was greeted initially with notable displeasure and opposition by key elements of the agricultural scientific, industrial, and policy communities. Later, following the 1980 presidential election and the arrival of the new administration, the report would fall into disfavor with USDA leadership as well. This was despite the fact that the report's definition of organic farming did not totally rule out the use of synthetic chemicals.

This reaction to a formal, scientific USDA report on organic farming was disheartening and dismaying to many within the organic agricultural community. They had hoped the report would finally and firmly establish the credibility and official acceptance of the role and importance of organic farming to all of agriculture. It soon became apparent, however, that such was not to be. The proponents of low-chemical production techniques had seriously underestimated the negative symbolism of organic farming, which had long since been dismissed by conventional agriculture as little more than a primitive, backward, nonproductive, unscientific technology suitable only for the nostalgic and disaffected back-to-the-landers of the 1970s.

In reaction to this newly perceived reality, advocates of organic agriculture made a conscious effort to identify and promulgate new language, new words, to describe the character and benefits of low-chemical agriculture. During this period, alternative agriculture spokespersons such as Wes Jackson and Eliot Coleman began to write about sustainable approaches. Robert Rodale, a major figure in organic agriculture, coined the word regenerative. Other leaders preferred ecological, or biological, or ecoagriculture. One even suggested "independent" farming. A small alternative farm organization in the Midwest chose to call itself the "Practical Farmers of Iowa." While some of these terms had been used prior to the 1980s, many were introduced for the first time in that period. It is our contention that this proliferation of terms was spawned consciously in an effort to find terminology for organic techniques that would be more acceptable to the conventional scientific, farmer, and policy communities. Such is the power and meaning of political symbolism (Edelman, 1964).

The Role of Symbolism in Sustainable Agricultural Policy

As noted above, throughout the 1960s and 1970s, critics of conventional agriculture, for the most part, bore the label organic—a word that by then carried with it highly negative symbolic meanings for most conventional agriculturalists. Even when critics used other terms, such as biological or ecoagriculture, they were often assigned the organic label by defenders of conventional, chemical-intensive technologies. Given the enormously negative symbolism of organic farming during that period, this strategy was a particularly effective way to discredit the critics of conventional agriculture. Because many of the technical principles of organic farming (e.g., crop rotations, minimal chemical use, mechanical cultivation) were associated with an earlier era in production agriculture, critics could easily discount its proponents as regressive and antiscientific pleaders for a more primitive and nonproductive form of agriculture.

In his well-known 1971 statement on organic farming, for example, former Secretary of Agriculture, Earl Butz, spoke for most of mainstream agriculture when he said: "We can go back to organic agriculture in this country if we must—we know how to do it. However, before we move in that direction, someone must decide which 50 million of our people will starve!" (Butz, 1971). Indeed, the symbolism of organic farming made it relatively easy for conventional agriculture to dismiss it.

Circumstances changed dramatically in the 1980s. By then, the concerns of alternative agriculturists were becoming widespread. The conventional agricultural community had, albeit begrudgingly, begun to recognize and acknowledge the adverse side effects of many conventional practices. Beyond these changes, meanwhile, advocates of organic farming had adopted the symbolism of sustainability, a far more promising label for advancing organic-type technologies and approaches.

This strategy had a profound impact on the policy process. Because of the symbolic power of sustainability, and organic farming's newfound association with it, critics of organic agriculture could no longer simply dismiss such technologies as regressive and unimportant. After all, these were the technologies, it was now claimed, that could lead to a sustainable agriculture—an agriculture of the future, not the past. The symbolism of low-chemical agriculture had been turned upside down. Its critics were suddenly placed in the awkward position of trying to discredit technologies that were now associated with the enormously powerful symbolism of sustainability. For these critics, new strategies would be required.

The decision to adopt sustainability as the symbolic vehicle for advancing low-chemical production technologies has had a number of unanticipated effects on the agricultural policy process. First, it has, as its proponents had hoped, greatly helped to legitimize and thus advance low-chemical, organic-type approaches within the conventional agriculture scientific and policy communities. Indeed, partly as a result of this new symbolism, many conventional scientists have helped in the process of legitimization. They have begun to do research on low-chemical approaches and speak out in behalf of such technologies. One need only peruse the most recent annual list of scientific and policy conferences on "sustainable" agriculture to find confirmation of this result. The recent and rapid proliferation of various sustainable agriculture centers and programs within the U.S. land grant university system provides additional evidence. Finally, the extraordinary response of many agricultural scientists to USDA's much-heralded Low-Input Sustainable

Agriculture (LISA) research and education program, first funded in 1988, is also noteworthy in this regard.

Second, and quite predictably, the power of sustainability as a symbol has drawn virtually every agricultural interest within its embrace. Fertilizer and pesticide technologies, as well as the products of biotechnology, now reside under the sacred temple of sustainability. The same is true for best management practices, integrated management practices, reduced tillage, improved manure management, and even corn–soybean rotations. Presumably, the latter qualifies for the mantle of sustainability because it reduces the corn rootworm problem and thus the need for insecticide use. The symbolic power of sustainability is truly intoxicating.

Third, at least for the time being, the symbolism of sustainability has made strange bedfellows among formerly quite different ideological camps and spokespersons. This, on the surface at least, has created the impression that all of agriculture is marching to the same drummer—a notion with a tantalizing and powerful symbolism of its own. As Edelman (1964:32) noted nearly three decades ago, "Emotional commitment to a symbol is associated with contentment and quiescence regarding problems that would otherwise arouse concern." After several decades of accelerating conflict over the future of American agriculture, sustainability has provided a comforting, although largely unanalyzed, symbolic refuge for an incredibly disparate array of agricultural interests.

Finally, the range of ideologies and technologies presently being advanced under the symbol of sustainability has direct consequences for policymakers and their staffs. On one hand, sustainability has, indeed, fostered a measure of good will among otherwise competing interests. It has drawn attention to the undeniable fact that all of agriculture—indeed all of society—has an urgent stake in the concept of sustainability. This realization has engendered a vigorous and broad-based dialogue—one often involving widely divergent views. In this way, the concept of sustainability has, to some degree, opened new pathways for discussion and even some limited consensus on needed future directions.

On the other hand, while this process may have brought a measure of civility to the sustainability debate, it has also blurred and confused the policy process. When virtually all proposals are justified on the grounds that they will contribute to agricultural sustainability, it becomes increasingly difficult for decision makers to distinguish among them. They cannot do so on objective grounds, because the agricultural community has yet to develop scientific criteria for and indicators of sustainability. Moreover, given the powerful and positive emotional symbolism of sustainability, it is politically difficult for both elected and appointed officials to dismiss proposals and programs bearing the sustainable label, regardless of their content. Finally, confusion surrounding the definition and character of sustainability creates an added burden for those in positions of political leadership who would attempt to initiate and fashion specific policy proposals.

SUSTAINABLE AGRICULTURE AND FARM POLICY IN THE 1980S

Every 5 years, the Congress writes a "farm bill" covering everything from commodity price support programs and food stamps to soil conservation and research and education. Between farm bills, agricultural policy is continually debated and shaped by the process of implementation and the appropriation of federal funds. As the decade of the 1980s unfolded, this multifaceted process was pervaded by, and in turn influenced, the emergence and meaning of the term "sustainable agriculture" and the awkwardness it created for both defenders and critics of conventional agriculture.

The Early Years

In the 1981 farm bill, the first one passed during the Reagan administration, sustainability as an embracing symbol for agriculture had not yet replaced the concept of organic farming. Indeed, proponents of low-chemical approaches clung to the organic label in a provision of the 1981 bill which called for "establishing integrated multidisciplinary organic farming research projects designed to foster the implementation of the major recommendations of the U.S. Department of Agriculture's *Report and Recommendations on Organic Farming,* July 1980" (U.S. Congress, 1981).

A year later, in direct response to this authorizing language, Congressman Jim Weaver of Oregon, laying the groundwork for the 1985 farm bill, introduced in the House of Representatives "The Organic Farming Act." Its purpose was "to facilitate and promote the scientific investigation and understanding of methods of organic farming and to assist family farmers and other producers to use methods of organic farming to complement conventional chemical-intensive methods of farming" (U.S. House of Representatives, 1982).

Meanwhile, Senate supporters of organic agriculture had already grown wary of using that label to advance the research and education agenda of USDA's organic farming report. Senator Patrick Leahy of Vermont, anxious to support such legislation but aware of mainstream agriculture's opposition to "organic farming," put a more politically acceptable label on the companion bill he introduced in the Senate. He called his bill the "Innovative Farming Act of 1982" (U.S. Senate, 1982). But its stated purpose, like that of the

Weaver bill, was to develop and disseminate sound information on the alternative of organic farming.

When neither bill passed, more and more supporters of organic farming came to believe that only by embracing a more palatable term could they hope to win significant policy support for the organic alternative to conventional farming. Defeat of the Weaver and Leahy bills also sent a strong signal to participants in the upcoming 1985 farm bill process that the traditional agriculture community, though ready to begin addressing the unanticipated ill effects of conventional farming, was vigorously opposed to major changes in the farm policy agenda and unwilling to share control of the agenda with critics of conventional farming.

The 1985 Farm Bill

Labeled the Food Security Act, the 1985 farm bill (U.S. Congress, 1985), more than previous ones, backed programs to reduce the adverse effects of conventional agriculture. It established a landmark Conservation Reserve Program to take highly erodible land out of the production of corn, soybeans, cotton, and other highly erosive row crops. It extended the sensitive concept of cross compliance to soil conservation, telling farmers that they would lose their eligibility for federal farm program benefits if they did not farm in ways to curb soil erosion on their cropland.

"Sodbuster" and "swampbuster" provisions also threatened farmers with a loss of program benefits if they converted range and wetlands to crops without following approved soil conservation plans.

The 1985 bill supported research and education on farming practices to conserve resources and protect the environment (Subtitle C, Title XIV). As a sign of the times, however, the title chosen for it, "Agricultural Productivity Research," carefully shunned use of terms such as sustainable, let alone organic. Ironically, the program authorized by this provision soon came to be widely known as the popular Low-Input Sustainable Agriculture (LISA) research and education program.

Between the 1985 and 1990 Farm Bills

Despite ambiguities in the 1985 farm bill, it opened doors to the agricultural appropriations process and gave new hope to proponents of low chemical approaches to sustainable agriculture. However, little new progress was made until 1988, when the Congress appropriated $3.9 million to launch the LISA program. The program had remained on the drawing board since its authorization in 1985 due to lack of new funds and the reluctance of the

Administration to redirect any of its traditional research funds for that purpose.

The appropriations language referred openly to the need for sound information on low-input methods that could reduce farmers' dependence on chemical fertilizers, pesticides, and other purchased inputs. Further, the launching of the LISA program sparked new interest in alternative agriculture research and education, both in the land-grant colleges of agriculture and the USDA. Also, as soon as LISA was funded, the USDA issued a Secretary's Memorandum "to state the Department's support for research and education programs and activities concerning 'alternative farming systems,' which is sometimes referred to as 'sustainable farming systems'" (USDA, 1988).

But words and symbols continued to foster ambiguity and uncertainty. Consider the choice of the term LISA. Although the exact origin of the term is not clear, the combining of "low-input" with "sustainable" apparently was an attempt to balance the apple-pie concept of sustainable agriculture with a more focused term, low input. Adding "low-input" also straddled a fence. It told defenders of conventional farming that business as usual was not the solution, and it told backers of organic and related persuasions that LISA's purpose was not to eliminate the use of all synthetic farm chemicals. But, from all indications, neither side of the issue was ever convinced.

The attention devoted to the sustainable issue, especially media attention, reached new heights in 1989 with the release of a long-awaited study report, *Alternative Agriculture,* by the National Research Council of the National Academy of Sciences (National Research Council, 1989). The report gave defenders of conventional agriculture a target for their resentment and frustration, and supporters of alternative, organic, and sustainable agriculture considerable ammunition. It all but guaranteed that sustainable agriculture would be a central issue in the upcoming 1990 farm bill debate.

THE 1990 FARM BILL: TURNING THE CORNER

Expectations were high as the 1990 farm bill debate unfolded. Since numerous and diverse interest groups were talking about sustainability by this time, it seemed reasonable to expect Congress to enact significant changes in farm policy to ensure the adoption of less-chemically intensive systems of production. But Congress found the issue of sustainability confusing and divisive. Congressional members and their staffs spent months fighting over issues such as research to reduce chemicals in farming, national standards for organic food production, and incentives for farmers to engage in crop rotations. Many sustainable agriculture advocatives were frustrated and disappointed. Senator Robert Kerrey of Nebraska spoke for many when he said, "My concern is that we are getting behind the curve on environmental issues related to agriculture . . . . We are fighting against the clock" (U.S. Senate, Congressional Record, 1990:S 10316).

In the end, the 1990 farm bill differed dramatically from farm bills of the past: it included a multitude of initiatives dealing with natural resources, environmental protection, and sustainable agriculture. But changes came at the margins. The lack of consensus on sustainability left most traditional farm support programs untouched.

Defining Sustainability

The definition and goals of sustainable agriculture were controversial topics from the start. The bill reported out of the Senate Agriculture Committee used the definition for sustainable agriculture developed by the National Research Council in their report *Alternative Agriculture:* agriculture that, among other things, reduces chemical use to the extent practicable (National Research Council, 1989).

Some Senators and Congressman, after hearing protests from commodity group organizations and chemical and fertilizer companies, became uncomfortable with this definition and preferred the definition promoted by the U.S. Department of Agriculture: agriculture that is environmentally, agronomically, and economically sound over long and short periods (USDA, 1990). As Senator Richard Lugar of Indiana conceded during debate on the Senate floor, the differences between the two definitions are "very subtle, but they are important" (U.S. Senate, Congressional Record, 1990:S 10313). Senator Lugar was right. The farm bill contained over 700 pages, included 21 titles, authorized the spending of billions of dollars, and covered commodity support payments, international trade, natural resource conservation, credit, marketing, research, and many other issues. Yet, the definition of sustainable agriculture—less than a paragraph of text—became one of the most significant and controversial issues of the 1990 farm bill. The definitional dispute fueled weeks of debate and, ultimately, a battle on the floor of the Senate that required every Senator to record his or her vote on which definition should become the law of the land (U.S. Senate, Congressional Record, Rollcall Vote No. 164, 1990).

Members of Congress were asked to join one of two sides in the sustainable agriculture debate. Trying to decide which side to join was not difficult: members readily and passionately joined their traditional allies in farm policy and pesticide regulation. But the substance of the debate itself was close to meaningless as members evoked symbols and emotions rather than scientific information to defend their chosen definition of sustainability.

On one side, Congresspersons argued that our national goal should be a reduction in the amount of pesticides and fertilizers used in agricultural systems. They based this argument on the belief that pesticide and fertilizer use has contributed significantly to environmental degradation and food safety problems. Further they argued that changes in government policies, along with targeted research, would bring about a food production system capable of sustaining a sizable reduction in pesticide and fertilizer use without a significant sacrifice in production or profitability.

On the other side, members argued that our national goal should be the efficient use of pesticides and fertilizers. They asserted that modern pesticides and fertilizers have significantly improved American agriculture and that their careful and efficient use only minimally contributes to problems of environmental degradation and food safety. In other words, how farm chemicals are used is more important than the quantity of farm chemicals used. Further, they believed that the concern over pesticide and fertilizer use is disproportionate with their real hazards.

The debate culminated on July 23, 1990, on the Senate floor when Senator Charles Grassley of Iowa offered an amendment to replace the National Research Council definition of sustainable agriculture with the USDA definition that made no mention of reduction of chemicals. Many senators were confused over the difference and their remarks during the debate reveal their failure to grasp the importance of establishing a clear definition for sustainable agriculture. Senator Tom Harkin of Iowa complained that "definitions are definitions," and urged his colleagues to move forward on sustainability no matter how defined (U.S. Senate, Congressional Record, 1990:S 10317).

While the Grassley amendment was defeated by a vote of 60 to 32, the victory was evanescent. Not only did a few Senators understand the details of the sustainable agriculture debate, unfortunately, the battle left each side more determined to win the next time around. The definitional dispute continued for several more months in the House and in the House–Senate conference committee before final passage of the law.

Research and Education

Despite all the time spent in debate, Congress could not agree on the definition of sustainable agriculture. In an effort to appease all sides, the 1990 farm bill included two definitions of sustainable agriculture, embodied in the research title. Very similar—some would argue duplicative—research programs were established under each definition and both programs now compete for limited federal research funds.

The sustainable agriculture research section of the 1990 farm bill was divided into three chapters. The first chapter retained the National Research Council definition of sustainable agriculture and expanded what was known as the USDA's Low-Input Sustainable Agriculture program or LISA. But it was given a less controversial title: "Best Utilization of Biological Applications" (BUBA). The BUBA chapter directed the Secretary of Agriculture to conduct research and extension programs that reduce use of toxic materials in production, improve low-input farm management, and promote crop, livestock, and enterprise diversification. The program requires that farmers be involved in the development, implementation, and evaluation of all projects receiving assistance. A Federal–State matching grant program was also established to encourage States to carry out sustainable agriculture programs and activities. The sum of $40,000,000 annually was authorized to be appropriated through the Cooperative State Research Service (CSRS) for this chapter (U.S. Congress, 1990, Sections 1621 and 1623).

While sustainable agriculture advocates were pleased with the BUBA chapter, they were unhappy with chapter two which they characterized as a "watered down" approach to sustainability. Chapter Two established an Integrated Management Systems program based on the USDA definition of sustainable agriculture. This chapter received strong support from several livestock organizations who did not want the concept of humane care included in the definition of sustainability. In Chapter Two, the Secretary is directed to establish a program to enhance research and dissemination of information related to farming operations, practices, and systems that optimize crop and livestock production that are environmentally sound. The Secretary is to encourage producers to adopt and develop individual, site-specific integrated crop management and resource management practices. The sum of $20,000,000 annually was authorized to be appropriated through the Extension Service (ES) for this chapter (U.S. Congress, 1990, Section 1627).

There was one point, however, on which everyone agreed: The Extension Service was not well-trained in sustainable agriculture nor was it disseminating sustainable agriculture information to the extent needed. Advocates of Chapters One and Two joined together in writing Chapter Three, which directed the Secretary to provide education and training for Extension Agents and for other professionals involved in the education and transfer of sustainable agriculture technical information. Such training would take place at regional training centers to provide intensive training for agriculture specialists. A competitive grants program was also authorized to provide funds for workshops to familiarize all other extension agents with basic knowledge on sustainable agriculture. All agriculture Extension Service agents were required to receive one of the two types of training in sustainable agriculture no later than 1995. The Secretary was also directed to develop technical guides, handbooks, and other educational material that describe farm production systems that foster sustainable agriculture production systems (U.S. Congress, 1990, Sections 1628 and 1629).

Incentives to Adopt Sustainable Agriculture

Most of the sustainable agriculture debate centered on research. But many sustainable agriculture advocates argued that while research was critical, it was equally important to begin establishing incentive programs to encourage farmers to adopt sustainable farming systems. They argued that there is much information and technology not used by farmers because farmers are either unaware of it, suspicious of it, or prevented by government programs from adopting it.

Several efforts were undertaken to fundamentally change the structure of farm programs in order to provide incentives for farmers to adopt alternative approaches to farming. The success of these efforts was limited because the agriculture committees have been notably conservative in this area and sensitive to the concerns of traditional farm constituencies. In the end, only two modest programs were authorized.

An Integrated Farm Management Program Option was designed to allow farmers to receive commodity price supports on their commodity base acreage while shifting from monoculture to practices involving rotation crops (U.S. Congress, 1990, Section 1451). The voluntary program was designed to help producers adopt integrated, multiyear, site-specific farm resource management plans. If enrolled in the program, base acres and yields on which program payments are based were not reduced as a result of planting a resource conserving crop as part of a resource conserving crop rotation. The Secretary was directed to enroll from 3 to 5 million acres annually in this program through 1995. It is important to note, however, that Congress considered this program to be an experimental "demonstration": it was not viewed as central to the farm price support system.

The farm bill established national standards for "organically grown" food (U.S. Congress, 1990, Title XXI). Many sustainable agriculture advocates supported national standards as a way to help eliminate consumer confusion, provide food alternatives, and give farmers market incentives to use fewer chemicals.

Implementation

So far, implementation of these programs and provisions has been spotty (Center for Resource Economics, 1992). The federal deficit and the claim that the goals of programs authorized in the farm bill are being addressed through traditional programs are reasons commonly given by the Administration and the Congress for failure to begin or to expand activities authorized in the bill. Another reason, often hidden, is pervasive skepticism or outright opposition among the traditional policy participants to some of the new programs, as well as to legitimate technical problems of devising workable rules and regulations.

The USDA has reluctantly implemented some of the sustainable agriculture provisions, perhaps to avoid adverse publicity. The Department's first response to the Integrated Farm Management Program Option was to issue regulations, since revised, that would have reduced the acreage of land eligible to participate in the program. In implementing the National Research Initiative (U.S. Congress, 1990, Section 1615), a competitive research grants program, the Department was required to consider sustainability as it solicited and evaluated proposals for funding of research projects. To begin to meet that requirement, representatives of sustainable agriculture were invited to identify farmers and fellow proponents who might serve on advisory and subject matter panels involved in identifying research needs and project approval criteria. Though well-intentioned, this step clearly treats sustainable agriculture, not as a central goal, but as an "additional consideration" imposed on business as usual.

Years after the 1990 farm bill became law, funding for the sustainable agriculture research provisions remains well below the $40 million authorized for it. Other provisions remain on the drawing board. The coalition of diverse interest groups that came together to push sustainable agriculture provisions in the 1990 farm bill faltered once it came time to lobby for appropriations and provide more than lip service to the goal of sustainability.

Overall, the 1990 farm bill represented a meager gain for sustainable agriculture. While there was confusion and debate over the definition of sustainable agriculture, members of Congress were united in their belief that many farmers lack the necessary information to adapt to an increasing number of pesticide and production restrictions designed for environmental and food safety protection. However, the fact that disagreement on the basic definition of sustainable agriculture persisted throughout the farm bill debate demonstrates that the future support for sustainable agriculture in the Congress is still far from secure.

Perhaps the old adage that knowledge is power holds especially true here. Until more science and education are available to refine the definition of sustainable agriculture, the political debate will continue to churn. As Senator Kerrey noted, the clock is ticking. Many analysts predict that if agreement is not reached soon within the agricultural community over the definition and future of sustainability, the environmental and consumer interests will succeed in enacting tough federal and state regulations, forcing farmers toward sustainability rather than guiding them there.

A STRATEGY FOR THE FUTURE

Having discussed the concept, ideology, and politics of agricultural sustainability, and how the farm policy process addressed sustainability through the 1980s, we turn now to a suggested strategy for the future. We do so convinced that sustainability must be agriculture's preeminent goal in the decades ahead. In its simplest terms, the strategy we recommend includes three fairly obvious steps. They are (1) agree on where we want to go, (2) determine the best path to sustainability, and (3) design and implement policies and actions needed to succeed.

Agreeing on Where We Want to Go

A majority of those who say they support the sustainability of agriculture today seem to have in mind an agriculture that, through time, will produce adequate supplies of food and fiber at a profit for farmers, conserve resources, protect the environment, support rural communities, and enhance human and animal health and safety. But lack of understanding and agreement about the combination and importance of different goals of sustainable agriculture remains a serious roadblock.

Some proponents of sustainability agree with Chuck Hassebrook, Center for Rural Affairs in Nebraska, that "It is not enough to have an agriculture that is environmentally sound if it destroys family farms and rural communities. Part of this agricultural vision is to be sustaining opportunity in rural communities" (Hassebrook, 1989). Patricia Allen and colleagues at the University of California in Santa Cruz believe that the goals of agricultural sustainability must also include social justice and equity for farmworkers and all people (Allen et al., 1991). At the opposite extreme, Timothy Crews and others (1991) at Cornell University reason that the concept must be limited to ecological sustainability, or physical "endurability." They fully agree that the economic profitability of sustainable practices, protection of the family farm, and social equity are all important and legitimate goals. But as all can vary from one time and place to another, they defy identification and measurement of the kind that applies to physical and biological phenomena. Therefore, it is better to address these other considerations separately. For example, they argue that "If an ecologically sustainable farming practice does not turn a profit, then we should look critically at our social structure to determine why, rather than use this economic indicator to judge the practice as unsustainable" (Crews et al., 1991:149).

Supporters of sustainability also differ in their perceptions of the scope of sustainability. Most speak only of promoting sustainable production practices, while others see sustainability as a yardstick to also judge other activities and functions that are part of the same total agricultural "system," such as the performance of production input industries, processing, and distribution of food.

So far, proponents of sustainability, anxious to avoid confrontation with defenders of conventional agriculture and thereby to gain wider acceptance for the movement, have been reluctant to dwell on differences in the way they define the concept. Unconsciously perhaps, they are lured to a definition that would offend no one and promise something for everyone. The definition put forth in the 1990 farm bill reads like a bid to do just that (U.S. Congress, 1990).

The tendency to seek a politically expedient definition is especially evident when discussions turn to the distinction between *low-input* and *organic* farming. While some proponents of sustainable agriculture believe that low-input farming means farming without any synthetic chemicals, most proponents of sustainability, aware of the negative symbolism of organic farming, have gone out of their way to explain that they never meant to eliminate the use of all chemicals. At least in the eyes of organic believers, the latter response has sanctioned a steady watering down of the meaning of sustainable farming.

By accepting a watered-down definition, proponents of sustainable agriculture, without realizing it, have also endorsed the companion view that sustainability can be achieved by fine-tuning conventional agriculture. According to this view, farmers can continue to use conventional farm chemicals, but they must do so prudently. Thus the case for organic farming has been virtually reduced to that of serving niche markets for chemical-free products. In this sense, regrettably, the potential role of organic systems in achieving overall sustainability in U.S. and world agriculture could be overlooked.

Another effect of the dilution of the meaning of agricultural sustainability—if not partly a cause of it—has been the expanding role of the traditional agricultural community in the quest for sustainability. The desire of proponents of sustainability to involve the traditional community has considerable logic. Land-grant colleges, the USDA, farm groups, and even agribusiness firms must be involved eventually. However, as long as supporters of sustainable agriculture hesitate to address differences in what they mean by sustainability, the traditional agriculture community is likely to fill the void with a definition that will be compatible with current mainstream agriculture. To vocal critics of conventional agriculture, this process is already underway (Buttel and Gillespie, 1988).

In our view, proponents of agricultural sustainability must forge a new and different agreement on the meaning of sustainable agriculture. Clearly, a measure of rational consensus regarding the scope, character, and requirements of sustainability would seem to be a prerequisite to the successful identification, development, and implementation of needed policy and research agendas.

We must agree on where we are going, or need to be, before arranging our travel plans. Viewed in this way, the most urgent research and policy needs revolve around mechanisms and strategies for achieving a measure of societal consensus about the essence of sustainable agriculture. Without such agreement, efforts to tinker with this or that policy alternative (for example, reducing target prices so that, presumably, farmers will apply less fertilizer) quickly become relegated to the purely political—an arena from which a truly sustainable agriculture is unlikely to emerge.

To illustrate further, unless a measure of consensus can be reached on the meaning of sustainable agriculture, how will scientists and policymakers develop criteria for measuring movement toward sustainability? How will they evaluate research programs and other policy proposals? How would such factors as the role of family farms, small rural communities, and rural aesthetics be measured and evaluated? The same questions can be raised with respect to virtually all segments of our food and agriculture system—farmers, farm suppliers, industrial processors and merchandisers, laborers and migrant workers, educators, advisors, wildlife advocates, environmentalists, retailers, media specialists, and, of course, consumers.

To address these critical questions, we must put science to work as a full partner in the sustainable movement. Doing so could help to overcome a related obstacle, namely, the politicization of the current dialogue and the tendency to think of the quest for sustainability as a contest between the "good guys" and the "bad buys."

Putting Science to Work

The thought of putting more science into the sustainable movement may sound at first like borrowing from the book of conventional agriculture. Science, to be sure, has been a major force in the development and implementation of conventional farming, which may help to explain why some supporters of sustainability became disillusioned with science. But the problem is not science. It is that science has not been developed and used effectively by scientists to ensure the sustainability of agriculture.

So far, the goals of sustainable agriculture espoused by many of its proponents reflect mainly a compelling set of values and feelings about what is right and wrong. But while movements may be catalyzed by values and feelings, few succeed unless they are also guided by facts. Science-based facts can be powerful allies by helping to explain the values and beliefs behind people's preferences, which in turn can illuminate their hopes for and interests in sustainable agriculture. Facts can inform the quest for sustainability by showing what is possible and what may be only a dream. Sustainability is not a magic formula waiting to be discovered. No one knows precisely or finally what farming practices and systems will turn out to be the most sustainable, and under what conditions. We have feelings and opinions about these things. We need more facts to go by.

Facts can expose the likely conflicts and trade-offs involved in the pursuit of sustainability, in addition to the complementarities. Full attainment of all of the goals implied in the sustainable ideal is improbable. Chances are that we will all have to sacrifice a little of one or more goals to make possible the attainment of others. As one analyst has put it, "Despite the logic of sustainable agriculture, there are inherent conflicts in the concept itself. These arise from trade-offs between production and conservation, resource allocation, short- and long-term objectives, differing participant perspectives, and incongruities in time and space. The primary purpose of a sustainable agriculture policy is to resolve these conflicts" (Wilken, 1991).

The incidence of conflicts and the necessity of trade-offs become more obvious as we extend the meaning of sustainable agriculture to include not only the goals of adequate and profitable food production, resource conservation, environmental protection, and health and safety, but also preservation of family farming and social equity. This is not to say that we should agree to limit the definition of sustainability to avoid conflicts between the different goals involved, but rather that we must understand what the conflicts may be, what trade-offs may be required in the quest for sustainability, and how they should be addressed in the policy process.

In short, we must begin now to replace the symbolism of sustainable agriculture with scientific facts if we are to encourage the development of public policies to foster sustainability. Without help from science, the sustainable movement could simply be chasing a mirage.

Overcoming the "Good Guys, Bad Guys" Way of Thinking

Discussions about sustainable agriculture all too often involve "preaching to the choir." While proponents of sustainability may disagree on ends and means, relatively speaking they try to talk the same language. Supporters of conventional agriculture do the same when among like-minded people. But when representatives of the two perspectives come together, as in conferences, task forces, and the like, it is our observation that self-interest prevails. The result is game playing and polarization. If only implicitly, participants are cast as "good guys" or "bad guys," depending on one's interests and perspective.

Putting science to work on the issue of sustainability can help to overcome these problems by establishing firmer ground for dialogue and thereby lessening polarization. After all, proponents of sustainable agriculture are not the only people who can and should benefit from the availability of scientific facts. Everyone with a stake in the future of agriculture, from farmers to fertilizer manufacturers and food processors, should enthusiastically support an expanded, even-handed role for science. However, we cannot rely solely on science.

Understanding the causes and consequences of polarization is an important first step. Too often, proponents of sustainability conclude that the chief barrier to attainment of a sustainable agriculture is opposition to the concept from established agricultural institutions; hence, sustainability will be achieved only when those "forces of darkness" are overcome. Proponents or defenders of conventional agriculture react defensively. They object to the insinuation that the agriculture they developed or grew up with is "bad." The resulting polarization impairs the development of sound agricultural policy.

Moreover, the "good-guys, bad-guys" perspective incorrectly implies that there are only two sides to the sustainability issue, and that wide agreement exists within each camp as to the meaning of the concept. As explained above, the truth is that many different views of sustainable agriculture are held by its proponents, as well as by its critics.

Finally, when sustainable proponents imply that agribusiness, commodity groups, the USDA, or colleges of agriculture are the "bad guys," they mistakenly impugn the growing number of scientists and others employed by those institutions who are now seriously engaged in the pursuit of a sustainable agriculture. Similarly, when defenders of conventional agriculture classify proponents of sustainability as uninformed "tree huggers," they too depress the quality of the debate.

Determining the Path to Sustainability

Agreeing on the meaning of sustainable agriculture is but the first step toward such a goal. The second step is to carefully examine and encourage the selection of the most promising path to sustainability. To perform this task, we must again call on science for help. This is now being done to some extent through the LISA program (USDA, May 1991), a growing number of university research projects, various farmer-based research initiatives, and even some private laboratories. But the dollars devoted to sustainable farming research are trivial when compared to those used to address agricultural problems from a conventional perspective.

Questions such as the following illustrate the kinds of policy-related issues that must be addressed through research if we are to determine the best path to sustainability, including the policies needed to encourage its adoption:

- *The symbolism of sustainable agriculture.* What meanings do scientists, farmers, politicians, and lay citizens attach to the concept of sustainable agriculture? What does the term sustainable mean to them? What modes and methods of production do they see as sustainable? What are the implications of such perceptions and beliefs for the development of public policies that can lead to a sustainable system? Does the symbolism of sustainable agriculture, on balance, facilitate or impede the adoption of sustainable policies?

- *The demography and sociology of sustainable agriculture.* How many sustainable farmers are there in the United States? How are such farms distributed geographically? What methods do they use? How do these methods vary from one region to another? Why? Are there sustainable farms in all regions? If not, why? What role does the farm family play in the success of sustainable farms?

- *The economics of sustainable agriculture.* What is the true cost of food under various production, manufacturing, processing, and distribution systems? For example, how do we calculate the social and human costs of low-paid farm and food processing workers? What about environmental and natural resource costs? Who pays for these externalities now? Who should pay? What is the total bill? Can such costs be objectively determined?

  What are the micro- and macroeconomic characteristics and implications of sustainable agriculture? Will there be yield declines? What are the likely consequences of yield declines and an altered product mix under sustainable systems? What will be the extent and nature of such declines under various levels of adoption? What effects will any such changes have on consumer prices, export trade, world hunger, land values, input costs, etc.?

- *Indicators of sustainability.* Soil scientists are beginning to develop objective soil quality indicators. Once established, such indicators can be used to judge the sustainability of various production and cropping systems in diverse climatic and soil conditions. Such information is crucial to the development of policy incentive reforms. Can similarly objective and measurable social and economic indicators of sustainability be developed? Is it possible, for example, to specify objectively the effects of farm size on soil erosion, wildlife numbers and diversity, rural aesthetics, and the quality of rural life? What kinds of farming systems and technologies foster rural revitalization?

- *Education, research, and technology transfer.* Are agricultural courses, curricula, and textbooks helping to achieve a more sustainable agriculture? How is sustainable agriculture being presented in college and university class rooms? What changes are needed? The same questions are equally relevant for primary and secondary schools.

What methods and organizational arrangements work best in stimulating farmer acceptance and adoption of sustainable methods? Farmer networks? LISA-type projects? Other?

What effect do our existing professional and university criteria for promotion, tenure, and other forms of academic and scientific recognition have on agricultural research? Do such systems impede sustainable agriculture research? How? Why? What institutional changes are needed to foster more interdisciplinary, systems-oriented research?

How important is the role of extension in fostering sustainable agriculture? What reforms are needed? In the case of both research and extension, how could additional resources help?

What is and/or should be the role of the Soil Conversion Service (SCS) and the Agricultural Stabilization and Conservation Service (ASCS) in sustainable agriculture? What about other agencies such as the Environmental Protection Agency (EPA) and the Interior Department's Fish and Wildlife Service? Are there ways to creatively link these agencies in extending knowledge about the character and benefits of sustainable agriculture?

- *Public policy development and implementation.* How do present agricultural commodity, trade, taxation, credit, marketing, research, conservation, and regulatory policies relate to the adoption of sustainable farming practices and systems? Do such policies facilitate or impede adoption? How? Why? What changes are needed in current policy? What are the likely consequences of various policy reforms? How much will such reforms cost? How will they be administered? How does international agricultural trade policy relate to sustainability? What are the variable effects of such policies on developed and developing nations?

What methods work to gain intergroup cooperation on sustainable agriculture goals? What are examples of cooperation of farm dealers and market representatives with farmers in working to lower farm chemical inputs? How can industry and retail leaders be brought into discussions of sustainability.

What, actually, are the real objections of commodity groups, major farm organizations and "conventional" farmers to research and/or demonstration of particular techniques associated with more "sustainable" methods—for example, systems of rotations, the use of biological pest control, etc.? Do they actually object, and if so, do their objections stem primarily from economic self-interest, or are they symbolic in nature? Do they object to these practices on technical grounds, or because of misunderstanding of what such techniques may represent or seek—a totally pure organic system, for example?

In addition to addressing these questions, it is essential that researchers also recognize and examine carefully societal values and the "rules of the game" that will affect the pursuit of agricultural sustainability. The question of how to define sustainable agriculture cannot be answered in isolation from the world in which its pursuit must occur. Stuart Hill of McGill University makes the point that without fundamental changes in how we think and live, the ideal of agricultural sustainability may simply be unattainable and its pursuit a cruel hoax—that is, unless we define it only as conventional agriculture with a kinder and gentler twist (Hill, 1985). We need to know this if the movement is to be adequately informed.

Among the values and beliefs that could be major barriers to attainment of a more profound definition of sustainability, we would cite (1) the widely held view that Nature is humanity's servant, out there to be conquered, (2) the belief that the pursuit of economic self-interest leads automatically to a better life for all in our society, (3) the companion view that seeking profit is a noble and open-ended objective, even if obtained in ways that may seem to be lacking in nobility, (4) the belief that the value of things produced and consumed today, from food to forests, is invariably greater than the value of those things in the distant future, and (5) the view that people have a right to do whatever they wish to or with their property.

Implementing Policies and Actions

The movement's plan must extend beyond agreement on the ends and the means of a sustainable agriculture. It must deal as well with the process of designing and implementing the policies and the actions of individual farmers that will help agriculture move in a sustainable direction.

To a great extent, traditional agricultural institutions—from landgrant colleges to farm organizations and Congressional agriculture committees—are not well positioned to guide the sustainable movement. Research organizations are tuned to scientific disciplines and the reductionistic approach to farming, while sustainable agriculture calls for whole-farm, holistic, and cross-disciplinary research. Most farm and commodity organizations, and Congressional committees, think more in terms of individual farm products and commodities than integrated farming systems. Therefore, other institutional arrangements will be required to further the pursuit of sustainability and, in the process, encourage changes in the mainstream institutions that will increase their compatibility with sustainable agriculture.

In this regard, we recommend establishing a policy consortium or other institutional arrangements to creatively link existing institutions and individuals capable of objectively analyzing the goals of sustainable agriculture and the policy reforms needed to help reach those goals. In much the same way that many physical and biological scientists have begun to think about and to do research on sustainable farming systems, a policy consortium, we believe, would catalyze a similarly impressive array of social scientists and other interested individuals, to evaluate and propose sustainable agriculture policy alternatives and their likely consequences.

No single organization has the resources or insight needed to address the knotty intellectual, political, economic, and social challenges posed by the concept and goals of agricultural sustainability. No single organization alone can or should attempt to rethink agricultural policy. Yet, it must be rethought and reformulated if we are ever to achieve a sustainable agriculture.

Additionally, we would propose the vigorous expansion of a "sustainable agriculture coalition" of farmers and supportive groups across the country. The formation of such a coalition is already underway. The challenge now is to encourage the expanded participation of others to ensure that the sustainable movement is national in scope. The combination of a policy consortium and a national, grassroots coalition would constitute a significant force for identifying and examining sustainable agriculture policy options, and for communicating sound information about the feasibility and the effectiveness of those options to guide and strengthen the agricultural policy process.

REFERENCES

Allen, P., D. Van Dusen, J. Lundy, and S. Gliessman. 1991. Integrating Social, Environmental, and Economic Issues in Sustainable Agriculture. *American Journal of Alternative Agriculture* 6:37.

Buttel, F. H., and G. W. Gillespie, Jr. 1988. *Agricultural Research and Development and the Appropriation of Progressive Symbols: Some Observations on the Politics of Ecological Agriculture.* Cornell University, Department of Rural Sociology, Bulletin No. 151.

Butz, E. 19791. Crisis or Challenge. *Nation's Agriculture,* July–August, 19.

Center for Resource Economics. 1992. Farm Bill 1990 Revisited. Washington, DC.

Crews, T. E., C. L. Mohler, and A. G. Power. 1991 Energetics and Ecosystem Integrity: The Defining Principles of Sustainable Agriculture. *American Journal of Alternative Agriculture* 6:146–149.

Edelman, M. 1964. *The Symbolic Uses of Politics.* Urbana: University of Illinois Press.

Harwood, R. R. 1990. A History of Sustainable Agriculture. In *Sustainable Agricultural Systems.* C. A. Edwards, R. Lal, P. Madden, R. H. Miller, and G. House (eds.). Ankeny, IA: Soil and Water Conservation Society, 3–19.

Hassebrook, C. 1989. *Biotechnology, Sustainable Agriculture, and the Family Farm. Biotechnology and Sustainable Agriculture: Policy Alternatives.* Report #1. Ithaca, NY: National Agricultural Biotechnology Council.

Hill, S. 1985. Redesigning the food system for sustainability. *Alternatives* 12(3/4): 32–36.

National Research Council, Board on Agriculture, Committee on the Role of Alternative Farming Methods in Modern Production Agriculture. 1989. *Alternative Agriculture.* Washington, DC: National Academy Press.

U.S. Congress. 1981. Agriculture and Food Act of 1981. Public Law 97-98. Washington, DC.

U.S. Congress. 1985. The Food Security Act of 1985. Public Law 99-198. Washington, DC.

U.S. Congress. 1990. Food, Agriculture, Conservation and Trade Act of 1990. Public Law 101-624. Washington, DC.

U.S. Department of Agriculture. 1980. *Report and Recommendations on Organic Farming.*

U.S. Department of Agriculture. 1988. Alternative farming systems. Secretary's memorandum 9600-1, January 19.

U.S. Department of Agriculture. 1990. 1990 Farm Bill, Proposal of the Administration. February. Washington, DC.

U.S. Department of Agriculture. 1991. 1991 Annual Progress Report. Sustainable Agriculture Research and Education Program. May.

U.S. House of Representatives. 1982. Organic Farming Act of 1982. H.R. 5618. 97th Congress, 2d session.

U.S. Senate. 1982. Innovative Farming Act of 1982. S.2485. 97th Congress, 2d session.

U.S. Senate. 1990. *Congressional Record,* Vol. 136, No. 95, July 23. Washington, DC: U.S. Government Printing Office.

Wilken, G. 1991. *Sustainable Agriculture Is the Solution, But What Is the Problem?* Board for International Food and Agricultural Development and Economic Cooperation, Washington, DC, Occasional Paper 14, April, p. iii.

Youngberg, I. G. 1984. Alternative Agriculture in the United States: Ideology, Politics, and Prospects. In *Alterations in Food Production,* pp. 107–135. D. F. Knorr and R. R. Watkins (eds.). New York: Van Nostrand.

Index

Fertilizers:
Central American example, 56–57
sustainable agriculture policy and, 300
Feyerabend, P., 179–180
Financial factors. *See* Economic factors
Flora, C. B., 268
Flora, J., 268
Food, southern hemisphere and, 244
Food crops, Central American example, 57–58
Food preparation and preservation, regenerative food systems and, 82–83
Food Research and Action Center (FRAC), 256, 261
Food Security Act of 1985, 302
Food Stamp program (U.S.), 253, 257, 259, 262, 264, 268–269, 271, 277
Food systems. *See also* Regenerative food systems
alternative food regimes, 213–233. *See also* Alternative food regimes
ecology and, 12
environment and, 6
hunger and, 251
land access and, 213–214
Fordism, 34, 35
Forster, T., 254, 278
Fossil fuels, uncertainties in resources of, regenerative food systems, 86–87
Foucault, M., 178
France, poverty in, 255
Francis, C. A., 142
Frank, A. G., 140
Fraser, N., 264
Freudenberger, C. D., 143
Friedland, W., 264, 270, 279, 280
Friedmann, H., 3, 4, 12, 152, 226
Fuentes, M., 140
Furley, P. A., 242

G
Gaard, G., 116–117
Gall, G. A. E., 90, 96
Gallopin, G. C., 205
Gender differences:
sexism, hunger and, 264
United States, 147–150
General Accounting Office (GAO), 84, 146, 255, 257, 263
General Agreement on Tariffs and Trade (GATT), 221
Genetics, regenerative food systems and, 87
Gershuny, G., 254, 278
Ghimire, K., 51, 53, 58, 59, 62
Giddens, A., 178
Gilbert, G. N., 21
Gillespie, G. W., Jr., 2, 25, 310
Gilligan, C., 117
Gillis, M., 62
Gips, T., 142–143
Global perspective:
developing countries and, 196
economics and, 195
food regimes and, 224–225

food system and, 12
historical developments and, 221–223
local ecosystems and, 237–238, 239–243
markets and, 220–221
regenerative food systems and, 85–89
sustainable agriculture and, 27
U.S. agriculture and, 151–152
Global warming. *See also* Environmental degradation
macroregional perspective on, 242
sociology of science and, 27, 28
threat of, 240
Goldman, B., 274
Goldschmidt, W., 94
Goodman, D. B., 6, 7, 53
Goodman, D. E., 238, 241
Goonatilake, S., 179
Grassley, C., 305
Graves, R., 215, 227
Greenhouse gases. *See* Global warming
Green movement. *See also* Ecology; Politics
goals of, 140
sociology of science and, 35
sustainable development and, 173
Green Revolution:
agroecology compared, 202
ecology and, 12
Greenstein, R., 255
Grindle, M. S., 229
Groundwater, 6, 8
Guha, R., 184
Guither, H., 262, 275
Guttmacher, S., 145

H
Hadwiger, D., 254
Hambridge, G., 271
Hamlin, C., 5, 9
Haney, W., 264, 280
Harding, S., 155
Harding, T., 270, 271
Harkin, T., 305
Hartshorn, G., 62
Harvey, D., 141
Harvey, N., 189
Harwood, R. R., 297, 780
Hassebrook, C., 261, 267, 274, 275, 309
Hawkins, A., 35
Hazardous wastes, developing countries and, 196
Health food industry, 228
Hecht, S., 48, 53, 69, 185
Heffernan, W. D., 95, 265, 279
Hinkle, M. K., 267
Historical tradition, epistemology, 180
Hite, J., 269
Hochner, A., 95
Holism, vegetarianism and, 112–116
Horowitz, M. M., 51
Household:
food systems, regenerative food systems and, 97–98